Toward a New Culture of the Material

Toward a New Culture of the Material

Edited by Frank Bauer, Yoonha Kim, Sabine Marienberg, Wolfgang Schäffner

DE GRUYTER

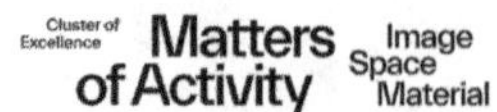 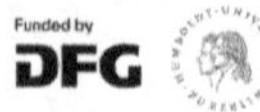

The editors acknowledge the support of the Cluster of Excellence "Matters of Activity. Image Space Material" funded by the Deutsche Forschungsgemeinschaft (DFG, German Research Foundation) under Germany's Excellence Strategy – EXC 2025 – 390648296.

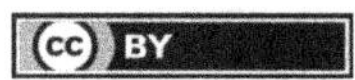

ISBN 978-3-11-071467-8
e-ISBN (PDF) 978-3-11-071488-3
DOI https://doi.org/10.1515/ 9783110714883

Library of Congress Control Number: 2024935272

Bibliographic information published by the Deutsche Nationalbibliothek
The Deutsche Nationalbibliothek lists this publication in the Deutsche Nationalbibliografie; detailed bibliographic data are available on the internet at http://dnb.dnb.de.

Editors: Frank Bauer, Yoonha Kim, Sabine Marienberg, Wolfgang Schäffner
Editorial staff: Elisabeth Obermeier
Student assistants: Pauline Bossauer, Anina Gröger, Julian Kutsche, Maxim Landau, Tim-Oliver Langner
Proofreading: Aaron Bogart
Image copyrights: Elisabeth Obermeier
Cover image: Internal structure of biofilm exposed to moisture. Photo: Michaela Eder, Max Planck Institute of Colloids and Interfaces, Potsdam
Cover design: Rüdiger Kern
Typesetting: 3w+p, Rimpar

www.degruyter.com

Contents

Growing Form

Indeterminacy, Ambiguity and Liveliness

Material Memories

Material Energies

Introduction

Toward a New Culture of the Material

Materials found on Earth occasionally have a history that predates our planet's existence. From this perspective, human material culture spans a relatively brief duration of 2.6 million years. In the early nineteenth century, Danish antiquarian Christian Jürgensen Thomsen classified the human being's own history and prehistory based on the use of materials, thus inventing the notion of Stone, Bronze and Iron Ages. At that time, and in contrast to these distant times, the dawn of industrialization initiated a period of excessive extraction of materials, notably coal and iron, that shaped the latest material age in just 200 years: Modern technology, as a strategy for dominating nature, is based on controllable substances—particularly iron, steel, concrete, glass and plastic—which all rely on materials to fuel the processing heat required to melt, move, and burn them. Only recently has it become evident that this ongoing age of destructive technological production is aggravated by the even more destructive impact of using its artifacts.

The actual ecological crisis is largely caused by the domination of nature combined with a disregard for the materials' own activity. As awareness of this situation grew, however, the analysis of the materials' usage in the biological realm showed a different picture: Long before human beings came into existence, bacteria, fungi, and plants developed practices that correspond to the materials' intrinsic activity rather than try to passivate it. Carbon, hydrocarbons, cellulose, and water are the highly active building materials of life and, as such, exhibit a sophisticated entanglement of matter, activity, and energy that may serve as a basis for the research and design of bio-inspired materials. Moreover, numerous non-industrial practices like fermenting, woodworking, weaving, and medicine, along with artistic endeavors, vividly illustrate an engagement with material activity. Although these practices have often faced challenges due to colonialism and extractivism, some of them have continued to this day, sometimes in new forms.

Our focus on active matter reverses the modern notion that raw materials (*Rohstoffe*) have to be made into suitable and obedient tools and means (*Werkstoffe*) that humans use. Taking the intrinsic activity of materials not as a failure or malfunction but as a dynamic sensorimotor structure offers a completely different starting point for designing artifacts.

It is the main objective of the Cluster of Excellence "Matters of Activity" to raise awareness of this fundamental change by combining the perspective of the humanities and their critical historical and theoretical approach with the perspective of the natural sciences and their new avenues of experimentation with the bio-technical principle of materials' activity, and above all with the transformative procedures of design. To achieve this, the Cluster has established dense interdisciplinary collaboration between more than forty disciplines, creating an integrative process for analyzing and designing with active materials.

Boundary Objects

Our research pursues two main goals: to pave the way for a new culture of the material in the digital age and to create a collaborative academic structure for interdisciplinary knowledge production that aims at its own "materialization" through the incorporation of architecture and design and the exchange with societal practices.

This interdisciplinary collaboration does not start out from a common language—still to be achieved—but from materials that constitute different but equally challenging features for all participating researchers. Active materials—such as cellulose and water—take the role of boundary objects in at least two fundamental respects: they serve as interfaces between and translate across diverse methodological approaches; and since they belong to multiple worlds simultaneously, their analysis integrates very different approaches, both natural and cultural. In the Cluster, material activity is analyzed by focusing on weaving, filtering, and cutting as elementary boundary practices. As procedures which are not simply applied to materials, they have to be considered as triggered by the materials themselves. The vascular structure of human organs, the cellulosic extracellular matrix of biofilms, the fibers of plant or fungi tissues, and the tessellated and elastic surfaces in plants and animals all raise questions of how artifacts fundamentally change when they are designed according to the principle of intrinsic agency. With these elements, we build up a material repository of high heterogeneity, a kind of *matériauthèque* with samples such as seed capsules, rubber, or wood, which represents a new world of soft mechanics, sensing tools, or self-acting structures.

Shaping

From the perspective of an interactive relationship between materials and their designers, users, or perceivers, form processes and object constitutions cannot be regarded as merely human-caused. Rather, we are dealing with a kind of multilayered mutual articulation in which materials, objects, and environmental properties are participants that are just as active as ourselves, animals, or plants. Shaping and being shaped by active matter leaves neither side unchanged. As a result, clear cut distinctions between action and perception, or doing and undergoing, become porous.

When studying atmospheric phenomena such as clouds or aerosols, engaging with the quasi-liveliness of our surroundings leads us to refine our practical and theoretical sensorium, and to design *with* active material rather than silence it. In the realm of neurosurgery, investigating the intricate dynamics between human and nonhuman agents within the surgical setting can enhance our comprehension of how the material world both forms and is formed by human activities. The surgical knife becomes an extension of both the surgeon's body and the patient's body. The act of performing and undergoing a cutting line converges within the instrument itself: shaping and sensing are two sides of one and the same process. Apparently, human senses have developed along with the tangible and comprehensible world. In order not to lose the feel for exploratory probing, one must remain open to invigorating contingencies and blurred boundaries of all kinds. The same applies, eventually, to processes of conceptualization in science. Scientific subject matters tend to be no less active than palpable ones—all the more so when they are tackled within an interdisciplinary framework. To acknowledge the tentative, vague, and not-yet-understood, or even to deliberately head towards it, means keeping concepts and terminologies in vibrant plasticity.

Collaborative Practices

As we withstand giving definite answers rooted in individual disciplines and dare to open a conversation on manifold collective questions, a new form of complexity arises. But what is instigated when we begin doing research in and for a messy world? Complex realities, as they occur in boundary objects, require correspondingly complex ways of dealing with them. This could mean entering into a "dance of agencies" which readily invokes how these practices are entangled between material, immaterial, and more-than-human actors. As a result, our path "toward a new culture of the material" may be found somewhere amongst more nonlinear or volatile approaches, and multiple modes of collaboration. They lead us to investigate the plasticity of hygromorphic willow as a responsive filament, conduct trials on programming auxetic surfaces with and from paper, and reveal catalyst values of filtering in cross-disciplinary environments. Such an endeavor is necessarily a collective one, growing as an organism of many parts between the humanities, sciences, and design towards more open futures.

Growth

But what happens when we move beyond predominant paradigms of control, exploitation, and domination? Maybe new common ground is constituted by shifting from the logics of efficiency and optimization towards more open-ended understandings of creation as growth, by embracing material indeterminacy, microbial activity, and behavior over time. The resulting models of how forms grow could mediate between bio-inspired takes on the whys and hows of "making," ranging from bacteria as collaborative environments of active structures to biofilms as agents that fold, buckle, and wrinkle into as-yet-unknown patterns and morphologies.

Such relations of matter and form are dynamic by nature, translated within processes of emergence, growth, and adaptation, resonating equivalently in realms of the nonliving, the living, and the conceptual. Inspired by morphological practices of nurturing, cultivation, and growth that are ubiquitous in nature, understandings of authorship entrenched in the world of standardized materiality and industrial production are abandoned in favor of more mutual and co-designed processes. In turn, and probably counterintuitively to some, growth can refer to both gaining and losing form, thus affording a perspective on the relation between structure and material that teaches us to conceive of moments of disturbance and crisis also in terms of their potential.

Indeterminacy

In view of various environmental and perceptional tipping points, our prevailing concern—and, in the case of aesthetic experience, also our playful tension—arises from the ambiguity and unpredictability of material-based form processes. In efforts to stabilize them, it becomes apparent that certain aspects persistently evade control. This phenomenon manifests itself in various ways: sea gates struggling against high tides, captions on photographs that fail to fully encapsulate the ambiguity of an image, or the human tendency to seek patterns in the fluidity of active materials. Within the anthropocentric paradigm, controllers and masters have benefited from extracting and stabilizing matter. Active materials, however, invite many more transformative interactions.

Humankind is embedded in larger open systems. Rather than seek control, we must learn to face uncertainty and partake in the continuous process of *poiesis*, which is particularly evident in transitional spaces. In these in-between realms where moments of equilibrium and stability are an anomaly, entities like the so-called duckrabbit encounter a landscape that is both terrestrial and aquatic. The potential of ambiguous imagery unfolds through relationships that entail mutual transformation. By paying critical and caring attention to such phenomena, one can merge and flow with transformative materials while imprinting their fluctuations as enduring memories.

Memory

The notion of memory as a space of storage and retrieval poses the problem of how a static reservoir can be suitable for recognizing something in a world of constant flux. In programming, memory is seen as an adaptive constructive dynamic that regulates its conduct on the basis of continuous recategorization and can only be detected while it is operating. In memory research, the difference between storage and operation has an equivalent in the distinction between explicit and implicit or declarative and procedural memory. Procedural memory is not located in a specific spot but *takes place*, be it in automated actions, sensorimotor recurrences, or in the so-called shape memory of materials. Material activity is rooted in both its memory and its potentiality, which are present as past and possible future states and processes. Here, potentiality can be understood as a possibility of becoming, a disposition to change with both an active and a passive side. While the active side is the ability to change something (or to change oneself), the passive one lies in the capacity to be changed and take on a different form. In preservation practices, the notion of material activity as a potential that can be heightened or halted takes on a further dimension. Wax, for example, due to its extreme malleability, stores the history of its processing down to the most unintentional traces. Inadvertent movements of the shaping hand and fingerprints are recorded as well as what can be regarded as planned processing. What if our focus on conservational efforts would shift from the fixation of states and results to this oscillation between activity and passivity, and to the transience and even decay of objects? In an Anthropocene dystopia in which the destruction of our habitat is imminent, this issue could be accentuated in yet another way: What should endure and be preserved in an imaginary museum when nothing remains of the world as we know it, including ourselves? Arguing with speculative archaeology, the "fossilization" of the greatest possible diversity, of delicate materials and fragile forms, could create a memory of precious and promising dynamics—a refuge in which memory and potentiality coincide.

Energies

Cultivating an attunement to the potentialities of a material-based and process-oriented memory necessitates developing a specific sensitivity—an ability to detect the latent energies in both the material and the symbolic realms. It calls for an acknowledgement of material activity as a new mode of symbolic material operation.

Felt, a fabric composed of densely interlocked animal hairs, functions as an effective insulator thanks to its physical properties. Furthermore, when worn, the vitality of the animal is transferred to the wearer. In this case, the symbol embodied by felt is not merely abstract but tangible and operational, influencing its interaction with its surroundings—its intertwining structure has enabled felt to even become a political symbol of democracy.

The structure of a material acts as an adaptive framework, allowing it to respond to environmental stimuli. This becomes particularly evident in biomaterials, where the interplay of internal interfaces and changes in free energy—influenced by humidity, entropy, and the shifting temperature patterns entailed by global industrialization—orchestrates their behavior. For instance, a seemingly inert sheet of paper undergoes immediate transformation upon contact with water, the latter acting as a catalyst, forging interactions at the molecular juncture that it forms with the cellulose fibers of the paper. Such phenomena emphasize the role that materials play as open adaptive systems. Intrinsic structures of active materials can be regarded as analog codes that not only symbolically represent but also physically execute their information at the same time. In recognizing this symbolic dimension of matter, the triadic relationship between material, energy, and information is redefined, heralding a paradigm shift in our engagement with material. The principle of active matter is a game changer—with the potential to cope with the challenges of the modern world and to conceptually and practically transform it from within.

Shape and Sensitivity

Sabine Marienberg

Matters of Vagueness and Articulation

> It is . . . easy to be certain. One has
> only to be sufficiently vague.[1]
>
> —C. S. Peirce

Abstract

Embraced by the arts but a nuisance to science, vagueness is often depicted not by its potential but by what it is lacking. As an inherent trait of language and embodied cognition, though, vagueness is the continual starting point of processes of articulation that allow for exploring and redrawing the boundaries between grasping and what is grasped. It is on inarticulate grounds, through vagabond thoughts and groping words, that forms as well as concepts come about.

Vagueness

Five grains of sand hardly make a heap. Five million of them do, provided they are arranged accordingly. If we add one grain of sand to five, the ensemble is still far from being heap-like. If *n* grains are not enough to form a heap, *n+1* grains won't do the trick either. From this assumption one might conclude that even by accumulating a million grains, one by one, a proper heap will never develop. Obviously, this is not the case. Eventually, at some indefinable point, if you were to ask someone what they see in front of them, they would answer without hesitation that it is a heap of sand.

This thought experiment is known as the sorites paradox, from the Greek *soros*, meaning "heap," and was devised by Eubulides of Miletus, a contemporary of Aristotle, in the fourth century before our time. It is the prime example of the philosophical problem of vagueness.[2] *Heap* is a vague term that provides for soft transitions and entails doubtful instances regarding its applicability. Other examples of such cases are *bald* or *blessed with a thick head of hair, early* or *late, cold* or *warm, large* or *small, still blue* or *already green.*

---

1 Charles Sanders Peirce, *Collected Papers of Charles Sanders Peirce* (CP), vols. 1–6, ed. Charles Hartshorne and Paul Weiss (Cambridge, MA: Harvard University Press, 1931–1935), vols. 7–8, ed. Arthur W. Burks (Cambridge, MA: Harvard University Press and the Belknap Press, 1958), vol. 4, *CP* 4.237.
2 For more details on the sorites paradox see for example Crispin Wright, *The Riddle of Vagueness: Selected Essays 1975–2020* (Oxford: Oxford University Press, 2021), chapters 2, 4, and 14; Dominic Hyde, "The Sorites Paradox," in *Vagueness: A Guide*, ed. Guiseppina Ronzitti (Dordrecht et al.: Springer, 2002), 1–17.

All these designations, like most if not all colloquial expressions, are vague. They do not allow one to specify exactly the point at which something can be considered to be a heap, warm, early, or green—and within a certain range they are enormously tolerant with respect to the conditions that determine when a predicate can be legitimately ascribed to something:[3] The difference between two or three million grains of sand can be easily ignored. Any temperature reading between 27° and 34° Celsius counts as warm in a weather forecast. In everyday life, the use of vague terms generally does not cause difficulties. They are handled with remarkable flexibility, and certainty about their meaning in a given situation is not affected by whether their attribution is logically permissible. However, for adherents of bivalent logic, in which every proposition has to be either true or false, when it comes to statements about the intension or extension of a term—that is, its internal content or the totality of cases to which it applies —such ill-defined boundaries are the subject of philosophical debates with a long and ongoing history.[4]

Of course, the above considerations are not seriously about the number of grains of sand or units of temperature. What is at stake here is nothing less than the question of how our predicates and concepts are formally defined and how they relate to reality.[5]

Etymological Nuances

In non-philosophical contexts, definitions and synonyms of *vague* have a rather negative overtone. They are consistently characterized by the absence of obviously more desirable qualities: *in*distinct, *un*determined, *im*precise, *un*specific, *not* clearly expressed, sensed, or felt, *not* sharply outlined—the deficient mode can hardly be overlooked.

However, the etymology of the term shows a different picture. In earlier usage, vagueness was associated less with a lack of precision than with a lack of consistency and stability, which was not necessarily seen as a disadvantage. The English term *vague* derives from French *vague*, which in the sixteenth century carried the meaning of

---

3 "What is involved in treating these examples as genuinely paradoxical is a certain *tolerance* in the concepts which they respectively involve, a notion of a degree of change too small to make any difference, as it were." Crispin Wright, *The Riddle of Vagueness*, 84.

4 For a short overview, see Geert Keil, "Vagheit," in *Handbuch Metaphysik*, ed. Markus Schrenk (Stuttgart: Metzler, 2017), 121–27; Philipp Stoellger, "Vagheit," in *Historisches Wörterbuch der Rhetorik*, vol. 10: Nachträge A–Z, ed. Gert Ueding (Berlin: De Gruyter, 2012), 1364–1377; Roy Sorensen, "Vagueness," in *The Stanford Encyclopedia of Philosophy*, ed. Edward N. Zalta, https://plato.stanford.edu/entries/vagueness/.

5 The question of whether vagueness is a feature of our symbolic access to the world or whether objects themselves can be vague has been intensely debated time and again. What follows is primarily concerned with the epistemic and representational aspect of vagueness regarding the means by which we conceptually and linguistically articulate particular objects, as was first elaborated in Bertrand Russell's seminal text on the topic. Bertrand Russell, "Vagueness," *Australasian Journal of Philosophy and Psychology* 1, no. 2 (1923): 84–92.

"empty, vacant" (stemming from Latin *vacuus*), or "wild" and "uncultivated" (stemming from Latin *vagus*, "rambling, strolling," or "wandering").[6]

Finally, a glance at the career of the term in aesthetics—especially in early modern Italy—shows that the connotation of an iridescent movability is anything but negative, quite the contrary. The Italian adjective *vago* and its corresponding noun *vaghezza* are conceptually equated with beauty, lightness, ease, and grace—qualities of things one takes pleasure in and objects of desire.[7] To be endowed with *vaghezza* can be a characteristic of faces and landscapes, of ways of acting and behaving, as well as of styles of expression—styles that are captivating in their variety and versatility and that can be attributed not only to painters or poets, but, surprisingly, also to scientists. In a letter to Leopoldo of Tuscany concerning the debate on the secondary light of the moon, Galileo Galilei notes that his way of writing might seem hyperbolic to certain people, namely, those who would prefer

> to see philosophical teachings confined to the narrowest possible spaces, so that they would always use that rigid and concise manner, stripped of any vagueness or ornamentation, which is proper to pure geometers, who do not utter a word that is not suggested to them by absolute necessity.[8]

Ornamentation here is not to be confused with mere embellishment. What Galileo is claiming is that a rhetorical approach to the world, including scientific topics, is capable of establishing a form of knowledge in its own right—a form of cognition that is in no way inferior to "pure" and straightforward scientific knowledge, but rather complements it.[9]

So, what is vague is fraught with a lack of clarity that may often be tolerable or even advantageous because it is more efficient for communication than overly precise specifications; but in critical cases it should be methodically remedied. Turned positively, however, vagueness can also be understood in terms of playful openness, as a freedom to approach the world afresh—beyond habits, beyond rigid rules and definitions.[10]

---

6 See Online Etymology Dictionary, "vague," https://www.etymonline.com/search?q=vague&ref=search-bar_searchhint.

7 See Istituto Giovanni Treccani, "vago," https://www.treccani.it/vocabolario/vago1/.

8 "[D]ubiterei che le mie parole, benché purissime e sincere, potessero apparire ad alcuno iperboliche o adulatorie: ad alcuno, dico, di quelli, che troppo laconicamente vorrebbero vedere, nei più angusti spazii che possibil fusse, ristretti i filosofici insegnamenti, sì che sempre si usasse quella rigida e concisa maniera, spogliata di qualsivoglia vaghezza ed ornamento, che è propria dei puri geometri, li quali né pure una parola proferiscono che dalla assoluta necessità non sia loro suggerita." Galileo Galilei, Lettera al Principe Leopoldo di Toscana (March 31, 1640), in Galileo Galilei, *Le Opere*, Edizione Nazionale, ed. Antonio Favaro, 20 vols., vol. 8 (Florence: G. Barbera Editore, 1933), 489–545, 491. Translation by the author.

9 See Gottfried Gabriel, "Logisches und analogisches Denken. Zum Verhältnis von wissenschaftlicher und ästhetischer Weltauffassung," in *Sprache und Denken/Language and Thought*, ed. Alex Burri (Berlin: De Gruyter, 1997), 370–84.

10 For a discussion of vagueness as an expressly valuable feature of language and thought, see, for example, Wolfram Hogrebe, "Indistinctness and Disunion," *Dialogue and Universalism* 3 (2016): 5–14; Nora

Shaping Concepts

One is particularly far from being ruled by habits—habits of acting, talking, or recognizing—when they have not yet been formed at all; that is, whenever one learns or begins to articulate something for the very first time. It is in such cases that the *acts* of differentiation and the *movements* of articulation come into view.

In a text on the essential role of the body for (re)constructing regularities, from fundamental spatial and temporal articulations to social and symbolic practices, Gunter Gebauer holds that "[m]ovements are the principle of the first creation of the world of man."[11] Born and placed into existing orders, we have to adapt to them if we want to act and communicate successfully. But at the same time, we also create and modify them through our own actions. This mutual form-giving takes place in an interaction between the plasticity of the human body and the malleability of its environment. The most striking example of such a two-sided plasticity is given by the touching, modeling, and indicating movements of the hand. The hand can take on the most diverse shapes, and it forms both itself and the world around us, "which makes it appear as the center of production of orders through the body . . . that leaves neither side unchanged."[12]

Embodied articulatory movements are initially vague and exploratory. Stable orientations and structured sequences of actions are developed and internalized over a long period of practice and differentiation, without having to be already grasped linguistically. Rather, explicit symbolic orders are rooted in these bodily learning processes[13]—and it is precisely the embodiment of language that is much lamented and has been subjected to attempts at methodical elimination when it comes to the dream of pure thought. To detach thinking from bodily experience, though, is to deprive it of its own origin and presuppositions. For accuracy, being embodied is both an obstacle and a prerequisite.[14]

Kluck, *Der Wert der Vagheit* (Berlin: De Gruyter, 2014); Kees van Deemter, *Not Exactly: In Praise of Vagueness* (Oxford: Oxford University Press, 2010).

11 "Bewegungen sind das Prinzip der ersten Schöpfung der Welt des Menschen." Gunter Gebauer, "Ordnung und Erinnerung. Menschliche Bewegung in der Perspektive der Historischen Anthropologie," in *Bewegung. Sozial- und kulturwissenschaftliche Konzepte*, ed. Gabriele Klein (Bielefeld: transcript Verlag, 2004), 23–41, 24. Translation by the author.

12 "Diese zweiseitige Plastizität, die sich darin ausdrückt, dass sie verschiedenste Gestalten anzunehmen und sowohl sich selbst als auch die Umweltdinge zu formen vermag, lässt die Hand als das Zentrum der Herstellung von Ordnungen durch den Körper erscheinen. Ihre Vermittlung zwischen den Dingen und dem Körper lässt beide Seiten nicht unverändert." Gunter Gebauer, "Ordnung und Erinnerung," 30. Translation by the author.

13 See Gunter Gebauer, "Hand und Gewißheit," in *Das Schwinden der Sinne*, ed. Dietmar Kamper and Christoph Wulf (Frankfurt: Suhrkamp, 1984), 234–60.

14 "Embodiment makes thought logically vague, but it also makes thought possible," writes John Michael Krois, "Image, Science and Embodiment: Or: Peirce as Image Scientist," in *Bildkörper und Körper-*

Precision and Flexibility

It was the program of analytic philosophy to conceive of the problems of thought as problems of language and to eliminate them either by constructing a conceptual language or by critically analyzing ordinary language. In his short text "On the Scientific Justification of a Concept Script," Gottlob Frege elaborates on the ideal of a system of symbols that leaves no room for vague and doubtful expressions. The image he chooses to illuminate the difference between embodied ordinary language and its purified logical sibling is precisely the malleability of the hand compared to the accuracy of a specialized tool:

> The shortcomings here stressed are caused by a certain softness and instability of language which, on the other hand, constitute the reason for its many-sided usefulness and potentiality for development. In this respect language can be compared to the hand which, despite its adaptability to the most diverse tasks, is still inadequate. We build ourselves artificial hands—tools for special purposes—which function more exactly than the hand is capable of doing. And how is this exactness possible? Through the very rigidity and inflexibility of the parts, the lack of which makes the hand so dexterous.[15]

The above passage is reminiscent of another comparison that figures in the *Concept Script* itself. Here, too, a physical access to the world is contrasted with an instrumental one. This time, however, not in relation to haptics, but to the visual sense:

> I believe that the relation of my concept-script to the language of life can be most clearly brought out if I compare it to the microscope's relation to the eye. Because of the range of its uses and the versatility with which it can adapt to the most diverse circumstances, the eye is far superior to the microscope. It is true that when considered as an optical instrument, it shows many imperfections, which ordinarily go unnoticed only as a result of its intimate connection with our mental life. However, as soon as scientific purposes require greater sharpness of discrimination, the eye proves to be insufficient. The microscope, on the other hand, is perfectly suited to precisely such purposes, but that is just why it is useless for all others.[16]

---

schema: Schriften zur Verkörperungstheorie ikonischer Formen, ed. Horst Bredekamp and Marion Lauschke (Berlin: De Gruyter, 2011) 195–209, 207.

15 Gottlob Frege, "On the Scientific Justification of a Concept-Script," trans. James M. Bartlett, *Mind* 73, no. 290 (April 1964): 155–60, 158. "Die hervorgehobenen Mängel haben ihren Grund in einer gewissen Weichheit und Veränderlichkeit der Sprache, die andererseits Bedingung ihrer Entwicklungsfähigkeit und vielseitigen Tauglichkeit ist. Die Sprache kann in dieser Hinsicht mit der Hand verglichen werden, die uns trotz ihrer Fähigkeit, sich den verschiedensten Aufgaben anzupassen, nicht genügt. Wir schaffen uns künstliche Hände, Werkzeuge für besondere Zwecke, die so genau arbeiten, wie die Hand es nicht vermöchte. Und wodurch wird diese Genauigkeit möglich? Durch eben die Starrheit, die Unveränderlichkeit der Teile, deren Mangel die Hand so vielseitig geschickt macht." Gottlob Frege, "Über die wissenschaftliche Berechtigung einer Begriffsschrift," in *Begriffsschrift und andere Aufsätze*, ed. Ignacio Angelelli (Hildesheim: Georg Olms Verlag, 1993 [1882]), 106–14, 110.

16 Gottlob Frege, *Conceptual Notation and Related Articles*, ed. and trans. Terrell Ward Bynum (Oxford: Clarendon, 1972), 105. "Das Verhältnis meiner Begriffsschrift zur Sprache des Lebens glaube ich am deut-

Apparently, what makes a good instrument and an excellent microscope is a lousy tool for practical life goals and for situations that do not require razor-sharp definitions and the precise spelling out of details. For in everyday life, our cognitions take place mostly in the form of a pre- or sub-semantic certainty that is sufficient for practical guidance. Wolfram Hogrebe calls such a tacitly given apprehension that is not built up along explicitly thought-out criteria "scenic understanding."[17] It allows the intuitive split-second evaluation of a situation in which we are involved. Instead of the propositional cognition of single semantic properties, what counts is a holistic and existentially meaningful hunch—is something for or against us, is it propitious, threatening, or indifferent?[18]

In his *Treatise on the Origin of Language*, Johann Gottfried Herder pictures an equally existential and inarticulate moment in which not just subject and predicate, but even agents and actions cannot be distinguished from one another. The first appearance of language takes the form of "Resounding verbs? Actions, and still nothing which acts there? Predicates, and still no subject? The. . . thought of the thing itself still hovered between the agent and the action."[19] As the cognitive linguist Michael Tomasello has shown, also in infant language acquisition, language is not learned by assembling elements that are understood in isolation, but by attending to larger, inarticulate scenes in which agent, patient, events, individual objects, and grammatical roles only subsequently become distinguishable from one another.[20]

---

lichsten machen zu können, wenn ich es mit dem des Mikroskops zum Auge vergleiche. Das Letztere hat durch den Umfang seiner Anwendbarkeit, durch die Beweglichkeit, mit der es sich den verschiedensten Umständen anzuschmiegen weiss, eine grosse Ueberlegenheit vor dem Mikroskop. Als optischer Apparat betrachtet, zeigt es freilich viele Unvollkommenheiten, die nur in Folge seiner innigen Verbindung mit dem geistigen Leben gewöhnlich unbeachtet bleiben. Sobald aber wissenschaftliche Zwecke grosse Anforderungen an die Schärfe von Unterscheidungen stellen, zeigt sich das Auge als ungenügend. Das Mikroskop hingegen ist gerade solchen Zwecken auf das vollkommenste angepasst, aber eben dadurch für alle andern unbrauchbar." Gottlob Frege, *Begriffsschrift. Eine der arithmetischen nachgebildete Formelsprache des reinen Denkens* (Halle: Verlag von Louis Nebert, 1879), V.

17 Wolfram Hogrebe, *Riskante Lebensnähe: Die szenische Existenz des Menschen* (Berlin: Akademie Verlag, 2009), see especially 50–58.

18 An example Hogrebe gives is a crime scene: "[S]omeone enters a dark cellar in the night, and there sees four armed figures at a table weakly lit by a swaying lamp. Unless this is a trusted group of conspirators to which this someone belongs, he/she will sense with lightning speed that the situation is 'risky'." Wolfram Hogrebe, "Indistinctness and Disunion," 13.

19 Johann Gottfried Herder, "Treatise on the Origin of Language," in *Philosophical Writings*, trans. and ed. Michael N. Forster (Cambridge, UK: Cambridge University Press, 2002), 65–164, 100. "Tönende Verba? Handlungen, und noch nichts, was handelt? Prädikate und noch kein Subjekt? . . . Der Gedanke an die Sache selbst schwebte noch zwischen dem Handelnden und der Handlung." Johann Gottfried Herder, *Abhandlung über den Ursprung der Sprache* [1772], in idem, *Werke in zehn Bänden*, vol. 1, *Frühe Schriften 1764–1772* ed. Ulrich Gaier (Frankfurt: Deutscher Klassiker Verlag, 1985), 695–810, 737.

20 Michael Tomasello, *The Cultural Origins of Human Cognition* (Cambridge, MA: Harvard University Press, 1999), 134–200.

Precisely individualized traits—be they of situations, of epistemic objects, or sentences—are singled out in all-encompassing scenic contexts with internally blurred boundaries between their compositional elements. Both the immediate comprehension of wholes and their articulation is deeply rooted in vague and continual embodied movements. The attempt to overcome the latter completely is paid for with the loss of flexible, versatile exploration and creativity that characterize original acts of determination, and lets the way in which situations concern us existentially fade into the background.

Family Resemblances

The later Wittgenstein sees vagueness not as a deficiency but as a necessary condition of language. According to his use theory of meaning developed in the *Philosophical Investigations*,[21] the introduction of concepts is mostly accomplished by giving examples of the use of terms in rule-based "language games." Since the number of possible usages is in principle open, the final determination of a concept cannot be achieved, which is why it remains intensionally vague. Moreover, its diverse extensions are not conceived as having one general characteristic in common, but as being interlaced in a braid of "family resemblances," as Wittgenstein explains with regard to various games: some are a matter of luck, others of skill; some you play with several people, and others on your own. Being cooperative, competitive, or entertaining is by no means a distinguishing feature of all of them and is not one of games alone. Soccer, hide-and-seek, chess, and solitaire are all instances of games, but they are neither organized in a hierarchical structure, nor do they all have a common property. Accordingly, the concept of *game* is also extensionally vague, since the single examples are perceived as belonging to the same family only through affinities and similarities:

> I can think of no better expression to characterize these similarities than "family resemblances"; for the various resemblances between members of a family: build, features, colour of eyes, gait, temperament, etc. etc. overlap and criss-cross in the same way. And I shall say: 'games' form a family.[22]

Like Frege, Wittgenstein also emphasizes the tool character of language.[23] However, he sees it not merely as an external instrument for the transmission of meaning. Just as the hand both shapes and is shaped by the objects and surroundings it deals with, language is the malleable active medium through which concepts are formed and devel-

---

21 Ludwig Wittgenstein, *Philosophische Untersuchungen / Philosophical Investigations*, 2nd edition, trans. G. E. M. Anscombe (Oxford: Blackwell, 1997). As is common in Wittgenstein research, passages from the *Philosophical Investigations* (PI) will be cited by paragraph numbers and *PI* I, and quotations from Part II by page numbers and *PI* II.

22 Ludwig Wittgenstein, *PI* I, 67.

23 "Look at the sentence as an instrument, and at the sense as its employment." "Sieh den Satz als Instrument an und seinen Sinn als seine Verwendung!" *PI* I, 421.

oped. Besides, Wittgenstein's demand to infer the meaning of concepts by observing their occurrences is explicitly tied to bodily participation and sense perception rather than to reflection: "Don't say: 'There must be something common, or they would not be called 'games'—but *look and see* whether there is anything common to all."[24] This view is most likely to be obtained by the unaided eye and not by peering down the microscope. Moreover, to *look and see* in a kind of groping, almost "haptic" way of gazing takes on the attitude of the hand and explores both what constitutes a family and the properties of its members. In this poietic, exploratory perspective, the traditional distinction between eye and hand—as the sensory modalities of theory and practice— is suspended.

If, in the vein of methodical culturalism,[25] we understand philosophical and scientific terminologies as stylizations of ordinary ways of talking,[26] and these, in turn, as rooted in practical handling of things, it becomes clear that the truth value of scientific statements derives from the certainty of action. The merely postulated essence of a general concept would, on the contrary, equate with a look through the microscope that no one takes, from an angle of view that does not concern us.[27]

Articulating Vagueness in Interdisciplinary Terminologies

Such a flexible approach can prove to be particularly fruitful when in interdisciplinary research we are concerned with developing and determining a common explanandum along with the linguistic means for capturing it. If we wish to know what the concepts of *matter, activity, symbol,* or *communication* are all about, we have to look at how the terms are actually used and to which cases they are applied. The different instances are likely to let the terminological boundaries appear vague simply by being so manifold.[28] Take, for example, the use of terms for actions and processes that could be understood

---

24 "Sag nicht: 'Es muß ihnen etwas gemeinsam sein, sonst hießen sie nicht 'Spiele" – sondern schau, ob ihnen allen etwas gemeinsam ist. – Denn, wenn du sie anschaust, wirst du zwar nicht etwas sehen, was allen gemeinsam wäre, aber du wirst Ähnlichkeiten, Verwandtschaften, sehen, und zwar eine ganze Reihe." Ludwig Wittgenstein, *PI* I, 66.

25 See Dirk Hartmann and Peter Janich, eds., *Methodischer Kulturalismus: Zwischen Naturalismus und Postmoderne* (Frankfurt: Suhrkamp, 1996); idem, *Die kulturalistische Wende: Zur Orientierung des philosophischen Selbstverständnisses* (Frankfurt: Suhrkamp, 1998).

26 Wittgenstein talks about guiding words to their "original home" by "bringing them back from their metaphysical to their everyday use." PI I, 116.

27 Referring to the "fleshless and skeletal entities" of mathematical propositions, which are the very opposite of vagueness, Peirce formulates the counterpart of what was quoted in the motto of this paper: "It is easy to speak with precision upon a general theme. Only, one must commonly surrender all ambition to be certain." Peirce, *CP* 4.237.

28 "[A] representation is *vague* when the relation of the representing system to the represented system is not one-one, but one-many." Russell, "Vagueness," 89.

as "symbolic" because they seem to be "communicative" in a way. Some of them include communicative intentions, others are "telling" or "showing," or display a kind of feedback behavior. The meaning of *communication* in its broadest sense is the "exchange or transmission of information," where the use of the term can comprise as diverse processes as talking, waving signal flags, annual rings of tree trunks, oxygen level changes in the blood flow, or intracellular small-molecule signaling between bacteria.[29] The activity of a pine cone opening and releasing its seeds under the influence of increasing humidity may show certain similarities to other responses to environmental conditions—like taking an umbrella for a walk when it's raining or when someone informs us that it will soon start to do so—but these are well-limited. In the face of a multitude of undeniable differences, how can we prevent the concept of communication from becoming intangible?

One might assume that obtaining precise terms is simply a matter of looking more closely, of narrowing the field of observation. A term would then become successively clearer the more we fade out larger contexts and concentrate only on a small subarea of its application. But both contextualizations and decontextualizations have their vagueness and exactness, and the more precisely the details are defined, the more blurred the overarching idea becomes. Grasping the borders of the latter is oftentimes rather a matter of taking a step back.[30] And the view of the all-encompassing context thus gained disintegrates when approaching the details again. Wilhelm von Humboldt captured the impossibility of determining the general character of language in the image of a cloud: From a distance, you view it as a whole, but when you dive into it, all its contours dissolve:

> If the description of the character of an individual or even of a nation is awkward enough, that of the character of a language is even more so. Whoever has attempted it will soon realize that when he is about to say something general, he becomes indeterminate, and when he wants to go into detail, the solid figures slip away, just as a cloud covering the summit of a mountain shows a solid figure from afar, but dissolves into mist as soon as one steps into it.[31]

---

29 It is, of course, highly problematic to fall into epistemological naturalism and not take the "communicative" anthropomorphization of trees and molecules for what it is, namely a heuristic metaphor.

30 "Who wishes to embrace contexts must do so from a distance, who desires detail, however, must draw closer. Who wants to bring out the whole must 'dim the lights'." Hogrebe, "Indistinctness and Disunion," 10.

31 "Wenn schon die Schilderung des Charakters eines Individuums oder gar einer Nation in Verlegenheit setzt, so thut dies noch mehr die des Charakters einer Sprache. Wer sie jemals versucht hat, wird bald inne werden, dass, wenn er etwas Allgemeines zu sagen im Begriff ist, er unbestimmt wird, und wenn er ins Einzelne eingehen will, die festen Gestalten ihm entschlüpfen, so wie eine Wolke, welche den Gipfel eines Berges deckt, wohl von fern eine feste Gestalt zeigt, aber in Nebel zerfliesst, so wie man in dieselbe hineintritt." Wilhelm von Humboldt, "Latium und Hellas oder Betrachtungen über das classische Alterthum," in idem, *Gesammelte Werke*, vol. 3, 1799 – 1818, ed. Albert Leitzmann (Berlin: B. Behr's Verlag, 1904) photomechanischer Nachdruck Walter de Gruyter & Co., Berlin 1968, 136 – 70, 167. Translation by the author.

The detailing of individual figures can even prevent us from coming upon analogies and affiliations at all, that is, it keeps us from identifying a family of usages or from tentatively outlining an area. Family resemblances are not given; they have to be inventively claimed.[32]

Turning back to the notion of communication, there is no one and only concept of it beyond the usage of the term in particular language games that all have their specific purposes. Within the framework of an assumed family resemblance, examples given in particular contexts—in the form of open-ended meanings that shift between the language games of different disciplines—broaden and modulate a concept. They do so by fanning out a multitude of nuances and aspects that can be perceived *as* aspects only by comparing them in the process of adumbrating a loosely organized network of resemblances,[33] including metaphorical ones. Metaphors forgive considerable vagueness and are still understood. And, what is more, they engender similarities rather than presuppose them.[34]

What we gain by the careful observation of uses and by tolerating notions with blurred edges is a plastic morphology of our common terms. This turns vagueness, a notorious scientific vice, into a virtue and fosters the discovery of both similarities and differences in conceptual wholes as they develop.[35] Embracing vagueness, welcoming metaphors as well as vagabond words and thoughts in interdisciplinary research, entails both the versatility of the hand and the pursuit of precision. It means playing with distances and a plurality of perspectives, and thus understanding the determination of a whole and the articulation of its manifold forms and figures as interrelated movements.

---

32 See Wittgenstein, *PI* I, 68: "For how is the concept of a game bounded? What still counts as a game and what no longer does? Can you give the boundary? No. You can draw one; for none has so far been drawn. (But that never troubled you before when you used the word 'game'.)" "Wie ist denn der Begriff des Spiels abgeschlossen? Was ist noch ein Spiel und was ist keines mehr? Kannst du die Grenzen angeben? Nein. Du kannst welche ziehen: denn es sind noch keine gezogen. (Aber das hat dich noch nie gestört, wenn du das Wort 'Spiel' angewendet hast."

33 "I contemplate a face, and then suddenly notice its likeness to another. I *see* that it has not changed; and yet I see it differently. I call this experience 'noticing an aspect'." "Ich betrachte ein Gesicht, auf einmal bemerke ich seine Ähnlichkeit mit einem andern. Ich *sehe*, daß es sich nicht geändert hat; und sehe es doch anders. Diese Erfahrung nenne ich 'das Bemerken eines Aspekts'." Wittgenstein, *PI* II, 193.

34 See Nelson Goodman, *Languages of Art: An Approach to a Theory of Symbols* (Indianapolis: The Bobbs-Merrill Company, 1968), 77: "Instead of metaphor reducing to simile, simile reduces to metaphor."

35 Or, as Hogrebe puts it: "Thoughts which bring out the whole even if they are somewhat vague (cognitiones obscurae vel confusae) are not bivalent in the usual sense—i.e. true or false—but, as a monovalent 'yes,' simply nice." Hogrebe, "Indistinctness and Disunion," 11.

Clemens Winkler

Atmosphere in the Making: Airborne Metabolic Pathways

> "The circular wind created a vertical air
> curtain in the rotunda of the Veterinary Anatomical
> Theatre with its thirteen-meter diameter, whereby a
> stable climate emerged in the center.
> . . . Because the cloud in the center was
> irreproducible, the bodily experiences could only be
> made deeply personal. A continuously forming
> atmospheric event connected specific energetic,
> material, and residual effects with something not
> clearly seen, a dream-like compassion of an
> extended 'we'." (page 8 of this article)

The collective curatorial project "Stretching Materialities: Hidden Activities in Objects and Spaces" 2021/22 brought an emerging cloud indoors. This enabled situated perspectives on becoming with water, energy, microbial life in the air, and its residues at a material and affective level. At a material level, these residues trigger questions about the physical composition and metabolic behavior of the cloud: How is the cloud formed? How is it accessible to whom, where and when? Moreover, at an affective level, how do we unintentionally deal with the cloud while constantly shaping our feelings within it?

Starting from the perspective of "in the making" to better understand the connections from trace gases and cloud formations to interactions with diverse actors, a participatory approach aims at new ways of perceiving compassion for human and non-human actors in increasingly precarious planetary environments. "In the making" explores the notion of dwelling within the intermediary realm and fostering corresponding in the design process, as demonstrated by Tim Ingold in his book on making.[1] Here, larger concerns across disciplines are added to open up the design process for ecological and future-oriented ways of making and participation in material practices and activities.[2] In a participatory opening for diverse nonhuman as well as human actors in the making, language in its complexity also plays an important role in thinking together implicit or explicit forms of being.[3] Consequently, language is regarded as

---

1 See Tim Ingold, *Bringing Things Back to Life: Creative Entanglements in a World of Materials*, NCRM Working Paper, Realities/Morgan Centre (University of Manchester 2008); see also Tim Ingold, *Making: Anthropology, Archaeology, Art and Architecture* (New York: Routledge, 2013).
2 Tim Ingold and Carolin Gatt, "From Description to Correspondence: Anthropology in Real Time," in *Design Anthropology: Theory and Practice*, ed. Wendy Gunn, Ton Otto, and Rachel Charlotte Smith (London: Bloomsbury, 2013), 141.
3 Joseph Vogl, "IV.2 Poetologie des Wissens," in *Grundthemen der Literaturwissenschaft: Poetik und Poetizität*, ed. Ralf Simon (Berlin: De Gruyter, 2018), 460–74, https://doi.org/10.1515/9783110410648–024.

sense-making in both ways: sensually investigating constructive moments in the dense materiality of a drifting cloud full of living entities and particulate matter, while also critically, semiotically, or poetically proofing the precise wordings through making anew. This is also about reading atmospheric processes in their indeterminate symbolic nature to then critically examine a potential farming of uses in working with atmospheres and climate, similar to synthetic biology working on synthetic forms of life between highly distributed systems and very fine modes of sensing. Accordingly, careful handling between *sensing* and *estimating* plays an important role in dealing with volatile materials, such as air, water vapor, and aerosols (dust, soot, bacteria, and spores).

The Cluster of Excellence "Matters of Activity" (MoA) provided a framework for such investigations into atmospheres in the making. For the aforementioned exhibition project, the Object Space Agency (OSA) research group conducted a six-month, in-situ, ongoing research process on "Stretching Materialities: Hidden Activities in Objects and Spaces" at the Veterinary Anatomical Theater (Tieranatomisches Theater, or TA T) in 2021/2022. In addition to coordinating the research group as a curatorial collective, my research was increasingly devoted to questioning the boundaries of various practices and figures of thought interwoven within them. Stretching the practices within the exhibition process encompassed questioning collecting and conservational practices to open up to ecological, technological, participatory, and political processes in contemporary exhibition contexts with visitors, exhibits, and the environment as active protagonists. The experiences gained in practice-led research ranged from methods on new bodily experiences, performative ways of notation, structural findings, and methods of fabulation, which allowed cloud beings, measurement types, and climate scenarios to emerge in negotiable ways.

The Act of Making Critically?

In the realm of making, there is an inherent potential for control, which presents its own challenges. To address these challenges, this article relates to the importance of focusing on the design process as part of embodied and situated knowledge, material agency, and culture.[4] The pursuit of adaptivity requires a critical and careful response to pressing issues such as climate change, extractivist capitalism, and cross-species living conditions that can be transmedially designed in an investigative world-making through material phenomena.[5] The element of control and activation through design

---

4 See Ingold, *Making*; Bruno Latour, *We Have Never Been Modern* (Cambridge, MA: Harvard University Press, 1992); Donna Haraway, "Situated Knowledges: The Science Question in Feminism and the Privilege of Partial Perspective," *Feminist Studies* 14, no. 3 (1988): 575–99, https://doi.org/10.2307/3178066.

5 Together with Léa Perraudin, I have explored the relationality of care, control and proximity and reflected between the discourses of the humanities and research through practice-led design in terms of their shared and contradictory modes of investigative world-making. See Léa Perraudin and Clemens

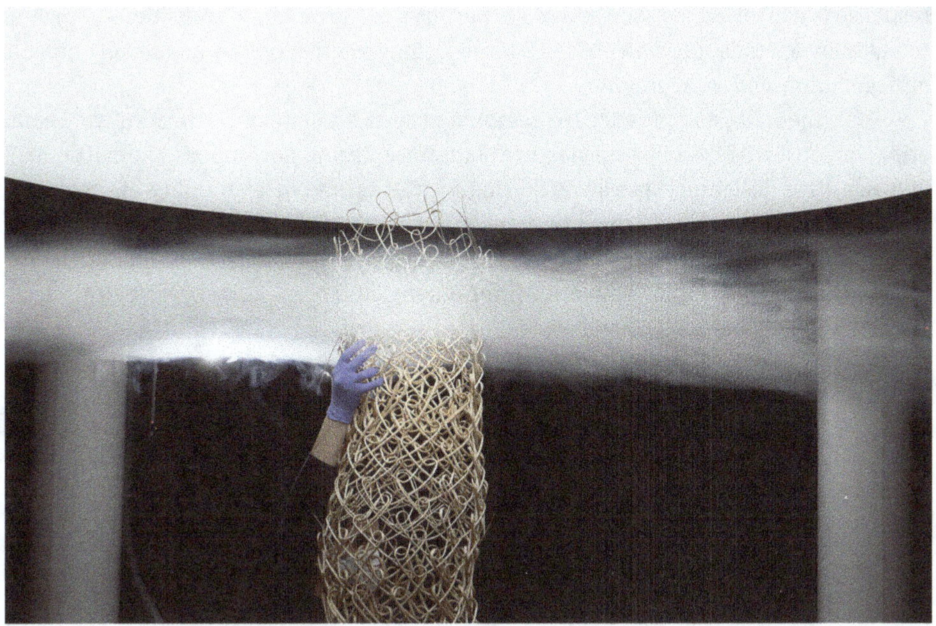

Fig. 1: The cloud at TA T responding to willow structures—with Natalija Miodragovic's Plektonik—Structural Textile. Credits: Michelle Mantel.

is critically examined, while design knowledge is seen as tacit, personal, and context-dependent.[6]

As in the act of space-time dispersion as well as the merging of culture with nature in Western society, this text transforms what Karen Barad expresses through intra-acting in posthumanist philosophy or Donna Haraway's famous claim of "staying with the trouble" of the planetary condition into staying in the damp atmosphere of a cloud, which became a critical symbol for postmodernity.[7] Here, matter as condensation of

Winkler, "Designing with Care? A Pending Question," in *Material Trajectories. Designing With Care?*, ed. Léa Perraudin, Clemens Winkler, Claudia Mareis, Matthias Held (Lüneburg: Meson Press, 2023) 15–29.
6 Claudia Mareis, Moritz Greiner-Petter, and Michael Renner, ed. *Critical by Design? Potentials and Limitations of Materialized Critique* (Bielefeld: transcript Verlag, 2022); see also Nigel Cross, *Designerly Ways of Knowing* (Basel: Birkhäuser, 2006).
7 See Karen Barad, *Meeting the Universe Halfway: Quantum Physics and the Entanglement of Matter and Meaning* (Durham, NC: Duke University Press, 2007), 133–39; Donna Haraway, *When Species Meet* (Minneapolis: University of Minnesota Press, 2008) and *Staying with the Trouble: Making Kin in the Chthulucene* (Durham, NC: Duke University Press, 2016); Anna Lowenhaupt Tsing, *The Mushroom at the End of the World: On the Possibility of Life in Capitalist Ruins* (Princeton, NJ: Princeton University Press, 2015); Astrida Neimanis, *Bodies of Water: Posthuman Feminist Phenomenology* (London: Bloomsbury Publishing, 2017); Ewa Plonowska Ziarek, "The Rhetoric of the Cloud: Celan and the Sublime," *Critical Inquiry* 28, no. 3 (2001): 634–71; Peter Sloterdijk, *The Art of Philosophy: Wisdom as a Practice* (New York: Colum-

response is conceived as a sensing of air particles between phase transitions, which itself already extends the "how-to" logic for designing and making by an ongoing process of negotiation and interpretation in actu.

Thus, the following practice-led research proposes a state of co-constitutive weathering, or intra-weathering, as a starting point for entering new modes of environmentally sensitive designing.[8] Intra-weathering focuses on the inseparability of the entangled relationship between environmental phenomena and their reciprocal bodily constitution in sensing and measuring porous material boundaries, as well as the debris and residue as fertile ground itself for entangled biological, meteorological, structure-building processes.[9]

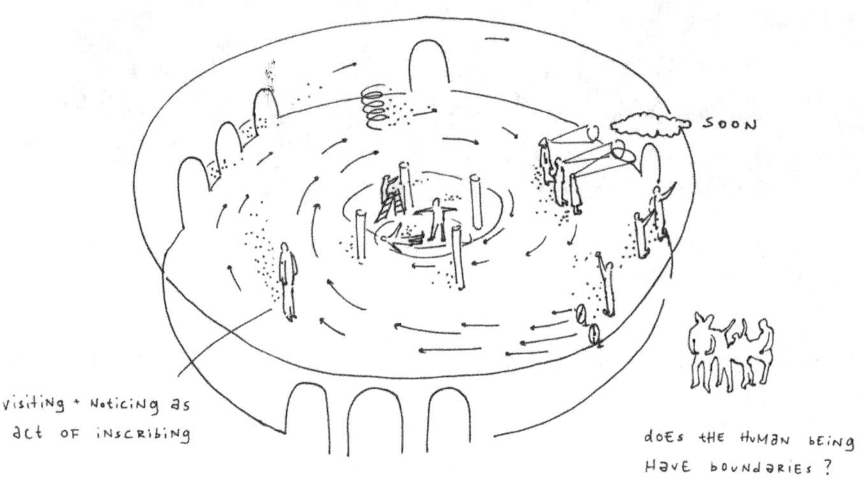

Fig. 2: Sketch of the exhibition space and possible ways of inscribing and loosening bodily boundaries.

Within this research practice, the political urge to make something is applied through the process of seeding and nucleating, which might bring us together with Tim Ingold's

bia University Press, 2009); and Tung-Hui Hu, *Prehistory of the Cloud* (Cambridge, MA: The MIT Press, 2015).

8 Yolanda Ariadne Collins, "Weathering Weather. Atmospheric Georgraphies of the Guiana Shield," in *Weathering: Ecologies of Exposure,* ed. Christoph F. E. Holzhey and Arnd Wedemeyer (Berlin: ICI Berlin Press, 2020),181–205.

9 Ted Krueger, "Microecologies of the Built Environment," in *The Routledge Companion to Biology in Art and Architecture,* ed. Charissa Terranova and Meredith Tromble (New York: Routledge, Taylor & Francis Group, 2017).

walk through "weather worlds" or Astrida Neimanis' "weather instructions."[10] In this exhibition context, event scores became an invitation for site-specific works, for sensing the impact of the self, and experimental spatial arrangements that follow towards an exploratory and participatory ground in practices of exhibition making, site-specific theater, and complicit knowledge production.

Making as understanding of self as well as inevitable third culture focuses in-between sciences and humanities on the emergence of *Gestaltung.*[11] From a critical perspective, it is essential to incorporate scientific methods and discursive views complementing the three perspectives of humanities, sciences, and architecture/design connected through various forms of making. By examining the modes of production in their scientific and cultural evolution, practices such as dismantling information, reassembling, and speculating offer critical insights into contemporary environment-human relationships throughout the disciplines.[12] As a result, three figures of practices evolved on a participatory exhibition stage with the indoor cloud: *enveloping* as designing new experiences, *mapping* through scientific tracking, and *fabulating* through conceptualizing and envisioning (fig. 3).

Designing Exhibitions as Investigative Research

A designing aspect of an atmosphere in the making is not carried out for a particular purpose from the beginning but rather in the process itself, where one's mind and body are exposed directly as barometers and thermostats standing in and shaping physical environments. Here, already before the actual process of an indoor cloud formation, there is a chance to witness ordinary material trajectories and metabolic pathways to open further important perspectives of making and designing as an intersectional, environmentally conscious and cautious practice of care.[13] It includes attentive experimentation and practical tinkering engaging with the material world in a non-

---

10 Tim Ingold, "Footprints through the Weather-World: Walking, Breathing, Knowing," *Journal of the Royal Anthropological Institute* 122 (2010): 121 – 39, 122, https://doi.org/10.1111/j.1467-9655.2010.01613.x; Astrida Neimanis and Jennifer Mae Hamilton, "Weather," in *Connectedness: An Incomplete Encyclopedia of Anthropocene*, ed. Marianne Krogh (Copenhagen: Strandberg Publishing, 2020).
11 Wolfgang Schäffner, "The Design Turn: Eine wissenschaftliche Revolution im Geiste der Gestaltung," in *Entwerfen—Wissen—Produzieren*, ed. Claudia Mareis, Gesche Joost, and Kora Kimpel (Bielefeld: transcript Verlag, 2010).
12 Helmut Kreuzer and Wolfgang Klein, ed. *Die zwei Kulturen: literarische und naturwissenschaftliche Intelligenz; C. P. Snows These in der Diskussion.* (München: Dt. Taschenbuchverlag, 1987); Claudia Mareis, *Design als Wissenskultur: Interferenzen zwischen Design- und Wissensdiskursen seit 1960* (Bielefeld: transcript Verlag, 2011).
13 Claudia Mareis and Nina Paim, ed., *Design Struggles: Intersecting Histories, Pedagogies, and Perspectives* (Amsterdam: Valiz, 2021).

normative form of ethical obligation.[14] In turn, this inquiry follows from the intra-weathering of objects and bodies against the backdrop of conservational practices in museums, theatre and archives as the potentiality of tracing clouds as an indicator for cautious practices. The making of certain atmospheres, specifying clouds, and experiences built upon my own previous projects in contexts such as laboratories, classrooms, volcanic sites, and high-altitude locations.[15]

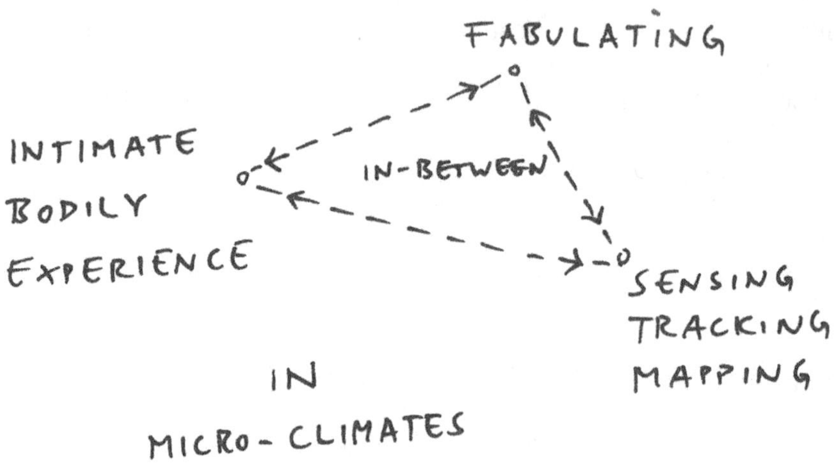

Fig. 3: Triangular diagram reflecting on emerging practices in the experimental situated exhibition context.

Why clouds as indoor atmospheres in the first place? Clouds are described as transformational states of water, particulate matter, and air that undergo phase transitions and precipitation processes. Given that clouds can only be seen as a visible condition of aerial events with their extreme peculiarities on the natural site, like abrupt phase shifts in so-called super-cooled conditions or turbulences that actually stabilize complex thermodynamic structural-functional relationships, for this research using wind, airborne dust, energy in- and output and vaporized water became the tools for testing and crafting. Consequently, the exhibition space in fig. 2 became a testbed for experiencing cloudiness, exploring the impact of changes in the physical atmosphere on human perception, and engaging with the natural and cultural history of creatures and other non-human bodies. The investigation of the atmosphere thus brought together mediating

---

14 María Puig de la Bellacasa, *Matters of Care: Speculative Ethics in More Than Human Worlds* (Minneapolis: University of Minnesota Press, 2017).

15 Clemens Winkler, "Per-Forming Clouds—Materielle Dynamiken als Kommunikationsmodelle," in *Matters of Communication—Formen und Materialitäten gestalteter Kommunikation*, ed. Sabine Foraita, Bianca Herlo, and Axel Vogelsang (Bielefeld: transcript Verlag, 2020).

qualities of various materials and the composing materials of media as environing media in climatic environments.[16] The aim was to experimentally sense and make with different forms of liveliness and stimulate interdisciplinary thinking around who or what might be materially active.

Figures of Practices

After situating a debate around critical making of and with atmospheres, the three disciplinary angles of sciences, humanities and designing particularly within a participatory exhibition project as a testbed for research, three figures of practices emerged. The first figure—*enveloping*—involves shaping the conceptual understanding of cloudiness by practical designing methods for sensorial experiences and new intimacies. The second approach—*mapping*—influences the understanding of cloudiness in situ by employing scientific methods of microbial air sampling and thermal investigations. The third approach—*fabulating*—attempts to understand cloudiness through discursive reflections and imagination, allowing for the emergence and disappearance of damp haptic atmospheres. As a curatorial collective, our research group focused on making aspects of the atmosphere inside the TA T, leading to attunement to specific air conditions and atmospheric qualities. Through experimentation, we engaged in critical ways of collecting, sampling, and mapping techniques, critically giving a voice to microbial companions, mineral or wet residues, bioindicators such as lichen, or even trace gas concentrations of carbon dioxide or sulfur oxide, which—with humidity—become weathering agents to certain surfaces in the indoor environment of TA T.

1. Enveloping. As conceptualized by Derek McCormack, atmospheric envelopments offer a productive framework for understanding the differential shaping and fabrication concerning atmospheric milieu, complicating the notion of inhabiting as a neutral passive act.[17] Understanding an active enveloping also fundamentally precludes complete control of atmospheric dynamics and questions how we engage in formless processes through the contingent capacity to sense and respond. The close link between atmosphere and technical control in this experimental project required critically examining the conditions for the emergence of particular microclimates in the exhibition space. Drawing on what anthropologist Timothy Choy calls "haling," this practice raised awareness of the changing indoor microclimate for which we are co-responsible, as participants, co-makers as well as so-called weather-ers.[18]

---

16 Adam Wickberg and Johan Gärdebo, ed., *Environing Media* (New York: Routledge, 2023).

17 Derek P. McCormack, "Elemental Infrastructures for Atmospheric Media: On Stratospheric Variations, Value and The Commons," *Environment and Planning D: Society and Space* 35, no. 3 (2017): 418–37, https://doi.org/10.1177/0263775816677292.

18 Clemens Winkler, Léa Perraudin, and Iva Rešetar, "The Body of Breath: Morphologies of Air Movement," in *Atem/ Breath*, ed. Linn Burchert and Iva Rešetar (Berlin: De Gruyter, 2021), 87–88.

The inscription of the atmospheric milieu and dimensions of weathering into the built site of the TA T occurs for reasons of preserving and exhibiting its "stabilized" or even passivated interior activities. The question arose concerning which institutional systems, tools, and infrastructures shape our relationships with atmospheric conditions' passivated materials and concerns of liveliness and activity in this practice of enveloping.[19] Why is it essential that exhibition sites take care of such dealings with ecological spheres or forms of contemporary response?[20]

The first step of investigation for this purpose was to provide a framework that enabled exploring certain airy conditions. This was set to bundle air in the rotunda, the former dissection area of the TA T with its thick masonry and small windows built in 1790 to keep the horse cadavers cool for the sake of anatomical research practices. In the exhibition context, two wind machines were installed to create a cyclonic motion of air with the eye of the storm in the center, where the former elevator for the horse corps was installed to bring bodies of research into the upper atrium. From here, in the exhibition, particles from visitors, objects, and built materials as contributors—inscribed unintentionally through residue—gathered in a stabilized climate. The circular wind created a vertical air curtain in the rotunda of the Veterinary Anatomical Theatre with its thirteen-meter diameter, whereby a stable climate emerged in the center. As the next step, a stable climate layering was installed in the center with a warm, dry air layer on top, a damp layer in the middle, and a colder, drier bottom layer. The creation of an indoor climate with diverse cables, relays, controllers, physical heating, cooling, and moistening engines looked different compared to how microbes or natural hygroscopic materials[21] would transform air quality.

As a result, a forming cloud could be observed as an accumulated object of condensation nuclei from the environment, where the leakage of objects, materials, and the occurring dust particles seemed to meet the condensation of water vapor and thermal transitions. In the context of this experimental design and collective exhibition work, the cloud was a differentiated process of "envelopments" and deeply interwoven with the objects, microbial organisms, visitors, researchers and cleaning staff of the exhibition (fig.4).

From a climatic and biological perspective, through breathing, body heat, and debris, the visitors were directly conversing with the installation: they tore holes into the cloud and changed its flow and structural differentiations through physical move-

---

19 Mark Jarzombek, "Haacke's Condensation Cube: The Machine in the Box and the Travails of Architecture," *Thresholds*, no. 30 (2019): 98–105, https://doi.org//10.1162/thld_a_00292.

20 See the Museum for Climate Action website, https://www.museumsforclimateaction.org.

21 Compared to a space full of olivine stones hygro-scopically digesting CO_2 in a so-called enhanced weathering process. Only the time scales are different to work with in a human life. James Temple, "How Green Sand Could Capture Billions of Tons of Carbon Dioxide," *MIT Technology Review* (2020), https://www.technologyreview.com/2020/06/22/1004218/how-green-sand-could-capture-billions-of-tons-of-carbon-dioxide/.

ment.[22] The cloud raises the question of where to find agency and a new way of being. A cloud is inherently a place for encounters of a material nature: human, ecological, social, and affective spheres of emergence.

Because the cloud in the center was irreproducible, the bodily experiences could only be made deeply personal. A continuously forming atmospheric event connected specific energetic, material and residual effects with something not clearly seen, a dream-like compassion of an extended 'we.' The cloud was intimately tied to notions of sublime mental worlds and alienated realities; however, at the same time, in this example it also becomes the result of technical operations between caretaking and extraction. How can we more actively shape our understanding by better experiencing the dialogue with various emerging bodies of air, as a metabolism of energy, material, work and exhaust through envelopment?

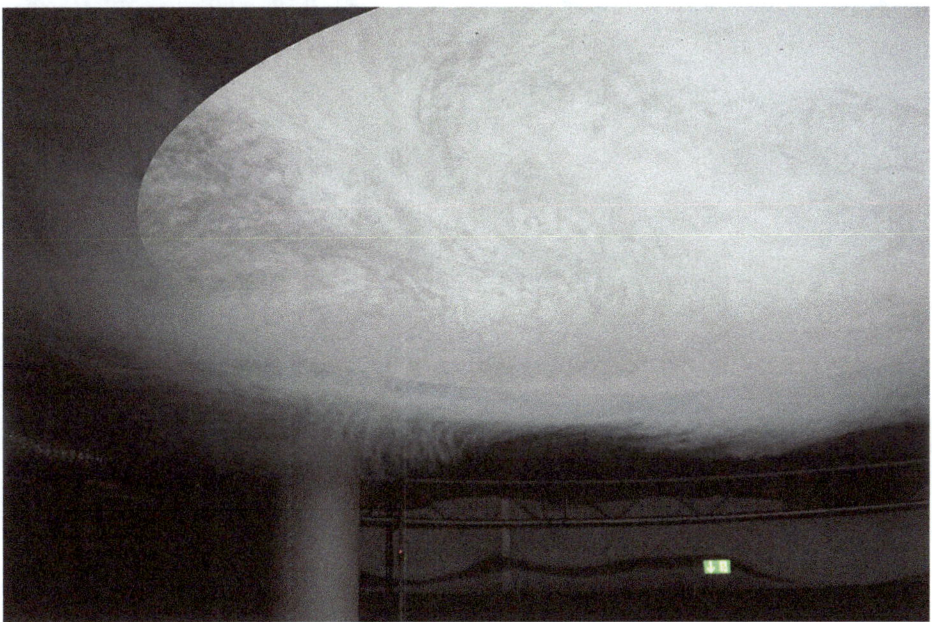

Fig. 4: Participatory cloud installation in the making.

2. Mapping. There are airborne particles—dust, debris, residue, and sediment—in every indoor environment. Outside, the nucleation of clouds and weather formation depends on the particles in the air. As residue, dust is carried within the space as a nucleator and occupies any surface level. Furthermore, it is the base for indoor micro-ecologies. How can we deal with dust? In this mapping practice, tools such as

---

22 Ingold, "Footprints through the Weather-World," 122.

air samplers helped to analyze and archive living and non-living air particles regarding participatory indoor events, such as tours and workshops (fig. 5).

In addition, the tracked number of visitors, taking dust samples weekly with the cleaning staff and settling dishes, and the pH value of outdoor filtrated water helped to complement the sampling concerning surrounding influencing factors. The experimental investigation of the practical questions of dealing with and living with the indeterminacy of residues and climatic tracing led to a testing series. For instance, a taxonomy of dust evolved when visitors brought in soil bacteria such as *Bacillus subtilis* through their shoes, and security staff unintentionally let pollen inside the building during the COVID-19 lockdown.

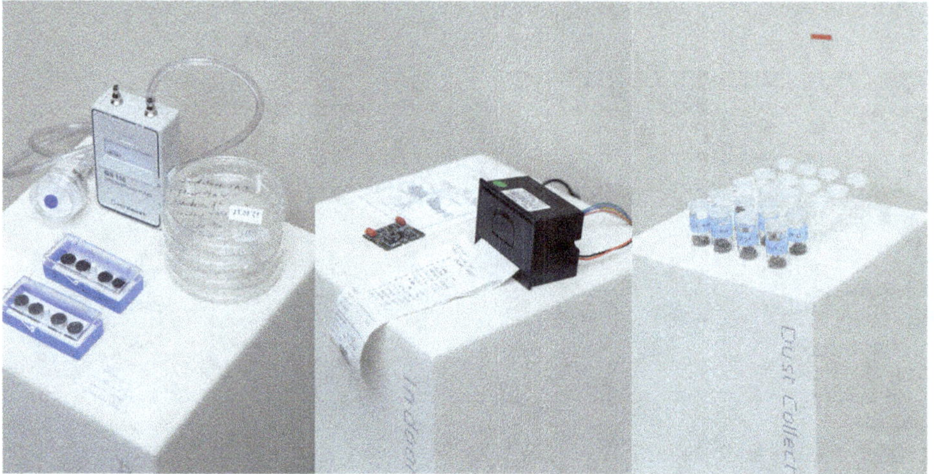

Fig. 5: *From left to right: settling dishes, microscopic plates with collections of biotic, abiotic, anthropogenic dust residues for SEM imaging; weekly dust samples collected with cleaning staff at Tieranatomisches Theater (TA T)*

As water is an immanent companion of airborne particles and almost any environment on our planet, it was included in our exhibition site at TA T for the purposes of cloud nucleation. The nearby river Kleine Panke recalled 200 years of penetration of the surrounding built environment, changing sediments, and micro-topographies on the site. By including the river- and rainwater into the exhibition, the soul of the Kleine Panke was brought metaphorically back into the "waterproof" building to make friends again with indoor air, microbiota, and overcome historical ghosts, thereby becoming a symbol of re-enactment and healing practices in the former slaughter hall (fig. 6). From here, making as caretaking was related to sensing and fabulation as well as measuring, mapping, and form-giving. Material residues were captured and made visible in new states through the addition of water, energy, and labor performed (fig. 7, fig. 8).

3. Fabulating. Due to experiencing different atmospheric material activities in the experimental exhibition space of the TA T, sampling and mapping certain conditions, the method of fabulating became useful to think of alternative treatments and changes in the situations between objects, humans, and spaces. Given that a new state for air assemblages was imaginable, a speculative archive of ideas and debates was tested based on a series of measurements and first physical experiences and brought onto the wall of the rotunda space at TA T. Like the airborne living and non-living particles mentioned earlier, they call for empathy regarding other strange airborne assemblages or microbial films to grow in what we consider to be harsh conditions. Further assemblages of skin cells, textile fibers, and lactobacilli were found through sampling in the space and brought imaginatively into undesirable and desirable climatic scenarios, where heavy metals or high levels of sulfur and carbon emissions surrounding the exhibition site could become nutritious for certain airborne species to metabolize into new forms.

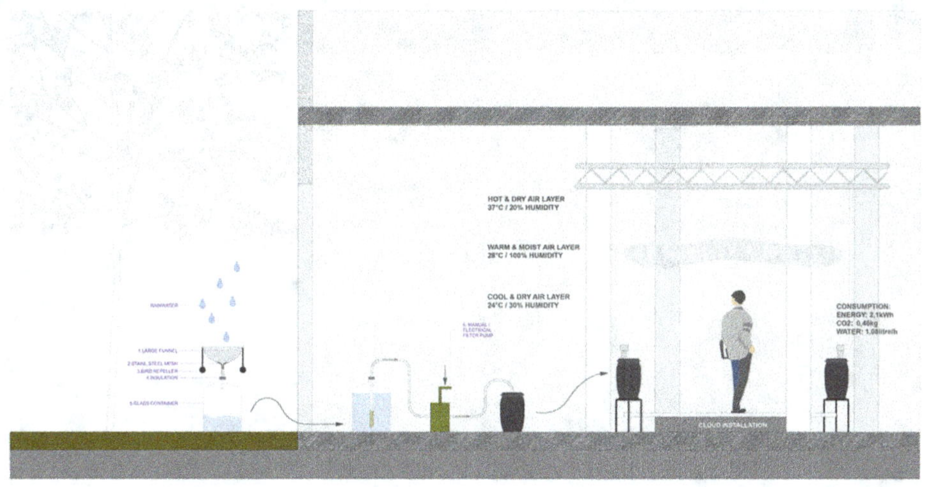

Fig. 6: *Environmental modes of making indoor atmosphere, from left to right: filtration diagram of the river Kleine Panke into the exhibition space with Dimitra Almpani-Lekka.*

The accommodated participatory formats, seminars, and workshops in the exhibition space were titled "Imagining Suspended Care-iers," asking the following questions for opening design processes through shared education: How can cross-species design offer more inclusive methods of climate action? What can we learn through the lens of air-traveling particles in the environment? How can we create access points to other species by designing material objects, exercising forms of embodiment, and short movie essays?

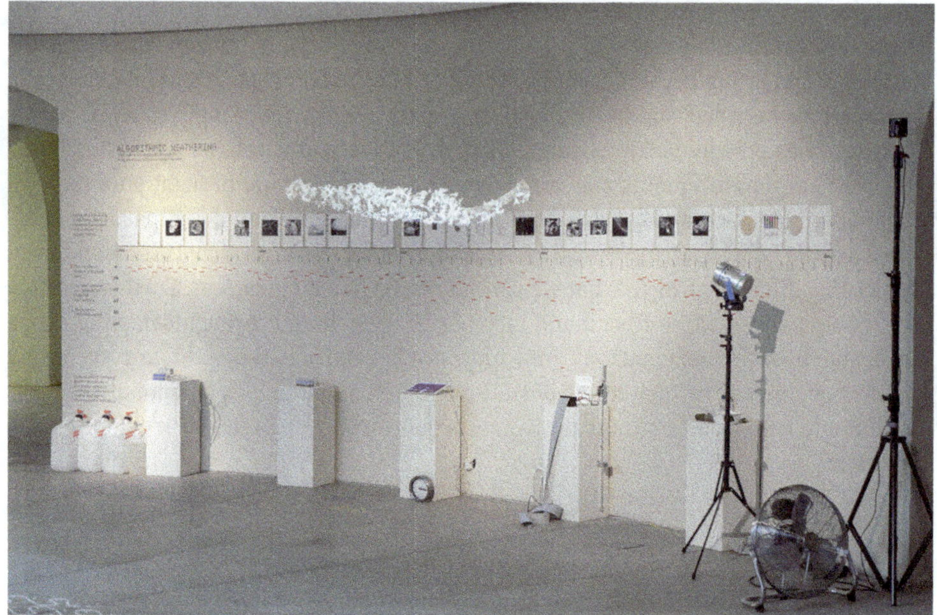

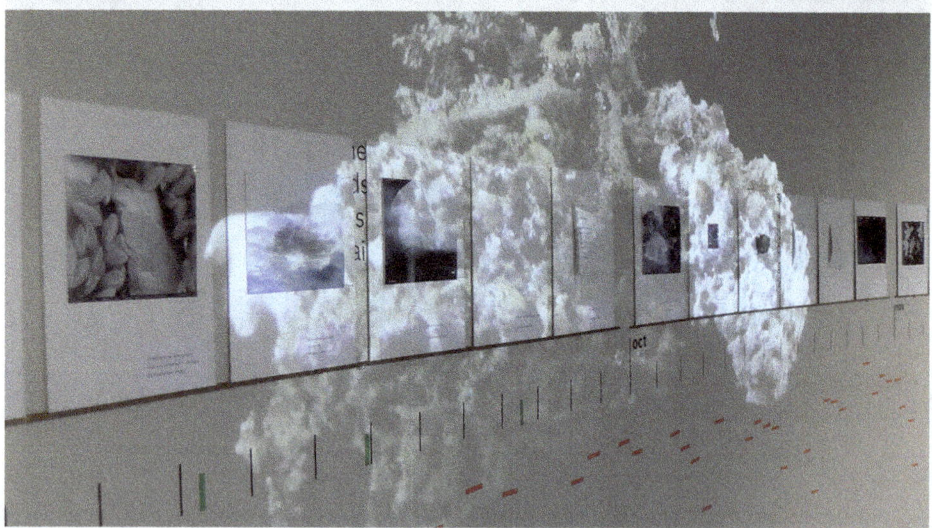

Fig. 7a and Fig. 7b: *Speculative archiving 1: Timeline of aerial samples of dust, microbiome, visitors, thermal conditions.*

Techniques such as templates for protocoling and event scores helped to follow up specific workshops, seminars, and public formats. As a co-designing practice and a cross-species designing approach, participants left notes and sketches in the exhibition space about individual air futures, geo-stories, and microclimate fictions. Drawing from the experiences with the indoor cloud, it called for re-attaching to bodily sensations of dampness, heat, and drought and changing the language towards probable modes of

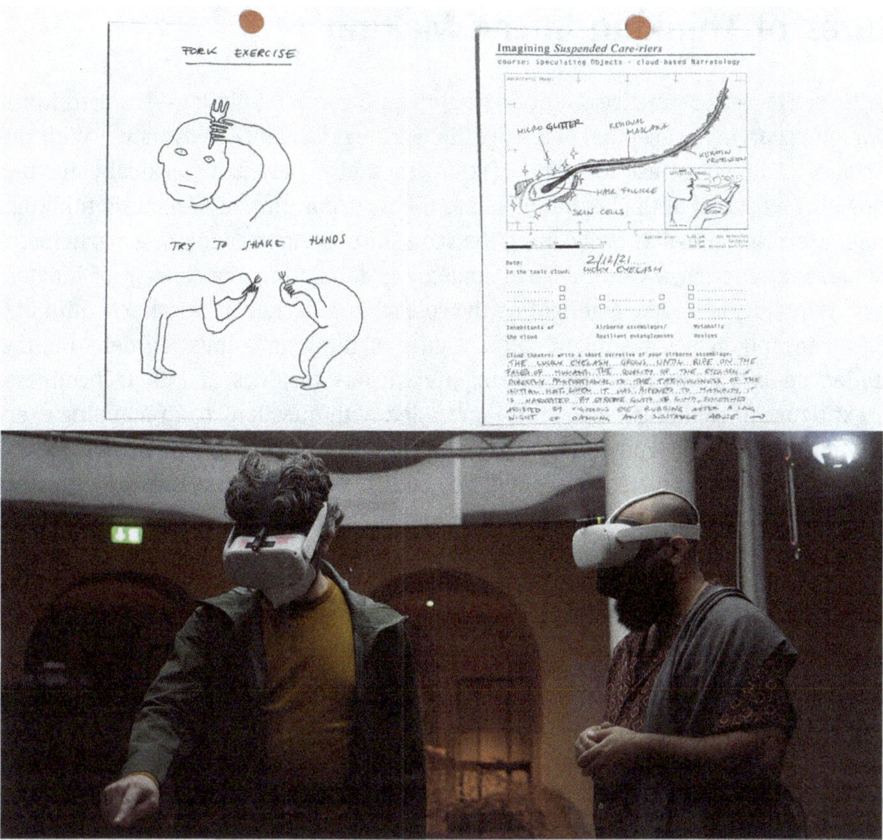

Fig 8a and Fig. 8b: *Speculative archiving 2: Participatory research methods on imaginary aerosols/embodiments, event score, VR experimentation.*

attunement[23] and material futures of acceptance, caretaking, and healing. As a figure of practice, fabulating thus enabled diving back into the other figures of enveloping and mapping.

---

23 Experiences and knowledge related to humidity, heat, and dust levels, as well as the dust sampling and classification process at TAT were effectively shared and transferred. As part of three lectureships, I had the opportunity to engage with institutions like the University of Buenos Aires in the "Master Open Design" program, the IDK design department at University of the Arts Berlin, and the MA program "Spiel und Objekt" at the University of Theatre Arts Ernst Busch Berlin.

Figures of Thinking in the Making

These three figures of practices—enveloping, mapping and fabulating—resulted in reflection on recurring conceptual challenges in a triangular dance of agency.[24] With the importance of language use for ways of knowing and experiencing critically, further investigating as stated at the beginning led to the incorporation of designerly thinking. Through exercises aimed at broadening the scope of making to encompass participatory exchange and larger socio-ecological contexts in fabulating, the concept of indeterminacy frequently emerged, highlighting the role of contingency in designerly thinking. Another concept—which we might call a "figure of thinking"—involved debating the exhibition space's cloudiness, considering its various residues and their liveliness. The next figure of thinking—scalability—grappled with questions from mapping exercises, which involved interpolating investments and investigations made during the designerly mock-up of the testing ground for bodily envelopments. This process also traced excavated energy and water resources outside the TA T, along with airborne microbes originating from plant pathogens, which were found far outside Europe but ended up in the exhibition space.

1. Indeterminacy. As earthlings, we are always in an atmosphere, whether in rough weather (extramural) or affordable passivated indoor climate (intramural).[25] We are immersed on a microscopic level in gas envelopes surrounding our planet. Therefore, we are in constant exchange with the environment, weathering and leaking into it and being affected by the atmosphere, forming it, and becoming informed. Embracing leakiness challenges notions of contamination and containment, blurring the boundaries between liveliness and non-liveliness.[26] How can we make new intimacies with ambiguous and indeterminate matter when we are always leaking in an atmosphere? What language can design practices offer to raise awareness of these relations? The epistemic potential of leaking is not a dematerialization or a drift into the diffuse, but—on the contrary—a concretization and disclosure of a profoundly material event. Otherwise, one

---

24 Performative dance as practice of epistemic acting between the lab and the environment in Andrew Pickering, "Being-in-the-Environment: A Performative Perspective," *Dans Natures Sciences Sociétés* 21, no. 1 (2013): 77–83, https://doi.org/10.1051/nss/2013067.

25 The atmosphere can be understood here metaphorically through language, there is "something in the air," expressing a vague qualitative something. The making of atmospheres from a psychological/sociological perspective follows Tellenbach and Boehme's latest socio-ecological framing of the concept of atmosphere as a question of who owns ecology when there can no longer be such a thing as mere nature. Federico Vercellone and Tedesco Salvatore, ed., *Glossary of Morphology: Lecture Notes in Morphogenesis* (Cham: Springer, 2020), 45–46. However, the handling of the indeterminacy of *atmosphere* shall be dedicated to microbes as an example to incorporate a nonhuman-centric view on matters of aerial activity.

26 "Materials—no matter how solid and inert they may seem—have the tendency to change sooner or later, to leave the place assigned to them, to mix, to form new connections or to dissolve." Tim Ingold, *Being Alive: Essays on Movement, Knowledge and Description* (New York: Routledge, 2011), 16.

would assume that materiality is an inert, passive, and resistant raw material for formal design. Therefore, contrary to a tendency towards dematerialization, working with air and fog involves a tightening of objects in the process of becoming. In this way, the physical world of hardware can be grasped through indeterminacy, continuously prompting us to ask: Where are we now? How do we locate ourselves now as leaking bodies? What language might help us to navigate anew through microclimates, bodily and mentally, in intimate and planetary haling?[27]

For designers, air is a design medium that is non-visual, preventing it from capturing our imaginations, despite triggering new forms of it. Therefore, some strategies are needed to bring the *atmosphere* into focus, to wrest something tangible from the vastness, invisibility, and complexity of this abstraction. What could constitute experiments on the materiality of the air that would help us to shape its imaginaries?

A *making-with-atmospheric condition* and a suitable language connecting theory and practice can be formed due to what Lucretius called nephological production as part of meteorology.[28]

Hippocrates—the son of Heraclides—held that humans were attacked by epidemic fevers when they inhaled air infected with dusty pollutants that are hostile to the human race. Further, in about 55 BC Lucretius called in the fields of meteorology and nephology—the study of clouds—to observe "the dance of dust particles on a sunbeam in a darkened room and concluded that their movement must result from a bombardment by innumerable, invisible, moving atoms in the air." This brilliant intuition enabled him to account for many interesting phenomena, including the origin of diseases.[29] The dust in its dance could be seen and understood as many forms, as "dust language." Developing a dust language, observing the shifting boundaries of people, objects, and environments was investigated in the practical part of *enveloping, mapping,* and *fabulating.*

This perspective still allows us to think of air as being dense and full of activity. Following Lucretius, it took another 1,500 years for scientists to discover the diversity and metabolic forces in airborne particles.[30] Indeed, it might shed new light on what

---

27 Cultural anthropologist Timothy Choy asks whether "we" are talking right now, grappling with implications. "We" depend on, share, and respire and also affect each other's atmospheric surroundings, including the air. "Ecological Reparation: Museum of Breathers," YouTube video, 11:55, uploaded on December 1, 2021, https://www.youtube.com/watch?v=CxpGrdXey50.

28 As a branch of meteorology, nephology (from the Greek word *néphos* for "cloud") was the study of the clouds. The British meteorologist Luke Howard established a cloud classification system in his 1865 paper *Essay on the Modification of Clouds.*

29 Maureen Lacey and Jon S. West, *The Air Spora: A Manual for Catching and Identifying Airborne Biological Particles* (Dordrecht: Springer, 2006), 16–17.

30 "The belief in 'spontaneous generation' of organisms causing decay and disease was held by many people and persisted for a couple of centuries. Micheli (1679–1737) was a botanist in Florence who, by putting spores of molds on slices of fruit, showed that they were 'seeds' of the molds. As some control slices became contaminated, he concluded that spores of molds were distributed through the air (Buller, 1915). In his letters to the Royal Society in 1680, Anton van Leeuwenhoek reported that he was able to

can be understood by "contamination" today in spaces such as cleanroom labs, bio labs, fermentation facilities, food plantations, and public institutions.

2. Liveliness. After seeing formation processes and anchor points in strolling inside of indeterminacy of the exhibition space in the Veterinary Anatomical Theatre, how do we investigate other-than-human species dealing vivaciously with orienting and acting in air matters and atmospheres? As recent human endeavors involving new technologies, the first air samplers thirteen kilometers above the ground are detecting airborne microbial activities rendering clouds as being biologically active.[31] It starts to prompt speculation about rendering harmful anthropogenic organic compounds carbon dioxide and methane not as hazards but rather as nutrients. Therefore, it is known that metabolic activities by airborne bacteria in suspension make clouds of the tropos- and atmosphere a lively place on our planet digesting in the same manner as on the litho-, hydro, and biosphere and living with—for instance—carbon, nitrogen, or sulfur cycles in the air. This perspective might be seen as a consequence of a contemporary attunement to atmospheric conditions.[32]

Computing these complex data models as a benchmark of digital processing is based on thermal layering, metabolic processes, and electromagnetic radiation to be mapped out in situ, from the ground, from inside, and from space. In the future, it is both expensive and uncertain whether we will be able to effectively study and understand the microorganisms present in the air, because the air has low density, significant temperature variations, fast movements, high volatility, and a considerable amount of radiation. These factors create technical challenges and stress that make it difficult and questionable to map the microbiome of the air and study its activities in a comprehensive manner.

We found fungal spores attached to grains of sand in the exhibition space by sampling and mapping. Another tiny actor within the exhibition was the bacteria *Pseudomonas syringae*, a plant pathogen that is also found in the upper troposphere and stratosphere. It cleverly deals with the oddity of phase transitions in water. On the ground,

---

see minute organisms with his handmade lenses; he later came to suppose that 'animalcules could be carried over by the wind, along with the bits of dust floating in the air' (Dobell, 1932)." Lacey and West, *The Air Spora*, 17.

31 Devices such as the Aircraft Bioaerosol Collector (ABC) system on the NASA C-20 A aircraft. David J. Smith et al., "Airborne Bacteria in Earth's Lower Stratosphere Resemble Taxa Detected in the Troposphere: Results from a New NASA Aircraft Bioaerosol Collector (ABC)," *Frontiers in Microbiology* 9 (2018), https://doi.org/10.3389/fmicb.2018.01752.

32 "[T]he global-scale transformation to which the anthropocenic moment is tracked is 'nowhere more evident than in the atmosphere' (Will Steffen, Paul J. Crutzen, and John R. McNeill, 2007, 616), . . . and this massively circulating designation of a human moment can be read not simply as the act of a sovereign and anthropocentric science but as a painstakingly developed attunement to an air condition." Timothy Choy and Jerry Zee, "Condition—Suspension," *Journal for Cultural Anthropology* (2015): 210–223: 217, https://doi.org/10.14506/ca30.2.04.

it likes to break off plant surfaces through ice crystallization to reach inner plant structures by releasing a protein as nuclei and therefore forcing water to freeze above 0 ° C.[33] If stronger winds catch it and lift it into the troposphere, it travels transcontinentally on a grain of dust. Finally, it gathers in the air and potentially controls the crystallization of water again into ice crystals to force so-called bioprecipitation and colonize new parts of the planetary environment, reflecting just one of many vibrant stories from the air and the exhibition space.

Again, other-than-human metabolic perspectives join approaches, questioning the human as a central force. However, for scientific investigation, other questions might be more pressing, such as who owns and profits by fostering metabolic processes in today's *atmosphere* by learning from airborne bacteria, seeing them as swarms of tiny metabolic machines. This was an important focus of the practices section on *mapping* and *fabulating.*

Connecting to high-tech solutions, it is even more interesting to note that 150 years ago, in 1854, Christian Gottfried Ehrenberg—the founder of the emerging field of aerobiology—called for the vitality of atmospheres and analyzed and classified collected dust particles as residues in his publication *Mikrogeologie.* He brought up categories of dust early on: biotic (pollen), mineral, and anthropogenic. Extraterrestrial dust is also measured today, which was also relevant for discussions in the exhibition space. Ehrenberg also presented three central principles of microbial activity: *growing* under suitable conditions, *inhibiting* during unsuitable situations such as on atmospheric flights in the stratosphere, and *decaying* due to community behavior or unsuitable conditions.[34] This is wonderful if one thinks of this as a potential for a designing-with approach to the liveliness of materiality overcoming passivating materials per se (fig. 9).

In her research project on microgeology and aerobiology at the Bundesanstalt für Materialforschung und -prüfung (BAM) for the Federal Ministry for Economic Affairs and Climate Action, our research colleague Anna Gorbushina described how dust collected by Charles Darwin was revived by her research team in 2008. Darwin collected residues from the sail of the Beagle while sailing near the Cape Verde Islands, and sent dust samples to Ehrenberg in 1846 to become part of the collection at the Museum für Naturkunde Berlin. Gorbushina and her team proved how inhibited spores at the Ehrenberg Collection of the museum were still able to grow while being attached to transatlantic dust particles 150 years later, forming biofilms with distinct colors of melanin and carotene. The latter pigments helped the spores to cope with high radiation levels in the stratosphere. Comparing this impressive microbial ability to humans being ex-

---

33 Ravindra Pandey et al., "Ice Nucleating Bacteria Control Order and Dynamics of Interfacial Water," *Sciences Advanced* 2, no. 4 (2016), https://doi.org/10.1126/sciadv.1501630.
34 Christian Gottried Ehrenberg, *Mikrogeologie: Das Erden und Felsen schaffende Wirken des unsichtbar kleinen selbständigen Lebens auf der Erde* (Leipzig: L. Voss), 1854–1856.

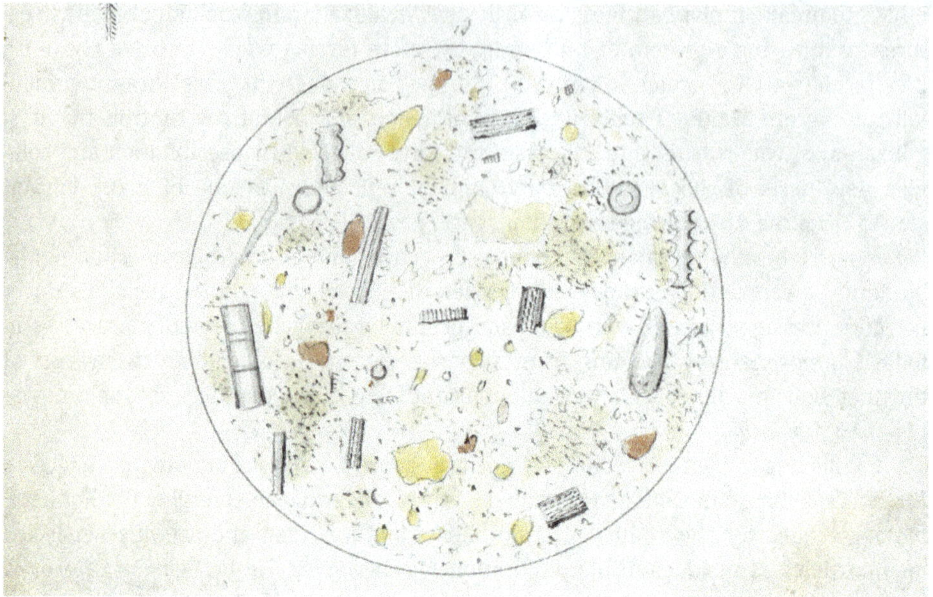

Fig 9: *Ehrenberg's illustration of dust collected by Charles Darwin on the Beagle* offshore Cabo Verde (at 17° 43N 26°W), January 1833.

posed to harsh—what we often call extreme—conditions hints at inhibition rather than growth or decay as an important principle of sustaining life.[35]

In comparison to satellite-sensing technologies to read out microbial journeys in the sky, Gorbushina and her research team further scratched local stone façades, e.g. of cultural heritages, to reveal colors of the biofilms woven by little microbes, depending on their flight duration and pathways. Melanin and carotene levels on the cell membrane can be seen as an analog approach to tracing microbial life besides satellite imaging techniques.

What are further adequate steps in tracing and understanding different forms of liveliness for and with exhibition sites in the making?

3. Scalability. The complication of such questions and the limits of scale lies at the heart of performing and understanding these processes of co-creation, requiring refined modes to account for our work as designers and scholars.[36] How does the contemporary human experience of smog and dust relate to various scales of airborne activ-

---

35 Calling extremophiles from a human perspective, surviving temperatures underneath -80° C, or high levels of UV radiation in stratospheric altitudes above 10 km; Anna A. Gorbushina et al., "Life in Darwin's Dust: Intercontinental Transport and Survival of Microbes in the Nineteenth Century," *Environmental Microbiology* 9, no. 12 (2007): 2911–2922, https://doi.org/10.1111/j.1462–2920.2007.01461.x.

36 Anna Lowenhaupt Tsing, "On Nonscalability: The Living World Is Not Amenable to Precision-Nested Scales," *Common Knowledge* 18, no. 3 (2012): 505–524.

ities interacting with climates, pollution, and exhaust? Based on Lucretius's thesis, nowadays researchers can investigate the extent to which one particle affects or connects to another particle through different scales. Nevertheless, it still allows us to think of air as dense and full of activity. What remains is a space filled with life. Focusing on air activity makes tracing different forms of liveliness across scales feasible, enabling us to understand and embody larger relationships and entanglements from the dust in front of us.

How can one scale of atmospheric phenomena connect to other scales, starting with soil moisture and vegetation at surface levels? How is this related to smaller scales of molecules of trace gases and the formation of dust particles up to the formation of cloud droplets, and even a larger scale of wind patterns, and planetary circulation patterns? In the framework of TA T, I showed how an activity is connected between the different scales of air: microbial activities, energetic states in a tangible cloud towards a better understanding of intercontinental weathers. Dust connected the nanoscopic world of molecules with the macroscopic world of human operations. What is the order of magnitude hereby?[37] Moreover, what are further references for measuring beyond what has been done in the cloud at TA T?
The exchange and travels across scales shift interfaces between human and environment, subject and object, and therefore the physical cloud as a condition of material and human activity in an experimental exhibition site. The direct involvement in certain conditions with other materials and species as companions in one atmosphere also bridges between scales into stratigraphic imaginaries.[38]

Concluding Remarks

In the open process of making, the TA T became an experimental laboratory for testing interactions of space, objects, humans and new technologies in the atmospheric milieu. The investigative design research incorporating scientific techniques and theatrical practices on various airborne nuclei can contribute to a proclaimed new culture of the material in this publication.

Here in the exhibition space of the Veterinary Anatomical Theatre, the ordinary material air is negotiated by various agents, vaporized and condensated, declared the object of empirical research, embracing what remains out of control. The demonstrated practical and theoretical figures of investigation opened the liveliness of making in atmospheric conditions, observations of clouds as boundary objects, and re-en-

---

37 Jens Soentgen and Armin Reller, *Staub—Spiegel der Umwelt* (Munich: Oekom Verlag, 2005), 124.
38 Astrida Neimanis, "Water a Queer Archiving of Feeling," in *Tidalectics*, ed. Stefanie Hessler (Cambridge, MA: MIT Press 2018), 191–195.

acting on residual leftovers as embodiments of knowledge production through empirical research.

In situated ways, certain atmospheric conditions in the cold indoor environment of the TA T—built for conservational purposes—led to various cloud formations, which carried particles of visitors, exhibits, and the building as nuclei for water condensation and energy exchange. Thus, atmosphere in the making led to understanding the exhibition side as a metabolism that was continuously negotiated by mapping, experiencing, and imagining.

In metabolic archiving, critical design epistemologies were formed not statically but performatively in the sense of constant exchange between figures of practice and thought. As a goal, iterative participatory formats within the indoor cloud were integrated to create new conditions for shifting from emotions such as fear or defeatism to acceptance, active listening, empathy, and caring.

This contribution ends with a call within "Matters of Activity" for transdisciplinary practices to think and feel various climatic conditions, forms of liveness and steadiness within ephemeral media and an invitation to observe and experience a walk through a body of air, vapor, and residual leftovers. The heightened sensitivity inside a cloud allowed us to imagine the powerful metabolic force of a cloud's microbiome, to comprehend entire ecosystems in their threat from humans even in the sky, which sooner or later flies back at us. What do we perceive? What do we associate with it?

The experimental, situated, and open archival approach, the cloud and winds in TA T, paired with the figures and forms of the practices described above, can be brought into the present as a site-specific theater for re-enactments of archiving. At the same time, it allowed for practical, experimentally guided pre-enactments with the forming material, energy, labor, and residue in metabolic exchange for further environments.

Acknowledgments:

I would like to thank my team members at the OSA research group for co-curating and the TA T for our testbed, as well as my family for spending many nights with me at TA T in the cloudy atmosphere. I thank Thomas Auer and Benjamin Maus for their constant technical advice in climate engineering, employing the importance of conduction, convection, and radiation as a base for further microbial testing; our microbiology department, especially Skander Hathroubi and Regine Hengge for sampling guidance; the Ehrenberg dust collection at Museum für Naturkunde Berlin for sharing their archiving and conservational practices; climate researchers Cornelia Auer from Potsdam Institute for Climate Impact Research and Tilo Arnhold from Leibniz Institute for Tropospheric Research for their important advice on the scalability of ecological pathways, aerosol research, and climate drivers; and anthropologist and museologist Sharon MacDonald for our fruitful mail correspondences during this contribution.

Maxime Le Calvé

"The Dance of the Knife." Ethnographic Reflections on Neurosurgical Experience and Agential Materialism

"A good butcher changes cleavers every year because of damage, a mediocre butcher changes cleavers every month because of breakage. I've had this cleaver for nineteen years now, and it has cut up thousands of oxen; yet its blade is as though it had newly come from the whetstone. . . . The joints have spaces in between, whereas the edge of the cleaver blade has no thickness. When that which has no thickness is put into that which has no space, there is ample room for moving the blade. This is why the edge of my cleaver is still as sharp as if it had newly come from the whetstone. . . . Even so, whenever I come to a knot, I see the difficulty to doing it. I am careful to remain alert, with my gaze steady. Moving slowly, I exert a very slight force, and the knot has come apart, like earth crumbling into the ground. Then I stand there with my cleaver, looking all around and pausing over the satisfaction in this. Then I clean off the cleaver and put it away."

The king said, "Excellent! Having heard the words of a butcher, I have found the way to nurture life."[1]

Introduction

Observing a neurosurgical team at work, I watch with the resident the tip of an instrument cutting out a tumor from a human brain.[2] The screen of the microscope is crisp, and the action is lively—the knife is moving within a microscopic site, with staggering stakes for the patient under the surgical covers, for the person holding the tool, for the team, and the institution around them. New to this field, my ethnographic mind is focused on the action, translating it to the movement of my own inscription devices. I have been hired to investigate the activity of the material pertaining to the practices of cutting in a neurosurgical context, at a world-class clinical research department. As I see the delicate nervous tissues in vivid colors being pushed around and aside speedily, I wonder at the number of human hours and human lives it took to get things moving so smoothly. The masterful healing cut brings to my mind a Taoist tale from a

---

1 Thomas F. Cleary, ed. and trans., *The Essential Tao: An Initiation into the Heart of Taoism through the Authentic Tao Te Ching and the Inner Teachings of Chuang-Tzu* (Edison, New Jersey: Castle Books, 1998), 48.

2 I would like to thank Joe Dumit, Laurence Douny and Yoonha Kim for their insightful remarks on this chapter, as well as Elisabeth Obermeier for her relentless editorial management. I am indebted to Joe in particular for steering me toward a deeper understanding to the work of Karen Barad. I am also grateful to Aaron Bogart for his thorough proofreading and reviewing work, and to Sabine Marienberg for coming to my rescue on a technical point during the revision. Thank you also to the team at Charité—Universitätsmedizin Berlin and especially for the warm welcome of Peter Vajkoczy, Thomas Picht, Anna L. Roethe, and Janett Grimm without whom this fieldwork would not have been the same.

book I like to read at night when I can't sleep. One of the stories of Chuang-Tzu, in part quoted above in the epigraph, introduces a master butcher whose instrument never gets blunt, as they have learned to cut around and with the bodies, throughout and alongside their lines. The knife is "dancing" within the animal body and the master is following it. The point of the story is hinting at a way of life which has merged with nature itself—it means in that context a life imbued with non-action, a way of life that doesn't impede the cycles and flows of the universe.[3] If the atmosphere in the operation room is quite relaxed, which is a cue to an arguably harmonious cosmic relation of the practitioner with the universe, the rest of the setup seems far from natural, and the heavy intervention that is happening would probably make a Taoist cringe. Or would it? The neurosurgeon at work, Peter Vajkoczy, likes to speak of humility in the neurosurgical practice. When he discusses and transmits his practice to others, he articulates action and nonaction in skillful ways: when there is a possibility of saving a patient with an intervention, deciding against the intervention is also an intervention. Stretching senses of the possible and bending the statistics of success is the baseline of his neurosurgical research. Vajkoczy is an influential proponent of considering neurosurgery as a dynamic and therefore daring field of investigation. Bringing the patient on-board is an act that requires charisma and a leap of faith, something that isn't purely medical skill. Or is it? Vajkoczy speaks of attracting talents to his team —and he knows the kind of teamwork that is necessary to take people to higher places. When people start speaking of a scientific practice as an art, we are at the limit of a delimited and theorized practice and we enter into an open area of reworlding and mystique.

Back on the fieldwork scene, I'm trying to gain critical distance facing the awe provoked by the surgical act; I survey the environing stage of the microscopic performance. Science and technology are everywhere I look: neatly ordered in category and procedures, the products and devices embed a hundred years of biomedical innovations, they are the infrastructures of what Byron J. Good has described as a reframing device for an emotional and human situation to be converted into a rational stage for biomedical operations.[4] The decision to operate on that patient was supported by a grid of criteria based on a review of studies with blind controls, itself published into a renowned peer-reviewed journal, tidying up a messy clinical reality into a workable decision process. A team of highly specialized professionals is at work, supporting a single person, and a single human hand, as the main channel of their definitive action. The sociologist Stefan Hirschauer argued that surgeons are sculpting the material to conform it to anatomical models.[5] Rachel Prentice complemented this view in the most elegant (ethnographic) manner: a lifetime of practice forms the hand of the sur-

---

3 I have to thank my colleague Shang Jing from the University of Tongji for this insightful remark.
4 Byron J. Good, *Medicine, Rationality, and Experience: An Anthropological Perspective*, The Lewis Henry Morgan Lectures 1990 (Cambridge ; New York: Cambridge University Press, 1994), 85.
5 Stefan Hirschauer, "The Manufacture of Bodies in Surgery," *Social Studies of Science* 21, no. 2 (May 1, 1991): 279 – 319, https://doi.org/10.1177/030631291021002005.

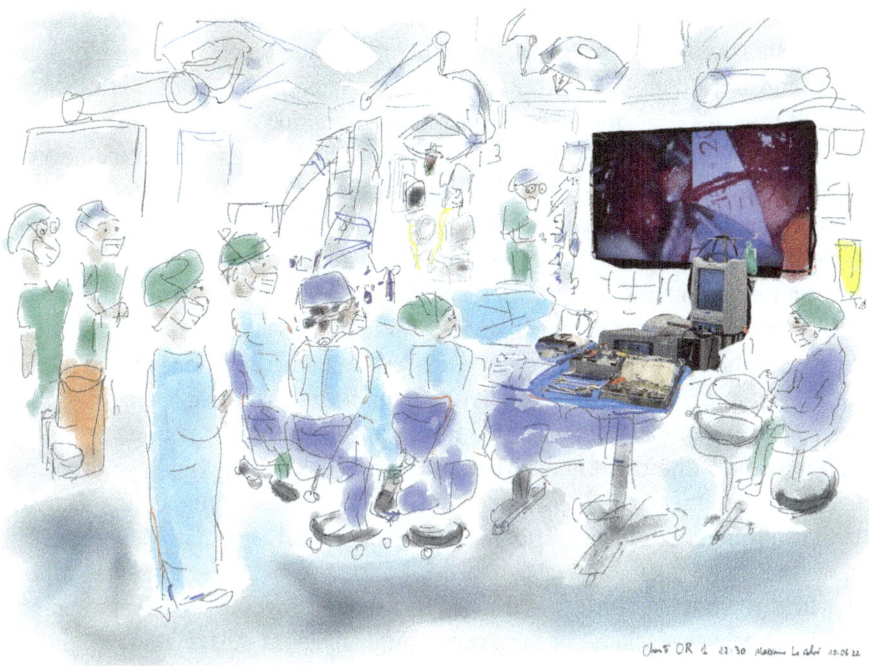

Fig 1: Peter Vajkoczy conducting vascular bypass surgery at Charité. Digital drawings by Maxime Le Calvé.

geon as much as the bodies of their patients.[6] And so does ethnographic practice: while I am sketching the scene, as I routinely do to capture the moment as an atmospheric note, the dancing tip of my digital brush catches my attention. The pen moves hectically as if by itself, like the knife of the surgeon, which brings me closer to the micro-scene broadcast on a big screen. I see no big consequences here for a wrong stroke, and certainly not the same kind of virtuosity at play, yet an improvised movement of note taking that has shaped me as much as it has shaped my work. It strikes me how the "middle voice" that arises in practice—not active, not passive, right in the middle—is bringing new dimensions of truth to light, and how making and knowing sometimes seem to flow.

With this text, I am sketching a direction for a larger project with an ethnographic study of the daily labor in the operation room. I attempt to pursue, in an essayistic mode, the idea of "cutting" as a micro-mystical practice: How do neurosurgeons stretch their senses and their sense of the world as they follow the cues of the material? I funnel the material through the lens of the practice of cutting as defined in two sources that I already mentioned above, one of Chinese origin and one from the Greeks, which I conceptually compare. I will also evoke the work of a number of scholars

---

6 Rachel Prentice, *Bodies in Formation: An Ethnography of Anatomy and Surgery Education*, Experimental Futures (Durham, NC: Duke University Press, 2013), 229–31.

who have written on the relation between the craftperson and the material—starting with Gilles Deleuze and Félix Guattari, over to Tim Ingold, Trevor Marchand, Lambros Malafouris and Don Idhe—on craft and on the interaction between the craftsperson and the stuff they are engaging with, as well as the extension of one's senses that happens through a technical device. I will use the notion that can be found in the new materialism literature and coined by Karen Barad in the context of her research on the history of experimental physics, that of "cutting together/apart." Language is tricky: it separates things into categories. Science has been caught in this trickiness, which extends through experimental apparatuses and discourses.[7] As the first section of this paper will show with a reading of two ancient parables, this seemingly antithetical notion of the cut, which bring things together, echoes far older reflections on the relation between subject and object. Insights into human embodied knowledge emerging from practices of cutting animal bodies can be found in the writing of two iconic authors from antiquity, one Chinese, Chuang-Tzu, and one Greek, Plato. The separation and the bond between them can be wonderfully emphasized when the link between them is the sharpness of a tool. Dichotomies are conceptual cuts that allow the classification of things into distinct categories.[8] They are central to modern scientific epistemology, which separates natural species, as Plato wrote, "carving nature at its joints," as one cuts a roasted chicken. As the French anthropologist André Haudricourt has noted, the way we handle animals and plants says a great deal about our relation to the world and to our fellow human beings.[9] Both Chuang-Tzu and Plato take on the same image of "cutting at the joints" of the animal body as a metaphor for the exemplary way of knowing. And yet, as we will see, their approaches are stunningly different. In a second step, I present a few graphic fieldnotes, sketches from my ethnographic survey of the neurosurgical cut, to illustrate some of the insights gained during my observation at the Charité Hospital among the neurosurgeons. The evocation of the fine skills of the craftsperson incites me to further the notion of skillful gesture as analytic in the context of digital mediations—exploring the premises of what I will call a 'stretching' of the master cutter's senses of the through training, scientific knowledge, institutions, and instruments.

---

7 Karen Barad, "Diffracting Diffraction: Cutting Together-Apart," *Parallax* 20, no. 3 (2014): 168–87, https://doi.org/10.1080/13534645.2014.927623.

8 As Joe Dumit pointed judiciously out to me, we should ask first if things are there before any cuts/ dichotomies/categories. Are humans the originators/senders of categories, or via Barad, is cutting together/apart a way of worlding worlds?

9 André-Georges Haudricourt, "Domestication des animaux, culture des plantes et traitement d'autrui," *Homme* 2, no. 1 (1962): 40–50, https://doi.org/10.3406/hom.1962.366448.

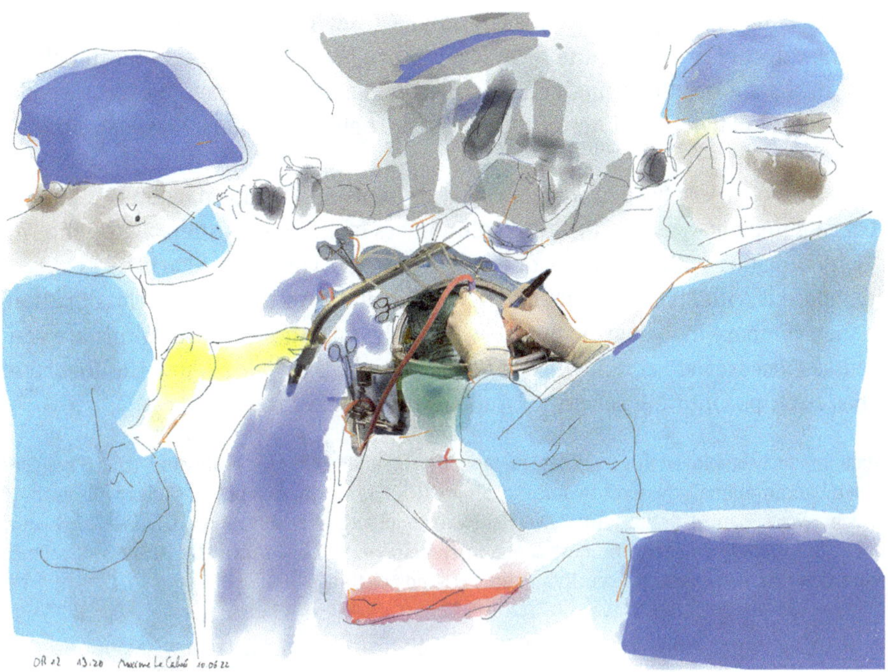

Fig 2: Surgeon operating with a resident at the binoculars.

I **Philosophical Butchery**

Let's start by reviewing and confronting two positions on cutting bodies from antique sources and their statement on the relation between knowledge and the cutting gesture. Both have to do with the idea of following lines in the structure of the body, therefore complying with a certain agency contained within them: a "sense" (direction) of the material that has to be acknowledged and "sensed" (perceived) by the cutter. Those two positions, however, differ starkly on the epistemic effect of the cutting action. On the one hand, a precursor to modern thought, Plato considers that dichotomies (meaning binary cuts) operated through language are justified by the preexistence of a structure of things in nature: the realist will "carve nature at its joints" to classify objects into categories according to certain characteristic behaviors or features, including their shape. On the other hand, the tale of the master butcher told by Chuang-Tzu speaks to the same action of cutting along the lines of the body, but with a very different conclusion: it is the sense of nature itself that is sharpened, rather than the knowledge or categories of the resulting pieces. The symptom of this progress is that the blade of the master will never go blunt, as the master butcher guides it or lets himself be guided by the knife in a sort of dance, always finding space within the tissues and, again, carving effortlessly at the joints. On the one hand, we have the philosophical butchery of cutting out species from each other. On the other hand, the unifying and integrating effect of mastering the technical gesture brings the subject and the object together in a state of inter-being and inter-becoming. Both texts agree on the fundamental idea of following the lines of the material. Their conclusions, however, couldn't be more different.

Plato speaks precisely of cutting nature at its joints. In the introduction to a book of 2011 bearing this quote as subtitle, the philosophers of science Matthew Slater and Andrea Borghoni start with the Taoist tale of the butcher who never sharpens his blade and juxtapose it with Plato's statement.[10] They are interested in scientific classification, which they compare to "scientific butchery," and the way these classifications are constructed as "discoveries" coming directly from nature. They tell us how the theory of Plato has been put into the service of a modernist ideal of science:

> Plato famously employed this "carving" metaphor as an analogy for the reality of Forms (Phaedrus 265e): like an animal, the world comes to us predivided. Ideally, our best theories will be those which "carve nature at its joints." While Plato employed this metaphor to convey his view about the reality of Forms, its most common contemporary use involves the success of science —particularly, its success in identifying distinct kinds of things.[11]

In the ensuing paper, they write at length on the paradigm of "discovery" as opposed to that of "invention": modern scientists have the ambition to "cut the chicken" (nature) in the most efficient way, which should be dictated by the chicken itself, as there aren't many ways of cutting it into distinguishable parts. The first example they take comes from ornithology: Ernst Mayr, long-time authority on the question of species, claimed that all people of the world recognize species of nature "as a western scientist."[12] While philosophers are trying to enforce the "skeletal structure" of nature, historians and philosophers of science are challenging the "naturalness" of natural kinds.[13] Indeed, when analytical philosophy takes over, the Taoist tale's insight is lost for good in the exploration of properties, categories, and essences of things and of what Slater and

---

10 Matthew H. Slater and Andrea Borghini, "Introduction: Lessons from the Scientific Butchery," in *Carving Nature at Its Joints*, ed. Joseph Keim Campbell, Michael O'Rourke, and Matthew H. Slater (Cambridge, MA: MIT Press, 2011), 1–3, https://doi.org/10.7551/mitpress/9780262015936.003.0001.
11 Slater and Borghini, "Introduction," 1.
12 Things are getting more complicated with the recent breach of speciation genomics, now that "the molecular basis of the splitting process" is within reach of the researchers, see Jochen B. W. Wolf, Johan Lindell, and Niclas Backström, "Speciation Genetics: Current Status and Evolving Approaches," *Philosophical Transactions of the Royal Society of London: Series B, Biological Sciences* 365, no. 1547 (2010): 1717–33, https://doi.org/10.1098/rstb.2010.0023.
13 Slater and Borghini, "Introduction," 2. See in particular David L. Hull, *Science as a Process: An Evolutionary Account of the Social and Conceptual Development of Science*, (Chicago: University of Chicago Press, 1990); Geoffrey C. Bowker and Susan Leigh Star, *Sorting Things Out: Classification and Its Consequences* (Cambridge, MA: MIT Press, 2008), 8.

Borghini call the "sortal" tradition—the modern scientific custom of separating things and species into distinct categories, which they consider a rebirth of essentialism.[14]

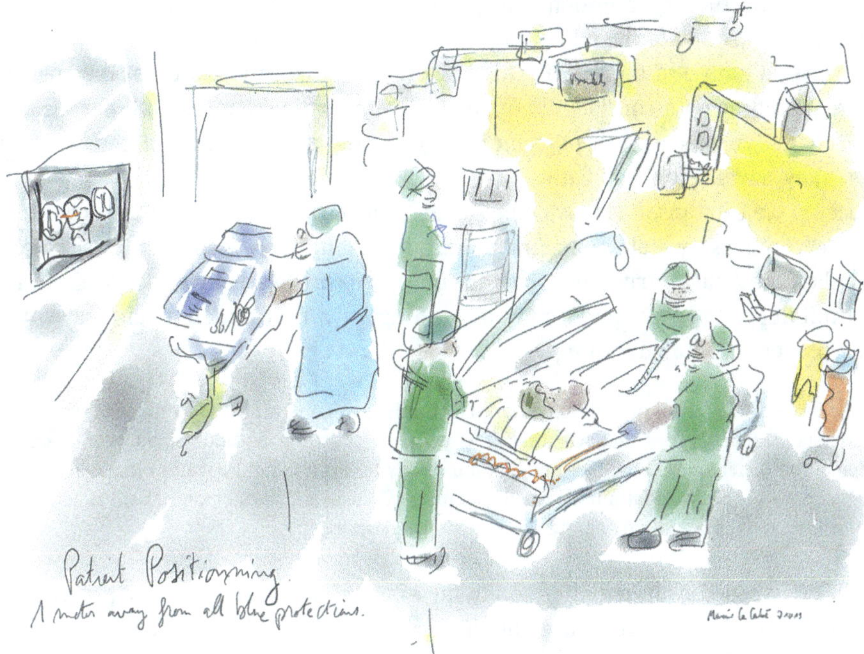

Fig. 3: Excerpt from fieldwork journal at the neurosurgery department: preparing an unconscious patient.

Returning to the scientific butchery essay of Slater and Borghoni with the Taoist story in mind, I couldn't rid myself of the impression that there is so much more to learn from the Chinese tale than just a quick parallel with the metaphor of Plato. What about the actual cutting and carving that scientists practice in their everyday work to make dichotomies and advance their knowledge? The "dance of the knife" that is captured so vividly in a few sentences by Chuang-Tzu points in other directions: What does it mean to follow the practical lines of the materials, so that our blade never becomes blunt, and the activity of the material, its resistance, reactions, immediate and in the longer term, are taken into account in the description and experience of the cutter? Moreover, the continuity existing along the line traced by the edge of the knife implies a relation between the cut, the cutter, and the hand/tool in between: the Taoist story highlights the fact that they are becoming one. This relation contrasts with the idealist notion that the different characters of the scene and the form that is being discovered have always existed and will keep on existing as essences whose "joints" would simply have to be distinguished. The parable of the master butcher is not

---

14 Slater and Borghini, "Introduction," 7.

about the dichotomy practice, which traces a line between already existing territories. To take the example of the neurosurgeon, this would be the ideal of an intervention planned on screen, with a 3D visual of the tumor appearing in its non-tumorous surroundings. The actual intervention, both in the case of the butcher and the neurosurgeon (and perhaps that of the ethnographer), is rather about letting the boundaries appear through a path and a practice. This is a process of co-constitution of the cut and the cutter, building on a consistently minded encounter that changes both parties. The Taoist Chuang-Tzu repeats it throughout his work: the practitioners get closer to nature itself as they advance on this path.

Because we can't productively embrace the Taoist practical philosophy of nature in the frame of this essay, I am appealing to Barad as an unlikely mediator. Physicist and philosopher, they have turned to the quantum physicist Niels Bohr to derive from his work a full-fledged process philosophy, advocating for a new understanding of nature and the process of knowing things.

> [T]he heart of the lesson of quantum physics (is that) we are a part of that nature that we seek to understand. Bohr argues that scientific practices must therefore be understood as interactions among component parts of nature and that our ability to understand the world hinges on our taking account of the fact that our knowledge making practices are social-material enactments that contribute to, and are a part of, the phenomena we describe.[15]

Barad has developed renewed attention to the effect of material-discursive practices in epistemic work. "Agential cuts" are operated by scientists as they construct categories and embed them into their apparatus. Inspired by the foundational reflections of Niels Bohr on the impossibility of separating scientific fact from scientific devices, the physicist-philosopher insists on the idea that we are always "meeting the universe halfway." In a move that the butcher of Chuang-Tzu would approve of, they read into this phenomenon "an invitation that is written into the very matter of all being and becoming."[16] The intensity of this encounter can be more or less sharp according to the practice and experience that is put into the technical gesture. Whereas Plato and the moderns do not let any sort of moral leak into their cutting story—although it is intensely moral as it polices expertise—this is not the case with the Taoist story, for which mastery stands for knowledge, practical wisdom, and inspiration for a way of life. The act of knowing relies on the production of a difference that is, as such, a "cutting together/apart," which is firstly engendered by the devices and the discourses that are embedding the practice.

---

15 Karen Barad, *Meeting the Universe Halfway: Quantum Physics and the Entanglement of Matter and Meaning* (Durham, NC: Duke University Press, 2007), 12.
16 Ibid., 12.

> Apparatuses enact agential cuts that produce determinate boundaries and properties of "entities" within phenomena, where "phenomena" are the ontological inseparability of agentially intra-acting components. That is, agential cuts are at once ontic and semantic.[17]

Barad calls this knowledge process a "mattering process." [18] This expression is a double entendre, hinting at the fact that this process is both giving matter its conceptual consistency and driving our sense of importance to a new, specific pole of what will matter to us in this newly forged circumstance. They describe the process of the production of knowledge in laboratories and elsewhere as a specific form of their "dance"—to quote them again: "Meaning and matter are more like interacting excitations of nonlinear fields—a dynamic, shifting dance we call science."[19]

These remarks on cutting will, hopefully, help me to explicate certain tensions that I have found in my ethnographic observations of the practice of neurosurgery: a cohabitation between the protocols and requirements of scientific experimentation, creativity, improvisation, and the attunement between the neurosurgical body and that of its patients. As they get their senses stretched by the practice, their relationship to the contingencies of the world is reshuffled and becomes unrecognizable—not in an exceptional event but in the daily grinding and drilling of clinical work.[20]

II Driving a Tipping Point through the Brains

The neurosurgical cut sits at the pinnacle of modern physiological science. The case of neurosurgical skill is extreme in many respects. Let us swiftly go through a review of a few ethnographic insights gained during fieldwork at the Department of Neurosurgery of the Charité Hospital in Berlin.[21] First, the "material" the surgeons are cutting is the flesh of a person: each central nervous system is embedded in an "ecology of relations," as Thomas Fuchs, and Gregory Bateson before him, have called it.[22] A mistake can have disastrous consequences on the life of that human being. While they are operating, they are aware of almost constantly standing at a tipping point between heal-

---

17 Ibid.

18 Ibid.

19 Ibid.

20 Although exceptional events also occur and are sometimes reported in the literature, see the opening scene of James R. Doty, *Into the Magic Shop: A Neurosurgeon's Quest to Discover the Mysteries of the Brain and the Secrets of the Heart* (New York: Avery, 2016).

21 My fieldwork focuses mostly on the perspective of the Image Guidance Lab, which gave me an overview of the current research on neurosurgical imaging and planning practices at this hospital. During this time, I could also observe several interventions, document experimental training, and shadow several members of the clinic, drawing graphic field notes.

22 Thomas Fuchs, *Ecology of the Brain: The Phenomenology and Biology of the Embodied Mind* (Oxford: Oxford University Press, 2018); Gregory Bateson, *Steps to an Ecology of Mind: Collected Essays in Anthropology, Psychiatry, Evolution, and Epistemology* (Northvale, N.J: Aronson, 1987).

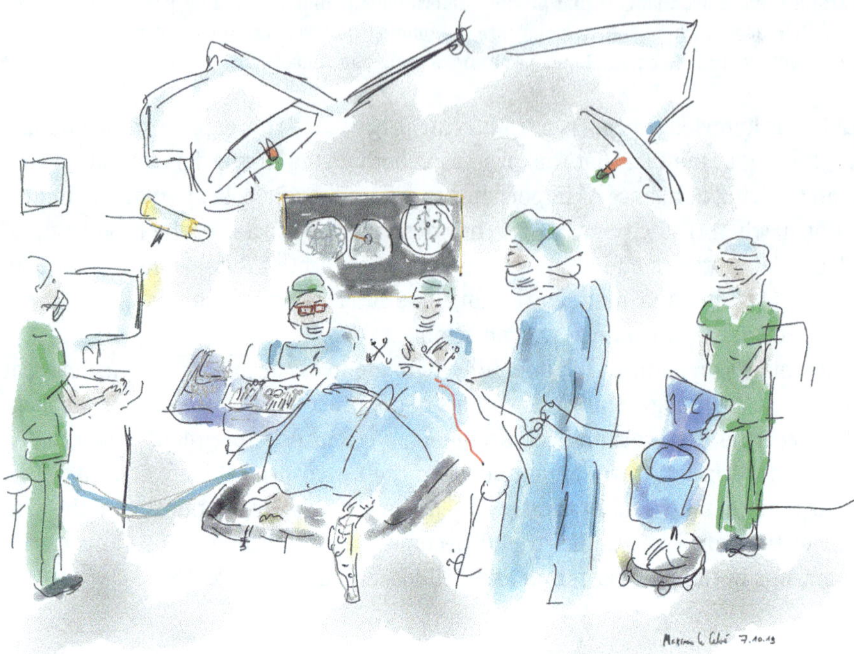

Fig. 4: Excerpt from fieldwork journal at the neurosurgery department: preparing an unconscious patient.

ing and harming, handling it with a delicate balance of caution and courage. Second, contrary to what I believed in the beginning (as well as most laypeople I present my research to), most neurosurgical operations are conducted by hand. The noninvasive techniques of robot-assisted surgery have, at the time of writing this chapter, very limited application in this domain, because of multiple difficulties related to the terrain itself. Brains are an instable and sometimes explosive material for the surgeons: The "brain shift" that occurs at the moment of opening the skull shuffles the tissues because of the difference of pressure. This scrambling is further aggravated as the operation moves on. It makes imaging devices incapable of rendering a precise cartography of the area of intervention, since things will move as the surgeon reaches the target. Imaging possibilities during the operation are making progress, but it requires interrupting the operation for a moment and to wheel out the patient in a machine, or to wheel in the machine to the patient. Handheld technologies are on their way, using ultrasounds, but real-time picturing seems to be out of reach.[23] Another cause

---

23 R. M. Comeau et al., "Intraoperative Ultrasound for Guidance and Tissue Shift Correction in Image-Guided Neurosurgery," *Medical Physics* 27, no. 4 (April 2000): 787–800, https://doi.org/10.1118/1.598942; David G. Gobbi et al., "Correlation of Pre-Operative MRI and Intra-Operative 3D Ultrasound to Measure Brain Tissue Shift," SPIE Proceedings: Medical Imaging 2000: Ultrasonic Imaging and Signal Processing 3982 (April 2000): 77–84, https://doi.org/10.1117/12.382260; Sean Jy-Shyang Chen et al., "Validation of a Hybrid Doppler Ultrasound Vessel-Based Registration Algorithm for Neurosurgery," *International Journal*

for the general use of noninvasive technologies in neurosurgery is the disastrous effect of bleedings—countermeasures to interrupt them in the delicate tissues are themselves a risky operation and make for the most stressful, tedious, and repetitive tasks of all. Third, the microscopic scale of the operation site is a miracle of human dexterity. These surgeons operate with the help of microscopes, which they fluidly pilot as prosthetics of their ocular organs. The precision of the tissue removal is of sub-millimeter accuracy when approaching the most critical areas, in particular during tumor resection within "eloquence areas." To access the operation site, they take "streets" that they have determined in advance, finding their way through the meanderings of the cerebral matter. And when they have to go through a part of the neocortex to access a tumor, they use a mini brain blaster device operating with ultrasound. Using the clinical equivalent of a vacuum cleaner, the assistant casually sucks out a person's liquified brain, and a wrong move at the wrong moment would have, again, calamitous consequences on the life of the patient, and on the career of the assistant. Fourth, they are engaged in teams that are enabling their tiny moves at the current limit of the hand-eye sensing scale, ensuring the stability of the life signals of the patient, the flawlessness of the aseptic environment, the continuous motion of instruments and products circulating around the room, the reanimation of the patients from earlier in the day, and the preparation of the patients next in line, while the surgeon remains steadily focused on the task, often at the workbench for more than ten hours in a row.[24] Fifth, they are inheriting and pushing forward their predecessors' tradition and experience that is incorporated in their training, in the way they frame anatomical problems, react to the unexpected, and in the way they hold their instruments.[25] Neurosurgery is a medical science on the move. The neurosurgeons of the team are researching as they are conducting their surgical practice, pushing the limits of the operable, and if they argue—controversy is the motor of any modern science —, clinical success will settle any discussion. They are pursuing that tipping point, chasing the next level of the possible cut, endlessly training and growing to the task. This slow and collective process of building up and stretching of the senses, generation after generation, brings them to find their bearings and to

---

of Computer Assisted Radiology and Surgery 7, no. 5 (September 2012): 667–85, https://doi.org/10.1007/s11548–012–0680-y.

24 In his recent biography, Peter Vajkoczy compares the process of resecting a veinous malformation to a legendary battle of man and fish, never quite over until it's over, *"Fisch raus, Fisch rein." Kopfarbeit: ein Gehirnchirurg über den schmalen Grat zwischen Leben und Tod* (Munich: Droemer, 2022). The sociologist Hirschauer has delivered an unmatched account of the distribution of tasks and attention in the surgical moment: Stefan Hirschauer, "The Manufacture of Bodies in Surgery," *Social Studies of Science* 21, no. 2 (1991): 279–319, https://doi.org/10.1177/030631291021002005.

25 The "Penfield dissector," for example, is an instrument that looks like a hook with a scoop, which was invented and named after one of the pioneers of the discipline, giving its descendants direct access to his gestural innovations incorporated into the tool. See Morenikeji Buraimoh et al., "Origins of Eponymous Instruments in Spine Surgery," *Journal of Neurosurgery: Spine* 29, no. 6 (2018): 696–703, http://doi.org/10.3171/2018.5.SPINE17981.

dealing with the increasing level of contingency that comes along with deeper, more complex, and invasive interventions into the brain material.

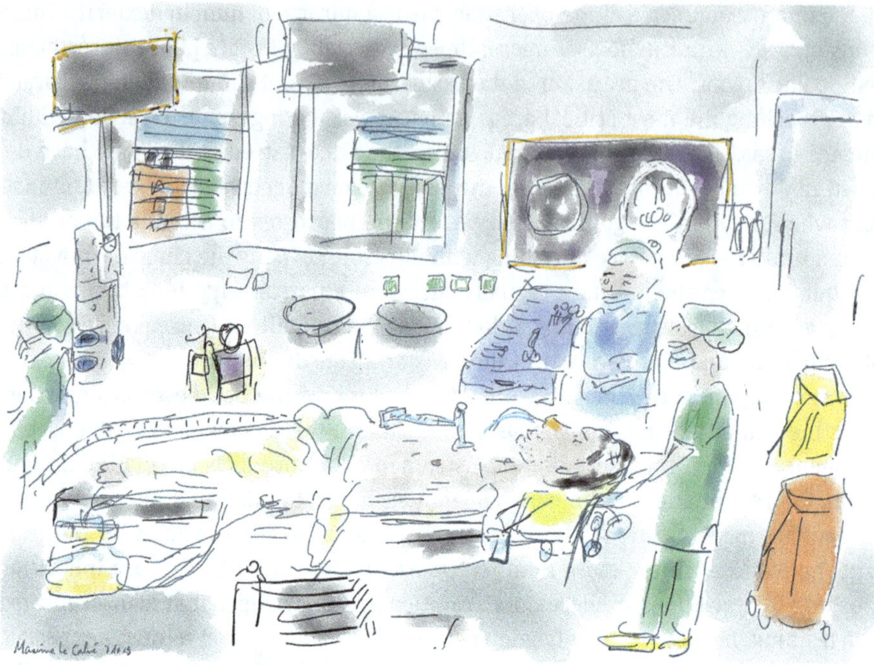

Fig. 5: Excerpt from fieldwork journal at the neurosurgery department: preparing an unconscious patient.

The evocation of the fine skills in the craftsperson's gesture leads me to refer to the contribution of Tim Ingold and other authors who have examined the question of thought in action, doing, and thinking, taking the notion of "skillful" gesture as analytic. Taking inspiration from archaeology, this literature focuses exclusively on low-tech environments, making them difficult to use as an exclusive framework to describe that sort of distributed skill.[26] I will refer to other thinkers in the field of technology studies, to comprehend this high-tech environment as a setup that is "stretching the senses": an extension of the human perception and intervention to more-than-human scales. My intention is to embrace elements of the literature considering technologies as prosthetics—including imaging and modeling—while keeping up with a phenomenological account of the neurosurgical practice. The hybrid digital/analog apparatus thus comes together in a single sensory experience as both the cutter and the cut are growing down

---

26 See also Lambros Malafouris, *How Things Shape the Mind: A Theory of Material Engagement* (Cambridge, MA: MIT Press, 2016). On the articulation between postphenomenology and the theory of material engagement, see Don Ihde and Lambros Malafouris, "Homo Faber Revisited: Postphenomenology and Material Engagement Theory," *Philosophy & Technology* 32, no. 2 (2019): 195–214, https://doi.org/10.1007/s13347–018–0321–7.

the scaling, reaching a state of micro-alertness to the contingencies of the environment.[27] This state is best grasped by conferring to the knife the status of an extension of both of the surgeon and the body. Thus, the touching is also touched within this device.[28] The knife is "dancing" the dance of the knife, following the impulses of both the material and the neurosurgeon. Perception is stretched, impulses are reverberated in the sensors of the magnetic resonance imaging devices, in the graphic processors, the architecture of the room, and the whole building. The many probings, trials and errors, multiple disciplines and design studios, the generations of medical specialists, and the career of each individual practitioner all are present within the encounter of the tool and the living tissue. None of them would exist without that instance. Cutting brings all these people and things together. Cutting is relating because the hand and the material become one at that precise moment: matter and activity are coproduced.[29]

III Stretching the Senses

Learning to dance with the knife is a matter of securing and fastening all possibilities, yet it is also a matter of embracing the instability as a fundamental element of this relation, and to ride it until mastery is achieved. This achievement is a tipping point. Another world becomes navigable and makes sense, not only in the visual domain but with all the senses. In the work of the anthropologist Tim Ingold, this sort of dance

---

27 This amplification of and coming-closer-to the object of attention is a learning process, a growing rather than a reducing (and reduction is what modern science would label it); on this point see in particular the anthropological argumentation of Hallam and Ingold in the introduction to Elizabeth Hallam and Tim Ingold, *Making and Growing: Anthropological Studies of Organisms and Artefacts* (London: Routledge, 2016), https://doi.org/10.4324/9781315593258.

28 From the perspective of the material engagement theory, see Lambros Malafouris and Maria-Danae Koukouti, "Where the Touching Is Touched: The Role of Haptic Attentive Unity in the Dialogue between Maker and Material," *Multimodality & Society* 2, no. 3 (September 1, 2022): 265–87, https://doi.org/10.1177/26349795221109231; and from the perspective of quantum physics and queer studies: Karen Barad, "On Touching—the Inhuman That Therefore I Am," *Differences* 23, no. 3 (December 1, 2012): 206–23, https://doi.org/10.1215/10407391–1892943.

29 Prehistorians have attributed to cutting and the making of cutting tools one of the major first steps of humans out of their animal condition, by entering the realm of the symbolic (chipping stones through systematic action), and hence developing frameworks of knowledge of trade and transmission. The production of cutting tools has also created an edge over other species with the introduction of tools in general. With the surplus that could be taken from the environment, specialization and further ensued, in a process that we used to call "civilization." Modern technologies and ecological disaster shouldn't be too easily lumped together however; from the perspective of quantum physics, see Karen Barad, "On Touching—the Inhuman That Therefore I Am," *Differences* 23, no. 3 (December 1, 2012): 206–23, https://doi.org/10.1215/10407391–1892943; and from the perspective of the material engagement theory in the field of anthropology, Lambros Malafouris and Maria-Danae Koukouti, "Where the Touching Is Touched: The Role of Haptic Attentive Unity in the Dialogue between Maker and Material," *Multimodality & Society* 2, no. 3 (September 1, 2022): 265–87, https://doi.org/10.1177/26349795221109231.

of "making" is a recurrent topic. In one of his most famous texts, he speaks of the act of weaving a basket as a way of feeling and *doing* with the willow branches, their resistance.[30] A correspondence with the world is established during this activity, as one understands in the most practical way that the basket is an assembly of forces, holding together because of the strength accumulated in the stems in a self-contained winding motion through the rhythmical action of the weaving hands. Ingold elaborates on this when he moves on to a description of another hands-on activity conducted with his students during his anthropology seminars: when making and flying a kite, one becomes slowly aware that the wind and the entire atmosphere is part of what we are composing and playing with, part of a perceivable and knowable world that the human senses have grown in correspondence with.[31] The kite becomes then an extension of the sensing self. More recently, as he was making and flying kites again in the company of Jennifer Clarke, Ingold reportedly went on to utter another brilliant remark: "[W]hen you are flying a kite, you are also flying in a way, although not like in a plane."[32] Feeling and sensing the air stream and its rhythms, the kite flyer extends and stretches their senses inward and upward, feeling through the wind streams and the intricacies of their moving boundaries.

The tool as a sensorial extension of the body is a theme that has been explored by many thinkers, such as Niels Bohr, Gregory Bateson, Maurice Merleau-Ponty, or Jean-Pierre Warnier. The most common figure to jam around this theme is the walking blind person.[33] Dis/ability studies scholars and blind phenomenologists have contested the use of this figure of the "blind man," which stands for a commonplace ability to synthetize the world through the tools we use.[34] As Niels Bohr noticed, the stick suddenly disappears as an object—a phenomenon that helped him to convey his understanding of the scientific apparatus and of the way the knower and their knowing devices, as well as the known, are intrinsically entangled. Stretching their canes in front of them, the blind sense the floor through the relation they establish along the staff,

---

30 Tim Ingold, *Making: Anthropology, Archaeology, Art and Architecture* (London: Routledge, 2013).

31 See among the many occurrences of the kite narrative in Ingold's oeuvre: Tim Ingold, "The Textility of Making," *Cambridge Journal of Economics* 34, no. 1 (2010): 95–96, https://doi.org/10.1093/cje/bep042; and Ingold, *Making*, 98–100.

32 Personal communication with Jennifer Clarke, November 2021.

33 Warnier and his working group "Matière à Penser" have conducted a vast number of studies on the relation between thinking and doing, often referring to the first thorough investigation of the body image authored by P. Schilder, *The Image and Appearance of the Human Body* (Oxford: Kegan Paul, 1935). See Urmila Mohan and Laurence Douny, ed., *The Material Subject: Rethinking Bodies and Objects in Motion* (York: Routledge, 2021).

34 Here, I take cues from a forthcoming paper by Asaf Bachrach and Joe Dumit, which Joe Dumit generously sent me along with his insightful comments: see Joel Michael Reynolds, "Merleau-Ponty, World-Creating Blindness, and the Phenomenology of Non-Normate Bodies," *Chiasmi International: Trilingual Studies Concerning Merleau-Ponty's Thought* 19 (2017): 419–34, https://doi.org/10.5840/chiasmi20171934; Jesse Workman, "Phenomenology and Blindness: Merleau-Ponty, Levinas, and an Alternative Metaphysical Vision," PhD diss. (University of Denver, 2016), https://digitalcommons.du.edu/etd/1210.

with swift movements that are prescient of the potential trajectories of their bodies, which "expresses the power we have of dilating our being in the world, or of altering our existence through incorporating new instruments."[35]

The American philosopher of technology Don Ihde has wrapped his mind around the problem of sensing and experiencing through technological devices for decades. One of the first and simplest examples that he has provided is a clinical examination tool.[36] The dental probe, which we all intimately know from the inquisitive metallic touch on the enamel of our teeth, does not only extend the fingers of the dentist to inaccessible corners of the buccal cavity, it also amplifies the sensation at the surface of the tooth. Through the sharpness and hardness of the metal and their quick probing actions, practitioners acquire a precise haptic image of the tooth, including texture, resistance, shape, and depth of cavities, all dimensions that are only made possible through this equipment. Just like the surgical blade, the dental probe is a historical artifact in which the dental knowledge comes together with the immanent senses of the dentist to produce her diagnostic and set her on a course of action. In this case, the senses aren't expanded in all directions, but rather, they are stretched into a precise direction, following the "pathway" of the handle. Moreover, senses become other, non-correspondent even, in enabling new forms of worlding to arise.[37] Does the sensitivity stop at the surface of the tooth, or does it go deeper? Like the kite flyers, the dentists and their probes are "becoming one" with the material, with its current dynamic state, they sense the internal structure of the dentin and the way it may degrade if the cavity lurks further. And so are the neurosurgeons as they build prosthetic scaffoldings around a fragilized spine, working with special alloy plates and screws, drilling into bones and preserving marrows.

At the risk of stretching my concept of stretching senses a bit far, I would like to include another means of accessing invisible and inaccessible realities. What about the magnetic resonance imaging devices used by the neurosurgeons to plan their navigation within the brain?[38] The technological development I traced during my early fieldwork at the university hospital is the translation of a technique used in neuroscience into a surgical planning method. Tractography is a practice that delineates the neural network. The team of Thomas Picht at Charité focused on speech-related tracks, in order to integrate the spatial information of the "fiber bundles" of the

---

35 Maurice Merleau-Ponty, *Phenomenology of Perception*, trans. Donald Landes (New York: Routledge, 2013), quoted by Dumit and Baschraff, forthcoming.

36 Don Ihde, *Experimental Phenomenology: Multistabilities*, 2nd ed. (Albany: State University of New York Press, 2012), 102 – 3.

37 See Workman, "Phenomenology and Blindness" and again my gratitude goes to Joe Dumit for his invaluable suggestions and hints for further developments.

38 Things become even more tricky when we look at the modeling practices, imported from linguistic computational neurosciences, that are now making their entrance into the domain of neurosurgical planning, putting sci-fi-grade claims to absolute control of the consequence of the cut, e.g., Friedemann Pulvermüller et al., "Biological Constraints on Neural Network Models of Cognitive Function," *Nature Reviews Neuroscience* 22, no. 8 (2021): 488 – 502, https://doi.org/10.1038/s41583 – 021 – 00473 – 5.

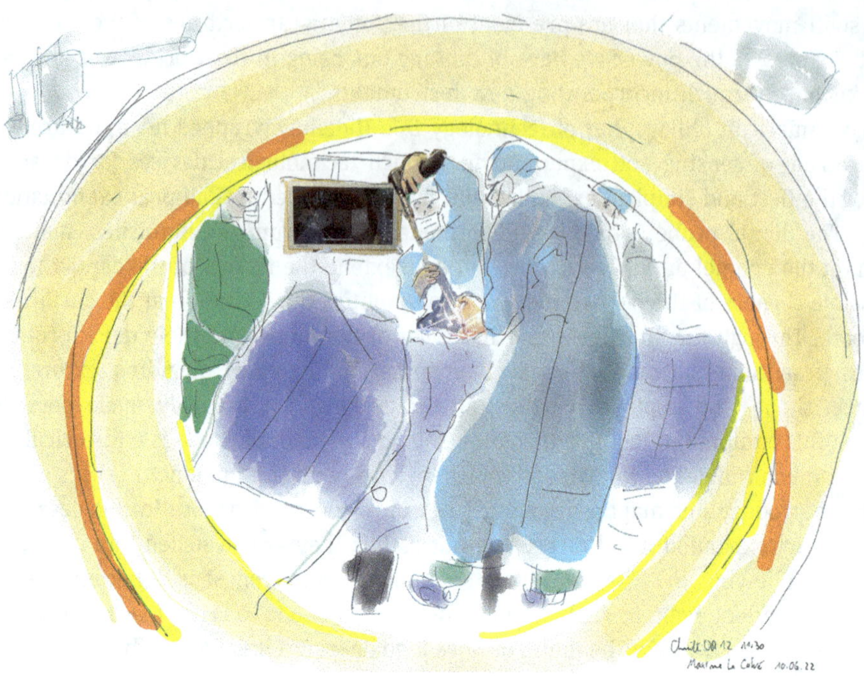

Fig. 6: Image-guided vertebra drilling as seen through intraoperative scanner.

brain into the navigation data.[39] The "tracks" and "hubs" they form outline a set of critical zones for the neurosurgeons: the brain can compensate for many disruptions, but the functions will "travel" to other places in the brain if, and only if, these pathways are preserved. The drama of a patient that can't move a limb or can't speak a word is a daunting threat in the daily job of the neurosurgeon, and the worst nightmare of the neurosurgical patient. This is why the novel technique of tractography is currently spreading, as one more factor that surgeons can be aware of—if they make themselves attentive to it. I want to claim that the new tractographic practices are a stretching of the senses as well.[40] Again, Don Ihde has provided some insight regarding what

---

39 Lucius Fekonja et al., "Manual for Clinical Language Tractography," *Acta Neurochirurgica* 161, no. 6 (June 2019): 1125–37, https://doi.org/10.1007/s00701–019–03899–0; Lucius S. Fekonja et al., "Detecting Corticospinal Tract Impairment in Tumor Patients with Fiber Density and Tensor-Based Metrics," *MedRxiv* (November 3, 2020), https://doi.org/10.1101/2020.10.28.20220293; Mehmet Salih Tuncer et al., "Towards a Tractography-Based Risk Stratification Model for Language Area Associated Gliomas," *NeuroImage: Clinical* 29 (January 1, 2021): 102541, https://doi.org/10.1016/j.nicl.2020.102541; for an ethnographic portrait of the principal map maker of this project, see Maxime Le Calvé, "Intersecting Cartographic Imperatives. Map-Making Practices of a Medical Artist in the Wake of the Computational Brain," *Kunstlicht* 41, nos. 2–3 (2020): 81–90.

40 The digital twin as a model from the engineering world stands for an all-encompassing simulation environment; the "discontents" of Sherry Turkle are appearing in the most splendid way. Sherry Turkle, ed., *Simulation and Its Discontents* (Cambridge, MA: MIT Press, 2009).

he called the "the understanding of science as an embodied technoscience, the instrument-embodied science of the contemporary world."[41] The work of Ihde aims to bring into the picture of the interpretative activity of science the "sensory interpretative activity" of its practitioners, an interpretative activity that is not related to words but images: a *"thing interpretation* through imaging instruments."[42]

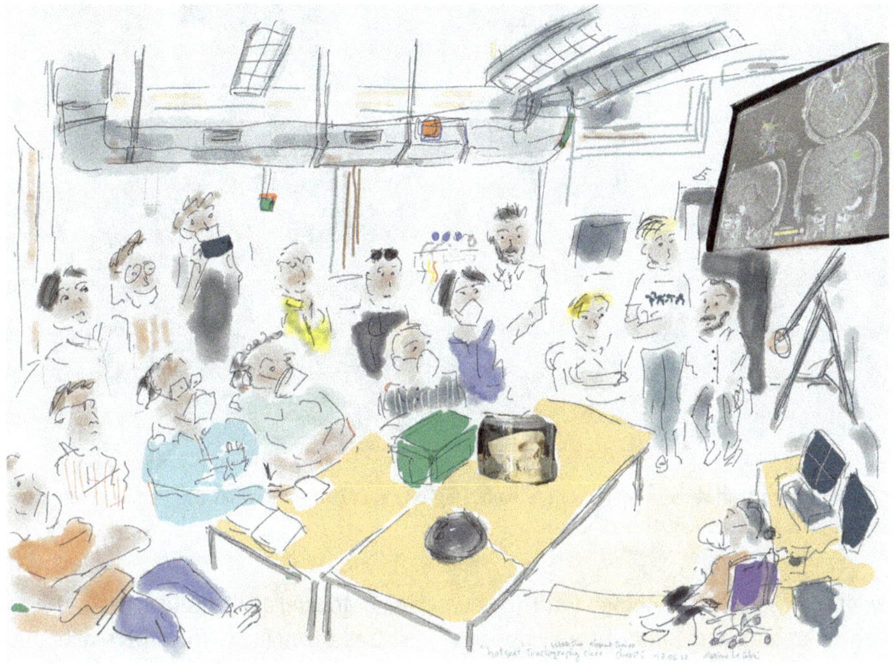

Fig. 7: Excerpt from fieldwork journal during a workshop on "eloquent tumors" at the Image Guidance Lab, Charité—Universitätsmedizin Berlin.

Under the umbrella concept of what they call "haptic creativity," anthropologists Natascha Myers and Joe Dumit have described a stream of experiments led by interdisciplinary teams of computer scientists, designers, and scientists to create a dynamic working environment into which geologists are invited to dive down into scalable models of their data.[43] Based on ideas developed by Myers, who worked on the embodied thinking of protein modelers, they show how the experience of data in virtual reality representation changes the way scientists relate to the material, triggering new insights.

---

41 Don Ihde, *Expanding Hermeneutics: Visualism in Science* (Evanston, IL: Northwestern University Press, 1998), 4.

42 Ibid., 8, emphasis in the original.

43 Natasha Myers and Joe Dumit, "Haptics," in *A Companion to the Anthropology of the Body and Embodiment,* ed. Frances E. Mascia-Lees (Chichester & Malden: Wiley-Blackwell, 2011), 239 – 61, https://doi.org10.1002/9781444340488.ch13.

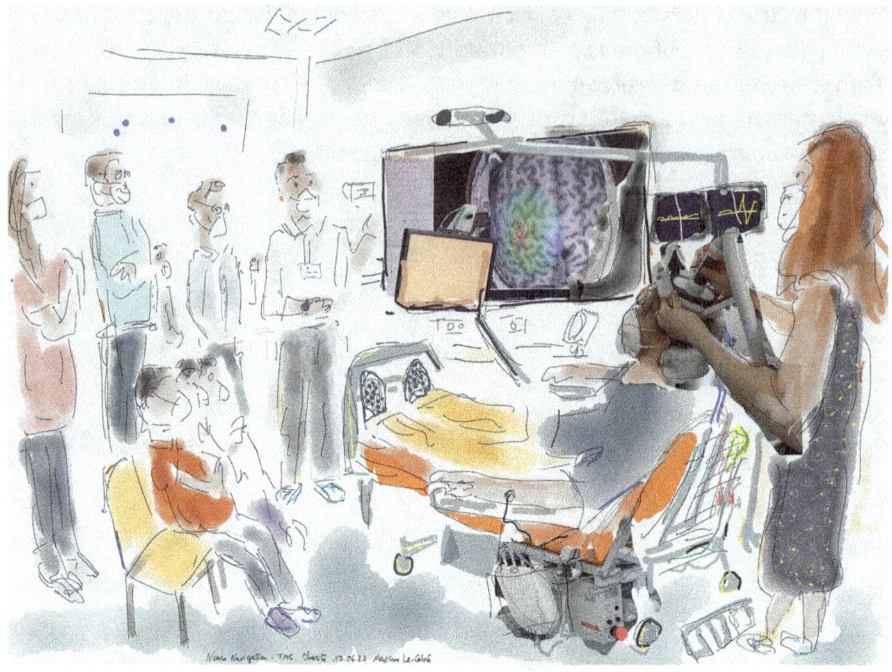

Fig. 8: Excerpt from fieldwork journal during a workshop on "eloquent tumors" at the Image Guidance Lab, Charité—Universitätsmedizin Berlin.

The scientists started "dancing" with the material: "Modelers *transduce* these affects through their body-work and propagate these gestures through performative articulations that excite others into action."[44] Today, virtual reality can include the scientists' bodies in the model. However, the routine "fly through" into brain image data sets conducted by the neurosurgeons at work is already a way of coming into and becoming alongside with the material. The probe, in this case, is a gigantic magnet that defines a large number of measurements, saved and processed by a computer, and rendered into series of images arranged in three directions, which can be displayed simultaneously to provide a spatial representation of the neurosurgical site to the practitioner.[45] For a few years, a fourth representation showed the image in 3D, making more legible certain complex spatial information, such as the visualization of the arrays of "tracks" running from distant parts of the brain through the white matter.[46]

---

44 Myers and Dumit, 240–41.

45 See Regula Valérie Burri, *Doing Images: Zur Praxis medizinischer Bilder* (Bielefeld: transcript Verlag, 2008), https://doi.org/10.14361/9783839408872; Silvia Casini, *Giving Bodies Back to Data: Image-Makers, Bricolage, and Reinvention in Magnetic Resonance Technology* (Cambridge, Massachusetts: The MIT Press, 2021).

46 Many attempts are being conducted (including by our team) to let the surgeons step in to the dataset using VR technologies: see the projects Brains Roads and TopoVox.

There are many ways to sense the lines of the material, and diverse equipment gives us access to different sets of boundary phenomena within the material. The master cutter's practical ideal—and ordeal—is to let the material guide the knife along its lines, tuning in to its demands, providing a balanced response to the question of the right form. This, in the light of Barad's thought, is how nature happens: a processual performance of knowledge that is intractably both representation and intervention, sensuous performance and bold invention.[47]

Conclusion

The idea of cutting along body lines can elicit two very different conclusions when translated to the metaphorical space offered by ancient philosophical tales: the idea of nature as a ready-to-discover ideal providing a set of "precut" facts, on the one hand, contrasts vividly with the phenomenological experience and the inherent perfectionism of its human practitioners on the other. The cutting scientist, like the Taoist butcher, experiences a form of growth as they move forward and contribute to their discipline and their team through their craft. This sensory and inventive process must be thought together with the rational aesthetics and the epistemic values of the field. Medical knowledge advances through a refinement of sensorial equipment, joining human bodies and metallic and magnetic instruments. Sustaining the form of elated working capacity of the neurosurgeon and of their teams, these dispositions and institutions are an interweaving of proprioceptive, technological, and social skills. Entering into a "mattering dance," they produce meaning and phenomena as a seamless unified working life-or-death experience for the physicians.

Through this reflection anchored in fieldwork, I attempted to challenge the common-sense modernist idea of the cut, and especially the breach introduced between the cutter and the thing being cut. I argue that the cut, as practiced in neurosurgery, can be considered as a response-able stretching of the senses toward materials and things, blending together elements of modeling, simulation, training, invention, and improvisation. The partitioners simultaneously cultivate the state of knowledge of the field and their relation to brains-as-material, and they grow individually and together, "becoming one" with a mattering process as they nervously sense their way through the nervous tissues. Through a lifelong process, the surgeon transmutes the nervous flesh into a familiar landscape—a labyrinthic, deadly playground that becomes more plastic and trusted as the gestures become more automatic and yet more defined and nuanced, anticipating myriads of contingencies and reading the constellations of

---

47 And certainly, also in the light of Alfred N. Whitehead, as in *The Concept of Nature*, see in particular *Thinking with Whitehead* by Isabelle Stengers for a comprehensive and generous guidance. Alfred North Whitehead, *The Concept of Nature* (Cambridge, UK: Cambridge at the University Press, 1920), http://archive.org/details/cu31924012068593; Isabelle Stengers, *Thinking with Whitehead: A Free and Wild Creation of Concepts*, trans. Michael Chase (Cambridge, MA: Harvard University Press, 2014).

signs that bode well or unwell for navigation that day. However, whether any surgeon will find "the way to nurture life" within their practice, like the king following the Taoist butcher's teachings, is a question that remains to be elucidated.

______

Collaborative Practices on Active Matters

Natalija Miodragović, Nelli Singer, Daniel Suarez, and Mohammad
Fardin Gholami

PLEKTONIK: Plasticity and Structural Textile

Abstract

The plasticity, the malleability of matter led to the evolution of application and aesthetic possibilities. Unlike timber wood, plant *filaments* such as willow and rattan are bendable, elastic, have hygromorphic behavior, and can achieve plasticity after water absorption. These *filaments* serve as design drivers for structural textiles, loop-based material systems for architectural production, yet they are not certified for construction. We assembled finite-length lignified plant fibers into (macroscopic) active structures that respond to mechanical stress and environmental humidity. Leveraging knitting techniques and traditional materials, our fabrication process challenges conventional practices. The formation of loops in textile production techniques is a negotiation between the elasticity of the material and the specific technique employed, while the properties of the yarn contribute to the textile's overall stability. These investigations are conducted within the context of interdisciplinary exhibition making. Change of intrinsic properties of material changes researchers and visitors—to adapt and explore new possibilities. Moreover, from a scientific point of view, we explore whether the loop of the preformed filament could be likened to a monomer found in polymeric materials. By bridging insights from both the nano and macro scales, we aim to transcend the limitations imposed by scale and uncover potential knowledge transfer between the realms. The exhibition provides a platform to evaluate the performance while fostering interdisciplinary innovation.

Introduction—Filaments and Droplets

In the pursuit of future architectures, this research aims to design architectural structures that operate as extensions of living systems by focusing on formable and tensioned *active yarns.*

The new culture of materials implies active individual and collective plasticity and the capacity to take form and to give form. "It is this very plasticity, the radical instability of the human, that is the basis of its massive impact," as Beatriz Colomina and Mark Wigley state, that is based on human interdependency with artifacts.[1]

---

1 Beatriz Colomina and Mark Wigley, *Are We Human? Notes on an Archaeology of Design* (Zurich: Lars Müller Publishers, 2016), 23.

This metamorphic and metabolic approach reappears on several levels in the project, which develops in an interdisciplinary environment through making, exhibiting, and textiling.

During the exhibition making process, research is informed by scientific insights at the cellular level, particularly regarding structures that define the boundaries between different entities or environments. For instance, it draws from research on the topology of swelling tissue in the outer mammal skin, where dead filamentous (threadlike) cells, corneocytes, respond to changes in humidity (see fig. 1B) as studied by Myfanwy Evans.[2] The spatial, complex tissues "exhibit variable permeability properties, actively mediating between different environments." This thermodynamic model is based on morphometric method: analyzing the form in change. To prototype these entangled looping helical filaments, rattan is selected as the material (fig. 1A).

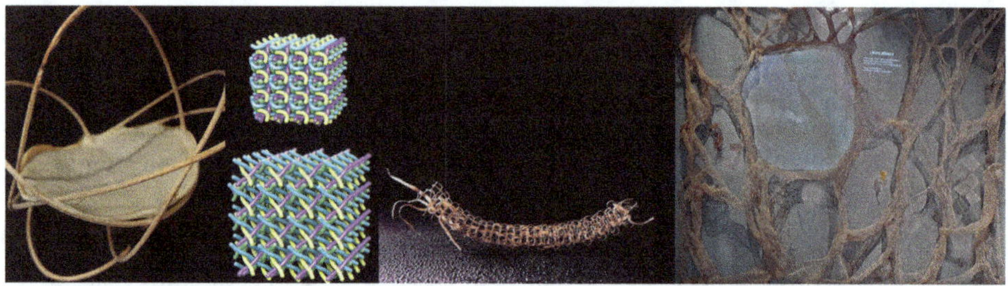

Fig. 1: Filaments and droplets: (1A) dried bacterial biofilm grown on the resilient rattan knot that can be read in its spatial state; (1B) swollen and subsequent drying states, from Myfanwy Evans and Roland Roth, "Shaping the Skin;" *Physical Review Letters* 112 (2014), 038102; (1C) knitted wood is actuated with humidity, Living Beings project, 2020; (D) warp-knitted scaffold showing elongation with open windows due to atmospheric water, Active Curtain Project.

The *Active Curtain* project in the exhibition *After Nature* (Humboldt Lab, Berlin 2020–23) featured a five-by-five meter netted structure that is an experimental setting that showcases interdisciplinary research on cellulose (fig. 1D). The exhibited projects demonstrate the controlled reversable movement of cellulose when exposed to water—in the project *Living Beings*, Nelli Singer contributed to the architectural scale with knitted wooden veneer. The research on bacterial cellulose addresses the public notion of cellulose not only as plant-based but also as microbial matter. The structure bridges architecture, design, microbiology, and material science, facilitating an *exchange* of insights into different operating scales and terminology among disciplines. The constituting spatial, tubular paths (fig. 2, left) are formed by the oversized loops (radius of five millimeters) and introduce the visitor to the scale of readable textile structure. Between construction and emergence, they form a structure reminiscent of the bacterial

---

2 Myfawny Evans and Roland Roth, "Shaping the Skin: The Interplay of Mesoscale Geometry and Corneocyte Swelling", *Physical Review Letters* 112, no. 3 (2014): 038102, https://doi,org/ 10.1103/PhysRevLett.112.038102.

filamentation and growth stance observed in the bio-spatial extracellular matrix in bacterial biofilm research conducted by Hengge's team.[3]

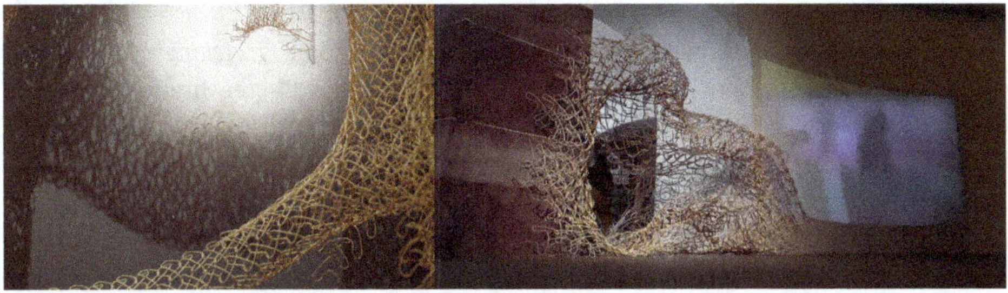

Fig. 2: Textile space mediating metabolic structures: (left) warp-knitted rattan tubular paths showcasing the Living Beings project, Active Curtain Project (right) tubular willow structure accessible also in VR, Stretching Materialities exhibition at TAT.

The long branches or *filaments,* malleable upon wetting, are preformed into stable looping segments entangled by warp knitting, an industrial knitting technique. Warp knitting is a loop forming process in which separate threads are interlocked with the neighboring parallel thread along the length of the fabric. The loops (or laps) may be open or closed. We translated the open lap path for manual knitting with pre-formed resilient fibers, rattan (and later willow). The size and spacing of the pre-formed segments are varied to determine the strength and stability of the structure. Using always the same 1.8 millimeter diameter, threads with the highest density were used at the top, followed by medium density (fig. 2, left) and lowest density as we moved downward, resulting in a loosening hanging structure with a total height of five meters. The two-meter-wide void area for a video projection of structure was bridged over (fig. 1D).

Material Landscapes

Unlike thick timber wood, plant fibers like willow twigs and rattan palm are not certified for construction in Europe and they are elastic and hygromorphic, achieving plasticity for the design of structural textiles after water absorption. Working with organic material encompasses a big picture, including landscapes, ecosystems, agroforestry, and the climate. Rattan palms, mostly liana, grow on other species (fig. 3, column 1), and are defined as non-timber forest products (NTFP), referring to biological materials

---

3 Christiane Sauer, Natalija Miodragović, Bastian Beyer, Iva Rešetar, Daniel Suárez, Nelli Singer, Regine Hengge, Skander Hathroubi, Michaela Eder, Cécile Bidan, exhibition project "Active Curtain," Humboldt Forum Berlin, since April 29, 2021, https://www.matters-of-activity.de/en/research/laboratories/49/humboldt-lab.

derived from forests excluding timber.[4] They are bendable and formable due to their filament structure.[5] The first Forest Stewardship Council (FSC) certified rattan plantations were established in Indonesia and Borneo in 2011.[6] Small-scale producers who collect rattan do not consider plantation grown rattan an NTFP.

In the second phase of research, and for the exhibition *Stretching Materialities*, 2021–22, more stable but formable willow (fig. 3, column 2) tree branches local to Europe were used. They are also categorized as NTFP due to their small diameter. Basketry plantations are rare and small, while SRC plantations for biomass[7] are expanding also due to willow's ability to grow on and remediate polluted soil[8] or for medicine.[9] Willow exhibits controlled growth and we can envision the preforming process to occur directly on plantations or construction sites. Its bast fibers show mechanical and thermal properties suitable for reinforcement, in addition to being biodegradable and low cost.[10] The applications of small diameter willow branches grown for biomass on an architectural scale is seen as a stance in circular economy and a step toward a post-waste society. The research aims to standardize and certify organic filaments for construction. Initial measurements are conducted. Moreover, standardizing non-uniform biological "rests," such as non-timber forest products for construction, can address the emerging need for standardization of non-uniform "human-made rests" such as textiles, plastics, and debris for construction.

Structural Textiles

Freshly harvested willow twigs exhibit remarkable flexibility, due to the high water content, and can be easily shaped into any forms. However, once the wood is harvested and dried, it becomes more rigid. We used the ancient and universal technique of

---

4 See "FAO's Forest Resource Assessment (2015) An international journal of forestry and forest industries," *Unasylva* 198, vol. 50 (1999).

5 Hanna M. Szczepanowska, "Deconstructing Rattan: Morphology of Biogenic Silica in Rattan and Its Impact on Preservation of Southeast Asian Art and Artifacts Made of Rattan," *Studies in Conservation* 63, no. 6 (August 18, 2018): 356–74, https://doi.org/10.1080/00393630.2017.1404693.

6 John Hontelez, EU Timber Regulation / Implementation Guide for Companies Trading FSC-certified Materials in the European Union, Revised Version (February 2018), https://ic.fsc.org/file-download.eu-timber-regulation-implementation-guide.a-13.pdf.

7 Ioannis Dimitriou and Dominik Rutz, *Sustainable Short Rotation Coppice: A Handbook* (Munich: WIP Renewable Energies, 2015), https://www.srcplus.eu/images/Handbook_SRCplus.pdf.

8 Yulia Kuzovkina and Martin Quigley, "Willows Beyond Wetlands: Uses of Salix L. Species for Environmental Projects", *Water, Air, & Soil Pollution* 162 (2005): 183–204, https://doi.org/10.1007/s11270–005–6272–5.

9 Jassem G. Mahdi, "Medicinal Potential of Willow: A Chemical Perspective of Aspirin Discovery," *Journal of Saudi Chemical Society* 14, no. 3 (July 1, 2010): 317–22, https://doi.org/10.1016/j.jscs.2010.04.010.

10 Oktae et al., "Characterization of Willow Bast Fibers (Salix Spp.) from Short-Rotation Plantation as Potential Reinforcement for Polymer Composites," *BioResources* 12, no. 2 (2017): 4270–82, https://doi.org/10.15376/biores.12.2.4270-4282.

bending dried wood[11] by applying water and for larger diameters also heat to initiate the wood plasticization. As a result, this research presents a material system for architecture production based on large-scale wood and non-wood warp-knitted structures, obtained through material plasticization and loop formation. The integration of form, active materiality, structure calibration, and bespoke fabrication methods call for a design process with a holistic approach.

Textile structures, in general, are inherently multilevel hierarchically assembled structures[12] and contain three interdependent levels at the macrolevel: fiber, yarn, and fabric. This research operates at these three levels, but we recognize the insight on the initial level of hierarchy—nano: *"Hierarchy is important because it allows materials to exhibit exceptional properties and diverse functionality through adapting structure across many different levels from nano to macro."*[13]

– At the fiber level, the project observes how moisture affects plant-based fiber. At a later stage, loop (pre-)formation, the study of the moisture conditions and hygroscopic behavior becomes a matter of particular interest and importance to this project.

– At the yarn level, the research recognized the loop as the basic building block of this material system. Thus, we explore loop formation processes based on wood bending techniques. The loop diameter (bending radius) is related to yarn thickness. Through wetting, the bending radius is reduced, and the filament "remembers" the form when dried. Hence, the wood is plasticized. In the course of our explorations, we have designed a collection of apparatuses to aid in the wetting, steaming, and bending of thin section wooden rods (fig. 3, columns 1 and 2, row C).[14] Willow stem and rattan filaments have finite lengths, therefore constraining those assemblies that demand longer threads or continuous arrangements. Here, the mono-material continuity is achieved by manually interlocking looped segments into chains (fig. 3, columns 1 and 2).

Moreover, we present methods of manufacturing continuous wooden yarns[15] using STFI's Kemafil® technology.[16] Kemafil yarns are hybrid yarns engineered to

---

11 Edward W. Berry, "Notes on the History of the Willows and Poplars," *The Plant World* 20, no. 1 (1917): 16–28.

12 Leslie Eadie, and Tushar K. Ghosh, "Biomimicry in Textiles: Past, Present and Potential. An Overview," *Journal of The Royal Society Interface* 8, no. 59 (2011): 761–75, https://doi.org/10.1098/rsif.2010.0487.

13 Jane Scott, "Hierarchy in Knitted Forms: Environmentally Responsive Textiles for Architecture," ACADIA 2013 (2013): 361–366.

14 Robert S. Wright, Brian H. Bond and Zhangjing Chen, "Steam Bending of Wood; Embellishments to an Ancient Technique," *BioResources* 8, no. 4 (2013): 4793–96, https://doi.org/10.15376/biores.8.4.4793-4796.

15 Natalija Miodragović, Daniel Suárez, Nelli Singer, Regine Hengge, Michaela Eder, Christiane Sauer, "Active Yarns for Structural Textiles," poster presentation, *Bioinspired Materials Conference Proceedings* (2022).

16 R. Arnold, A. M. Bartl and E. Hufnagl, "Production of Cord and Narrow Fabric Products with Kemafil Technology," Band und Flechtindustrie 31 (1994): 48–52.

fulfill different performance and functional requirements.[17] The not-continuous textile and non-textile material is sheathed with selected thread, resulting in yarn with thickness up to sixty millimeters (fig. 3, column 3). Hence by programming the yarn, the coding complexity expands from geometrical to more complex behaviors and applications. These wooden yarns are also directly processed as meandering weft material on a coarse warp-knitting machine (fig. 3, column 3, row C)[18] to produce architectural surfaces.

– At the fabric level, we find the textile structure formation process. Knitting offers a multitude of tectonic arrangements. The project exploits the possibilities of warp-knitting arrangements to organize discrete programmed loops of wood material. A variety of two-dimensional warp and weft assemblies have been studied. However, the most representative one is the tricot structure.[19] After wetting, the wood and non-wood loops hold their shape while remaining partially within their elastic limit. They behave similarly to plates with out-of-plane bending resistance. As in any knitting technique, the loops interlock with their neighbors, hold their position, and create a hinged plate shell structure.

Wood and non-wood warp-knit structures lend themselves well to doubly-curved forms. The surface's positive or negative curvature is achieved through a simple hand gesture when making the textile structure. Depending on the direction in which one loop interlocks with its neighbor, i.e., from above or underneath, the loop will hinge in one order, leading to a convex or concave surface (fig. 4, right) that can be efficiently coded using a binary code, therefore programming the surface into a particular form. The spacer textile, produced also with warp-knitting technique, are part of the research (fig. 3, column 2D), the fiber properties can determine the properties of the fabric.

In fact, a broader definition that describes textiles as flexible products made primarily of polymeric (natural or human-made) fibers, is more appropriate today. However, as Julian Vincent noted, "there is a huge potential to obtain new or unusual combinations of material functions/properties by structuring a given material, rather than by changing its chemical composition."[20] In fact, textile fiber assemblies can readily provide an ideal test bed for this concept.

---

17 Hugh Gong, *Specialist Yarn and Fabric Structures: Developments and Applications* (Amsterdam: Elsevier, 2011).

18 Wright, Bond, and Chen, "Steam Bending of Wood," 4793–96.

19 Yordan Kyosev, *Warp Knitted Fabrics Construction* (Boca Raton, FL: CRC Press, 2019), https://doi.org/10.1201/9780429094699.

20 Julian Vincent, "Biomimetic Materials," *Journal of Materials Research* 23 (December 2008): 3140–47, https://doi.org/10.1557/JMR.2008.0380.

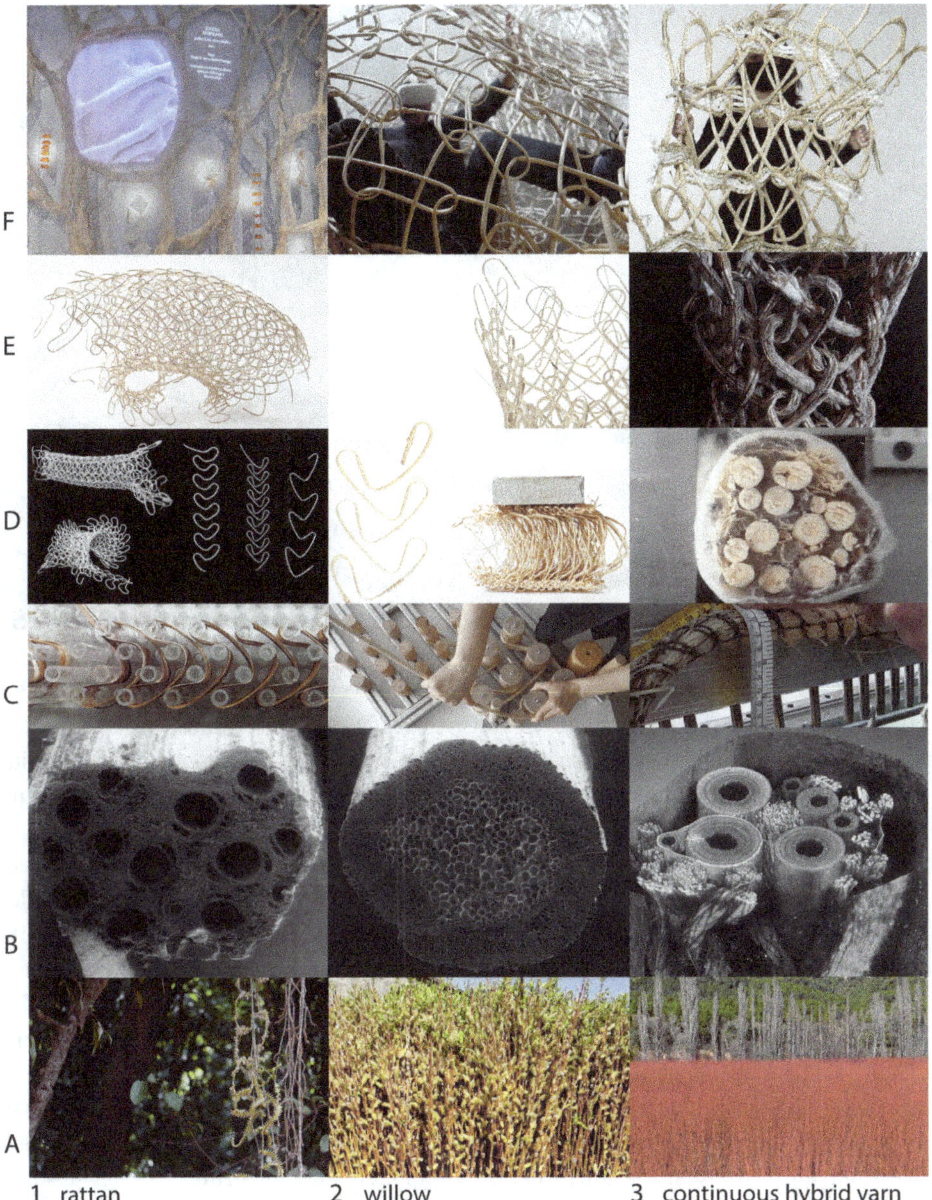

Fig. 3: Material landscapes, from agroforestry and fiber to spatial textile spatial structure: column 1, rattan; column 2, willow; and column 3, continuous hybrid yarn from finite branches.

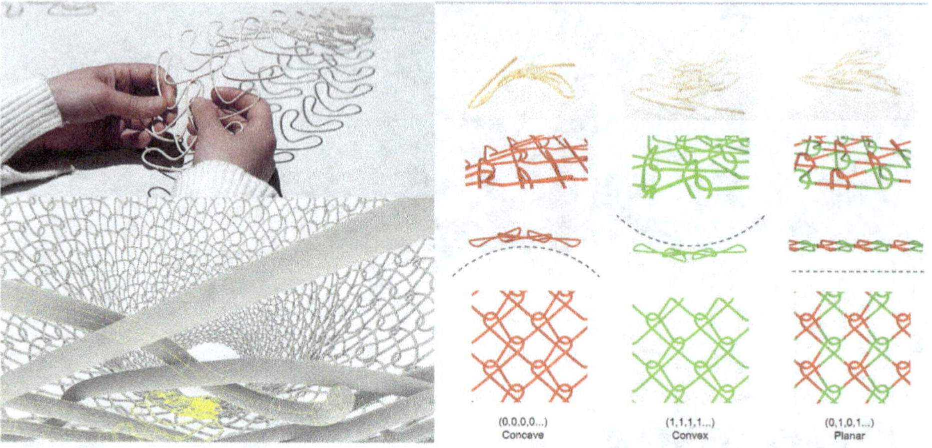

Fig. 4: Structural textile: interlocking order of loops determines the form concave and convex.

Polymeric Nature and Assembled Knitted Structures

While dealing with knitted structural textiles, we were repeatedly reminded of the other synthetic or natural structures consisting of linkage of many repeating units. An observer will notice the repeating units, as reminders of a polymeric material where chemical units known as monomers are chained together to create large chains of macromolecules.

We observed two major parameters affecting the activity of knitted wooden structures. These are, namely, the "humidity uptake" and the "actuation specific knitting design." In principle, the "humidity uptake" is a property of the matter that exists at its smallest scales. More specifically, this is in relation to the hydrophilicity of the molecular structure of the matter. In the case of wood as the subject matter here, the cellulose-based structure is the main hydrophilic structure that comprises 40 to 50 percent of the wood's fiber weight (fig. 5A). This water-friendly polymeric structure consists of repeating units of β-glucose, which are simple monosaccharides, forming polymeric chains and consequently cellulose fibers and cellular walls within plants.[21] The collective emergent activity of these hydrophilic components demonstrates a water absorption capacity that is dependent on the environmental water content (humidity).[22] The

---

21 Yoshiki Horikawa, "Structural Diversity of Natural Cellulose and Related Applications Using Delignified Wood," *Journal of Wood Science* 68, no. 54 (2022), https://doi.org/10.1186/s10086-022-02061-2; Jifu Wang, Daihui Zhang, and Fuxiang Chu, "Wood-Derived Functional Polymeric Materials," *Advanced Materials* 33 (2021), 2001135, https://doi.org/10.1002/adma.202001135.

22 Matthew J. Harrington et al., "Origami-like Unfolding of Hydro-actuated Ice Plant Seed Capsules," *Nature Communications* 2, no. 33 (2011), https://doi.org/10.1038/ncomms1336.

humidity uptake from the environment is then the main trigger for the process of structural expansions at cellular levels and consequently structural dynamics, or conformational changes of the structures.

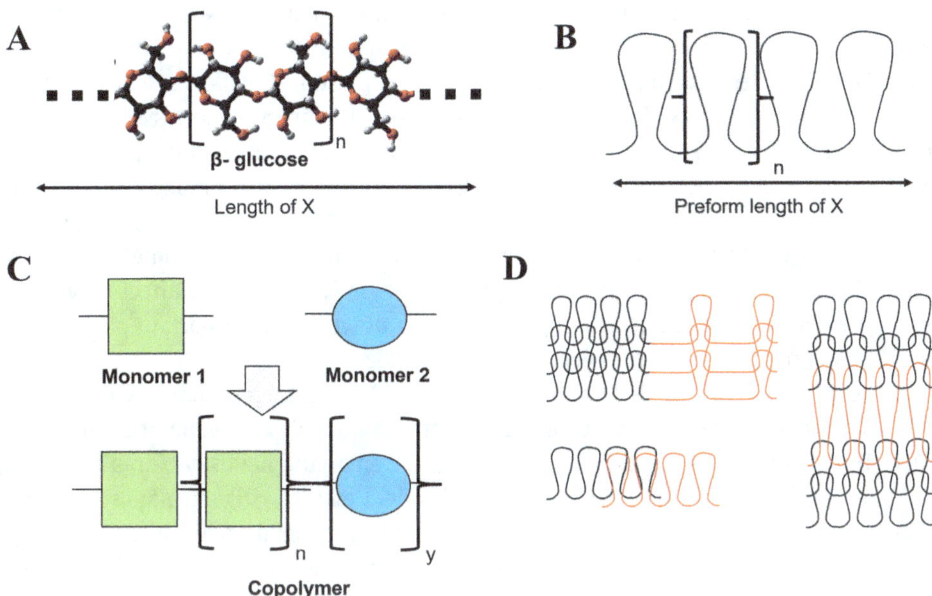

Fig. 5: (5A) the polymeric structure of the cellulose as the polymeric building block of the wood; (5B) repeating loop structure of the preformed wood before knitting and assembling into larger structures; (5C) co-polymeric structure based on two different monomers; (5D) co-knitted preforms of different geometries for specific abilities and activities.

On the other hand, the concept of "actuation specific knitting design" involves a deliberate framework of bending designed to imbue the structure with active properties. The architect or designer envisions and constructs the loops and their arrangements to control the extent and direction of the structure's response to environmental stimuli. Through the knitting/bending process, units are interconnected and chained, allowing for the incorporation of diverse geometries, sizes, and humidity uptakes, thus enabling controlled collective emergent activities, namely here "movements." This knitting process mirrors the assembly of preformed material strands and evokes the analogy of polymerization, where smaller polymer chains undergo growth to form longer chains capable of both withstanding stress and or dynamic behavior under the stimuli.

In this discussion, we highlight two key similarities between polymeric structures and knitted wooden structures that we observed to be independent of scale. These similarities are as follows: 1) Knitting process involves preforms of different geometry being combined (figs. 5C, D). 2) An activity of structures is emerged when triggered by the environmental water interaction (fig. 6). To underscore this point further, figures

5C and D exemplify the resemblance between a copolymer system comprising two different monomers and a co-knitted preform structure.

We propose that, similar to copolymer structures designed by polymer engineers, chemists, and physicists, knitted wooden structures, crafted by textile designers, also harness the amalgamation of various loop geometries, materials, and sizes to create dynamic frameworks of activity, whether it is load bearing, aesthetic, or stimuli-responsive.[23] We describe this as a "bidirectional inspiration junction". Recent experimental demonstration of the polycatenane systems as a direct demonstration of polymerization of interlocked loops of matter by Laura Hart et al. showed the possibility of achieving the interlocked assembly at nanoscale.[24] These structures indeed show multiple emerging modes of activity and vibrations as a result of stimulations and environmental changes. Furthermore, as a more classical example, the water content around a natural polymer chain, such as various proteins or DNA, is known to affect polymeric structure conformations and folding thereof, both of which are important parameters in the functionality of these molecules at their scales.[25]

The preform knitting and subsequent assembly into larger, complex structures with specific properties and functionalities captivate not only the realm of macrostructures but also serve as a source of inspiration for emerging nanostructures based on polymeric chains. A prime example lies in the field of DNA origami nanostructures, which has garnered considerable attention in the past three decades as a captivating avenue for nanoscale construction, manipulation, and the demonstration of geometrical control. This design perspective has captivated natural scientists from diverse disciplines such as chemistry, biology, and physics, empowering them to attain precise control over matter at the nanoscale. Given the existence of tools and knowledge for creating loops and self-assembly structures at the nano scale, these techniques hold significant potential for driving the synthesis of nanomaterials through a "top-down design" approach. This opens up new frontiers in material synthesis, leveraging the synergy between macrostructure and nanostructure design principles. Since the mechanical and structural investigation of the designed macrostructures such as preform knitted structure discussed here are easier and less costly, transfer of those new designs with specific emergent activities to the nano scale would be interesting and rewarding case studies.

We propose that fostering a new culture of materials involves not only the bottom-up design of materials but also the integration of "macroscale designers," such as textile designers and architects, in a collaborative, multidisciplinary environment alongside natural scientists. By adopting a top-down strategy or inspiration from structural

---

23 Nelli Singer, "Living Beings," master's thesis (Weißensee Kunsthoschschule Berlin, 2020 Master Studio Christiane Sauer, 2020).

24 Laura F. Hart et al., "Material Properties and Applications of Mechanically Interlocked Polymers," *Nature Reviews Materials* 6, (2021): 508–30, https://doi.org/10.1038/s41578-021-00278-z.

25 Marie-Claire Bellissent-Funel et al., "Water Determines the Structure and Dynamics of Proteins," *Chemical Reviews* 116, no. 13 (2016): 7673–97, https://doi.org/10.1021/acs.chemrev.5b00664.

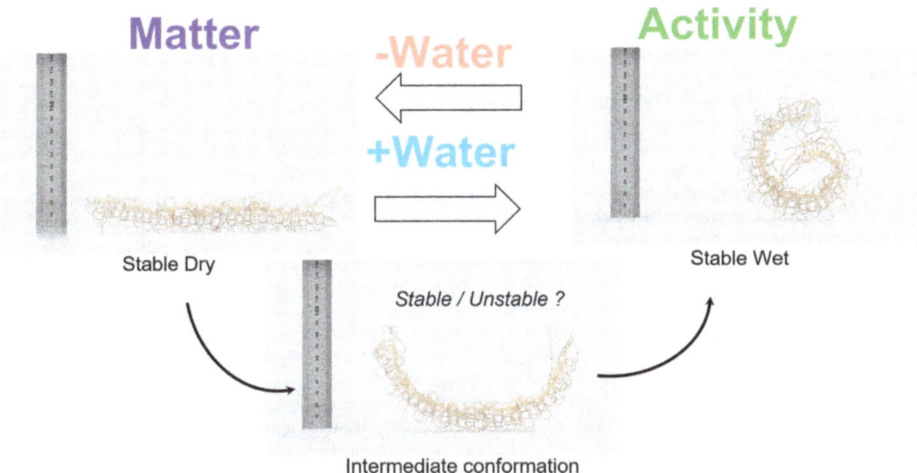

Fig. 6: Conformation balance observed in knitted wooden structures. The intermediate conformation may or may not be stable depending on the design and the humidity of the environment.

designers at macroscale, novel nanostructures can be achieved, potentially giving rise to entirely new materials and their respective emerging activities. These nanostructures will possess tangible macroscale counterparts that can be structurally modified and tested. An excellent example of this concept is the physical unification and interlocking process, which leads to the emergence of active matter, reflecting the latest rethinking in material science. Moreover, this approach encourages innovative methodologies in multidisciplinary cooperation. Additionally, morphometric-based approaches hold promise for architectural design, expanding the realm of design practices. For example, the physical unification and interlocking process creates an emergence of behavior that not only demonstrates *active matter* according to the latest rethinking, but also new methodologies in multidisciplinary cooperation that are material- and application-oriented.[26]

Whether at the building block scale of the source material or within the intricate locks and loops of knitted and assembled structures, the significance of multidisciplinary approaches cannot be disregarded. Furthermore, the plasticity of the material unveils new insights into loop-based textile techniques, where stable loops serve as initial units. Linking water molecules, environmental water, fibers, and circular economy principles to bioremediation supports the argument for a multiscale approach. Thus, this approach investigates a departure from traditional hierarchical approaches, transitioning from fiber and yarn to fabric, toward embracing multiscale modeling.

26 Mohammad Fardin Gholami et al., "Rethinking Active Matter: Current Developments in Active Materials," in *Active Materials*, ed. Peter Fratzl, Michael Friedman, Karin Krauthausen, and Wolfgang Schäffner (Berlin: De Gruyter, 2021), 191–222, https://doi.org/10.1515/9783110562064-011.

Felix Rasehorn, Hanna Wiesener, and Thomas Ness

The Role of Filtering in Design Processes: Creating Stabilities in Volatile Contexts by Tentative Prototyping

Abstract

This essay emphasizes the role of friction and filtering in design processes. We will explore filtering dimensions as an integral part and a concise moment in the production of prototypes. At the beginning of this essay, we will focus on the topic of plasticity and instability, taking a closer look at how these two terms are entangled in a nonlinear design process. From there, we aim to discuss the productiveness of nonlinearity in the design process and how the designer acts as embodied filter in exchange with the material that forms the prototype. We will look into how digital material can be manipulated with similar principles, suggesting that the instability in the moment of prototyping is arguably a method of approaching problems (*Probehandeln*). In elaborating these terms, we will introduce excerpts of practice-based research projects to contextualize the topic of prototyping with digital materials, the role of materiality in the design process, and the designer as a filtering dimension. The presented practical work will help us to conclude how creativity arises from instability, demonstrating that tentative prototyping is a productive method in design and collaborative work.

Nonlinearity

Product development and product design are nonlinear processes, meaning there is no straight pathway from an idea to a final product. The idea is not a crystalline nor a diamond that in a second step is poured into a material shape, the prototype.[1] Rather, this process can be described as subtractive, where the idea is metaphorically carved out of a solid block of material. Here, we describe a process that goes from rough to fine, the idea being the blurry shape (or the solid gemstone) that is able to inhabit multitudes of possibilities, shapes; while the prototype, in the process of iteration, matures and concretizes until its final shape is reached. In this process, the material, the designer, and the team of specialists actively filter decisions, degrees of probabilities, and ideas. In this process step, a multitude of changes in perspectives are necessary. The designer is confronted with specific contexts, material behaviors, and socio-economic

---

[1] See Julian Adenauer and Jörg Petruschat, *Prototype! Physical, Virtual, Hybrid, Smart* (Berlin: form +zweck, 2012).

patterns, just to name a few. It can be observed that this confrontation with the real-world causes of instability is of a kind that puts the stable concept in shaking momentum. The process of prototyping can therefore be described as the endeavor to once again balance out demands, with the goal to fit idea and material in a form or interaction. As every product is part of economic and material cycles, it influences more than on a personal level, reflection on parts of society bearing witness to abundance or lack. For the designer, it is indispensable to go to these linked disciplines for research purposes and to seek exchange with experts even at the risk of being called an amateur scientist.

Prototypes

Prototypes and prototyping play an important role in product development. Even when the product is still not more than a vague outline, designers are capable of focusing on certain snippets or features of the upcoming product. These snippets are the basis of the repeated prototyping cycle. On the one hand, the material form of the prototype enables concrete interactive feedback, as Julian Adenauer and Jörg Petruschat described in detail in their book *Prototype!*[2] On the other hand, the prototype in its provision still has sufficient plasticity and fuzziness and, depending on its context, it provokes and enables certain actions. A prototype can help focus on one specific domain of a question, it can communicate certain aspects of a situation or clarify false assumptions. Depending on the design requirements, temporary stabilities are created during prototyping, be it through the material, form, proportions or functionality. If one assigns prototypes to the models, then presumably referring to models that are binding thoughts and bear constructive potentials. Just like a prototype when a model is created, assumptions are fixed in the material.

Exploration Phase

We will leave the overarching description of prototyping to go back to the beginning of the design process, the so-called exploration phase, a moment where instability becomes most obvious. This phase embraces experimental arrangements and material studies, the self-activity of the material is assumed and appreciated, material activity is traced, the designer being the observer. The architect Günter Behnisch describes this approach as listening (Ger. *Lauschen*).[3] By listening to the manifold possibilities of the material, it will most likely reveal itself or parts of its potential. Designers are

---

2 See ibid.

3 See Elisabeth Spieker, "Günter Behnisch—die Entwicklung des architektonischen Werkes: Gebäude, Gedanken und Interpretationen," PhD diss. (University of Stuttgart, 2005).

listening to the material through and with the method of prototyping. What the designer hears, however, is obviously determined by his or her individual skillset, knowledge, and cultural imprint, but also the brief, context, and environment of the experiment. The designer's action oscillates between grasping and conscious tracing to fix and bind explored moments into the material.

Digital Technologies

But how to explore new information technologies? How can we employ the method of material exploration in the domain of digital technologies and complex or abstract systems? Now, the materiality we work with is equally important as working with, e.g., wood, but still, we cannot just take off to the workshop and fiddle around with it. To explore its modalities and possibilities we need a different approach and starting point. Anthony Dunne and Fiona Raby have successfully described and classified the variety and similarities of such approaches in their book *Speculative Everything.*[4] Using specific design projects, they lay out the variety of methods and topics and the unifying principles and rules of designing with prototypes. In this essay, we will take a similar approach of demonstrating some of the presented ideas embedded in a concrete design project as a case study. Because testing objects and prototypes creates a layer of spatial and temporary disruptive critique, a moment of engagement, that is hard to achieve with a text-based publication, but rather experienceable. In teaching and in our own design process we transfer the method of testing prototypes (*Probehandeln*) in the early exploration phase. *Probehandeln* does not mean to focus on usability, but to try to create a moment of engagement determined by a certain materiality, geometry, and interaction properties. The goal is to playfully create prototypes that lure users into a flow and thereby generate insights about specific behavioral patterns and unconscious reactions. These prototypes often employ friction as a productive moment. Especially useful when reflecting on or questioning social standards, everyday routines, or behavioral norms. Furthermore, designers produce *speculative prototypes* to explore technology and interactions in a performative act. The adjective "speculative" includes setting up scenarios to reenact and thereby explore a specific interaction or aspect of a routine, thus creating a better understanding of possible futures.

Case Study

The following part of the essay is an introduction into three excerpts of the ongoing practice-based doctoral thesis *Tessellated Skins and Shells: Designing Biology-Informed*

---

4 Anthony Dunne and Fiona Raby, *Speculative Everything: Design, Fiction, and Social Dreaming* (Cambridge, MA: MIT Press, 2013).

Wearables for Context-Sensitive Protection and Support by Felix Rasehorn. Three specific moments of this practice-based research are briefly presented to situate and contextualize the discussed filtering layers.

Tessellation Archive—Prototyping with Digital Materials

The first excerpt aims to depict the design approach of prototyping with digital technologies and abstract materials. It is part of an interdisciplinary project situated in the Tessellated Material Systems (TMS) research group. The group consists of researchers from the Cluster of Excellence "Matters of Activity" (MoA). Coming from morphology, engineering, material science, and design, members of the group are commonly interested in the relation between form and function in TMS. The group has collected over 120 specimens of natural TMS, and developed a taxonomy that prioritizes formal similarities over genetic relatedness. The compiled dataset can be appreciated as a research outcome, while from the perspective of a designer, it can also be perceived as a material. That change of perspective transforms the process of defining the taxonomy into a moment of tentative prototyping. It might not be equally productive for all members of the group. Researchers from the natural sciences in particular are most professionally trained in defining taxonomies, tapping into their domain-specific methodologies. The change in perspective, on the other hand, recalls the method of a designer, collaborating with other disciplines and applying design methods to productively contribute to a common effort. Through methodologically treating the dataset as material, the development of categories equals the construction of analogue filters. If the dataset is treated as material and the category understood as the filter, the equation is missing a user. So, who is this material and its shape made for? Answering this question leads to the creation of a persona, this is analogue to the process of writing a paper having a specific peer group in mind. In this project, the persona was a fictional group of architects, designers, and other creatives, interested in finding inspiration in nature's structural systems. Creating this persona implied that the material, the dataset, needed to be made accessible in a way that enabled the persona to draw inspiration and information from it. Through the approach of tentative prototyping, the vision of an interactive web-based archive, accessible for interdisciplinary researchers was manifested. Experienceable prototypes became containers for ideas and arguments in the discussion. Fostered by drawings (fig. 1), interactive prototypes (click dummies), and visualizations (fig. 2), the group could collaboratively decide, agree, and improve the archive. In exploring how a website unpacks to engage people with the presented dataset, an interactive drawing board became the tool to store suggestions, alterations, and ideas (fig. 2). It allowed the visualization and comparison of approaches and helped to evaluate one over the other. After one year of iteration, the tessellation archive has been published (fig. 3). It is a publication that allows practitioners to distinguish form from function, and thus recognizes the unbiased potential of formal tessellation patterns.

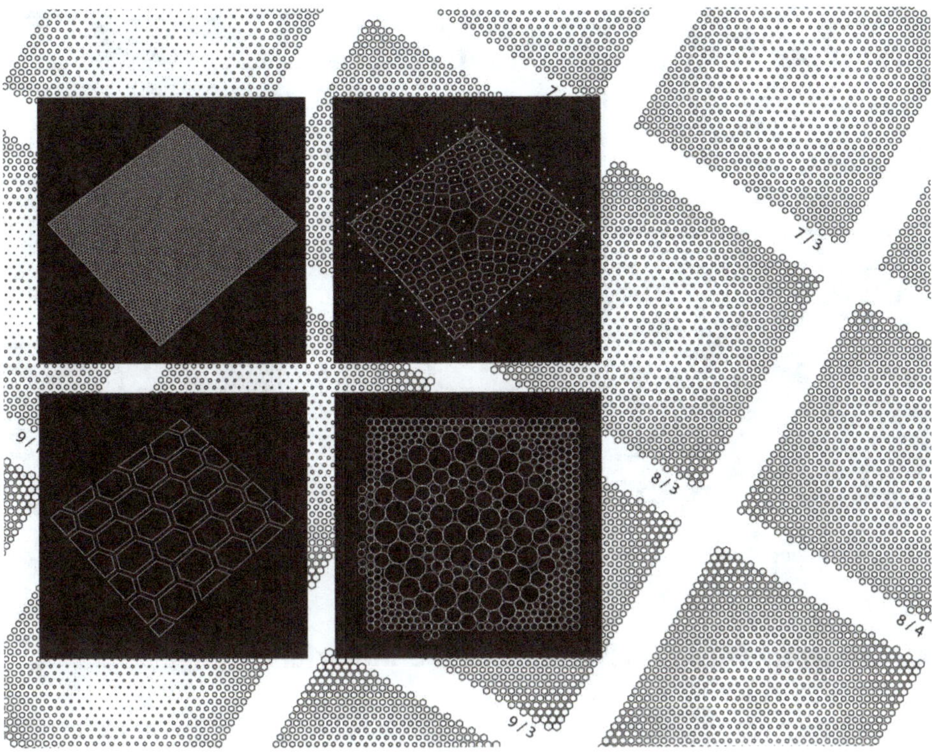

Fig. 1: Drawings of computational models of four categories describing TMS.

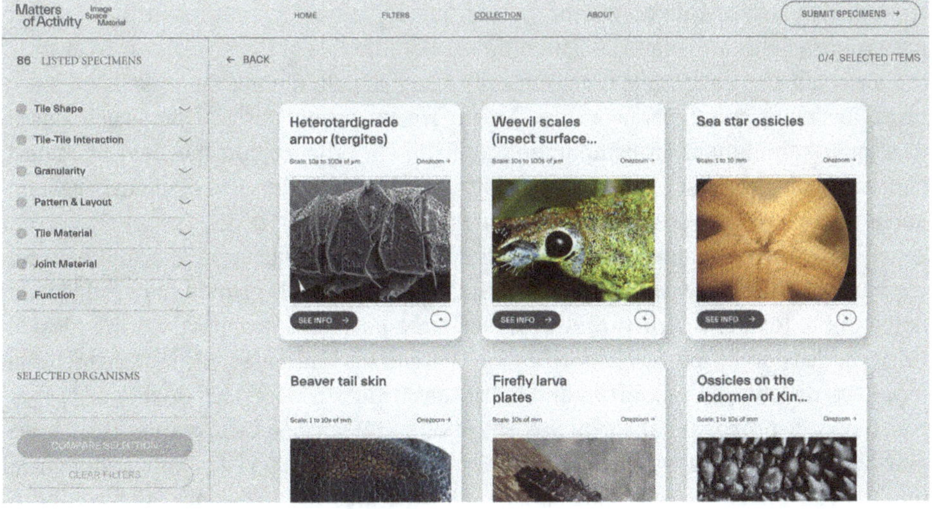

Fig. 2: Digital iterations of prototypes for the interactive archive.

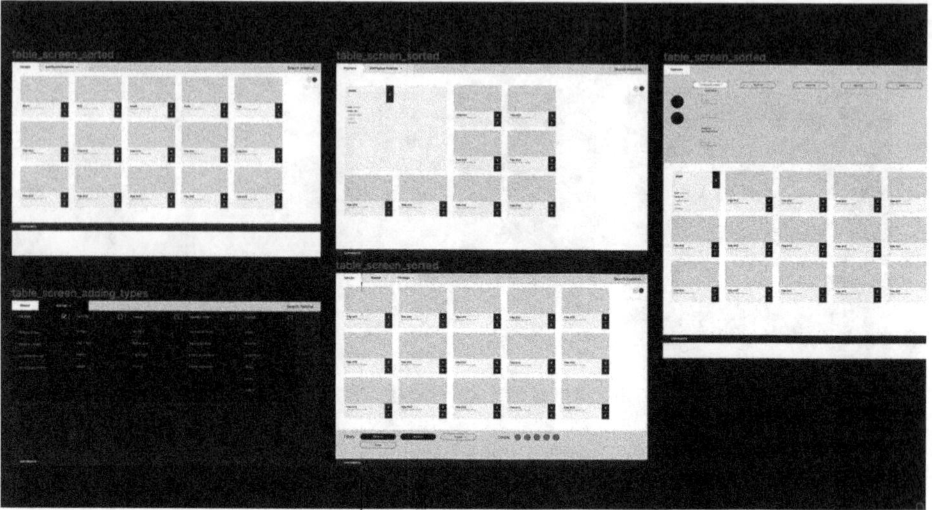

Fig. 3: Screenshot of the "collection" tab of the interactive TMS archive.

Materials as Filters—Listening to the Material

The second excerpt of this doctoral research focuses on the ability of materials to function as filters, not literally, but methodologically. This study aims to replicate the strategy of hierarchical structure that can be observed in natural systems. Natural TMS consist of gaps filled with fibrous interfacing membrane that connect solid tiles. This structural duality presumably leads to multifunctionality. How to transfer the observed natural principles into materialization processes? The role of the designer is to abstract the observed natural phenomena. Instead of digitally simulating the tessellated structures, analogue prototypes were developed. Working with physical materials aids in overcoming limitations in computation and thus in tapping into the field of material-based analogies. In the process of prototyping, mechanically rigid elements are laminated to pre-stretched textiles using 3D printing. The selected textile (jersey 94 percent cotton, 6 percent elastane) therein simulates the soft interfacing membrane between the hard plates as observed in natural systems (fig. 4). The 3D-printed material is laminated to the textile, it structurally represents the hard tiles, trapping surface tension. Once the tension of the fabric is released, those areas laminated with 3D-printed material are structurally reinforced and resist the shrinking force. This duality of properties allows disproportional shrinkage, resulting in 3D surface deformation (fig. 5). Primarily, this study seeks to explore analogies between materialization and natural hierarchical systems. Simultaneously, it demonstrates the relevance of listening to the material properties in the prototyping process. As materials come equipped with specific properties, they can transform into active players in a prototyping process. Therefore, their ability to filter possibilities and formal potentials is recognized.

While materials are usually regarded as passive matter, designers attribute them activity and agency, thus they become aids in order to prove or disprove assumptions.

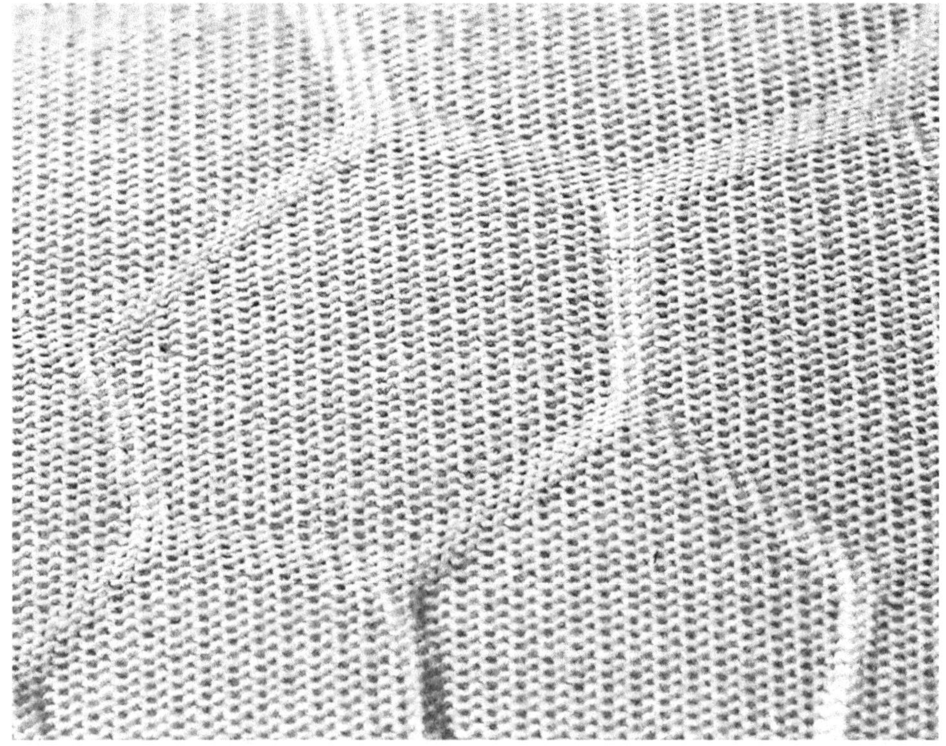

Fig. 4: Close-up of laminated jersey, showing the structural difference between the interface and joints.

Designers as Filters—Perception and Communication

Designers themselves can function as filters; unconsciously in discussions with others or consciously with the aim of changing the direction of a prototype. The third excerpt dives into the design process itself taking a closer look at the collaborative design project *embrace2* developed by Silke Hofmann. Her work is grounded in the extensive and sensitive research in the realm of breast cancer affected women and their specific clothing needs. The project *embrace2* focuses on the product development of alternative underwear individually designed for women affected by breast cancer after mastectomy. In this collaborative design project, designers from the field of fashion, textile, product, and computational design worked together with the wearer as co-designer. The collaborative process peaked in situations that are called fittings (fig. 6), an established methodology in fashion design. A fitting is in itself a technical process in which all participants come together to evaluate the prototype on the wearer's body. Special-

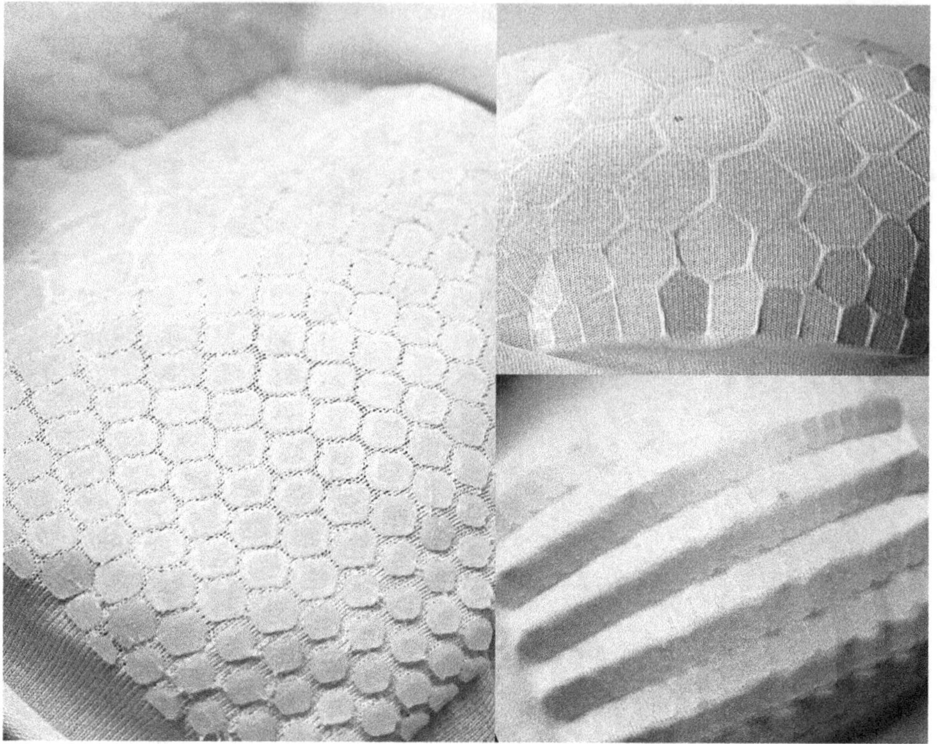

Fig. 5: Iterations of computational geometries 3D printed on pre-stretched textile.

ists from different disciplines with different vocabularies, ideas, and experiences came together to evaluate the respective work. Thus, the fittings were situations in which individual competencies stepped back and common interests were negotiated, discussed, and evaluated. These communication and development processes were intensively negotiated *in situ* on the physical prototype (fig. 7). In these situations, the designers functioned as filters that looked at the prototype from their specific field of expertise, contributing to a larger body of work. In this joint development step, the ideas of each specialist were exchanged, not primarily verbally but directly on the prototype itself (fig. 8), allowing to filter extremely efficiently and at the same time particularly finely what was technically feasible, aesthetically desirable, and functionally durable. To this day, the developed prototype contains all details and decisions of the discussion. Each collaborator embodies the specific filter to unpack this information which he or she turned into actions that have led to the final prototype.

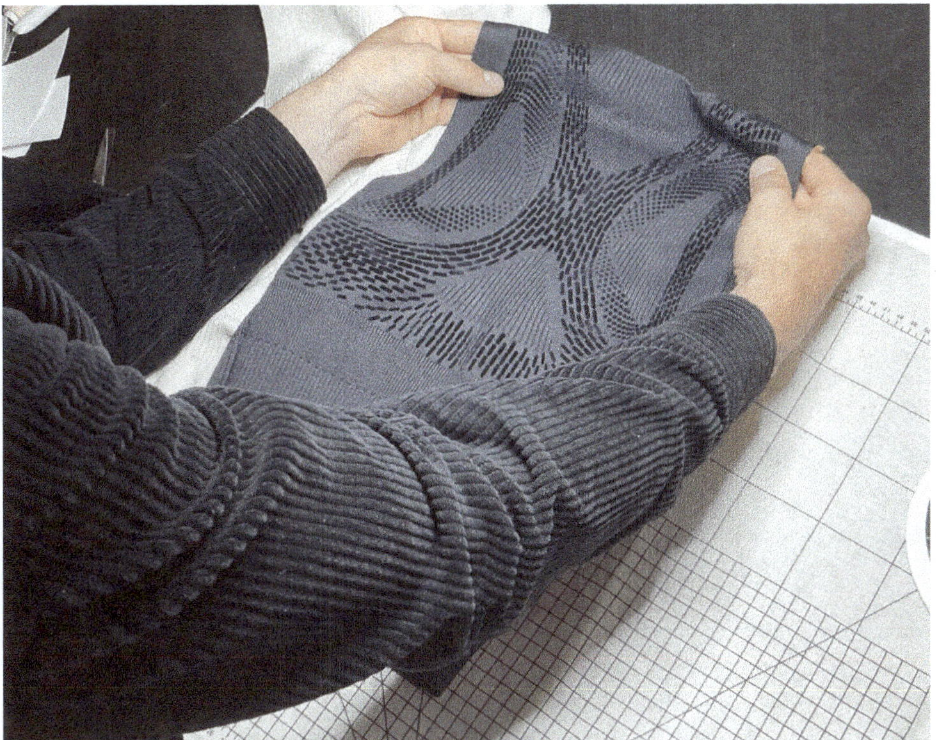

Fig. 6: Preparation for prototype fitting.

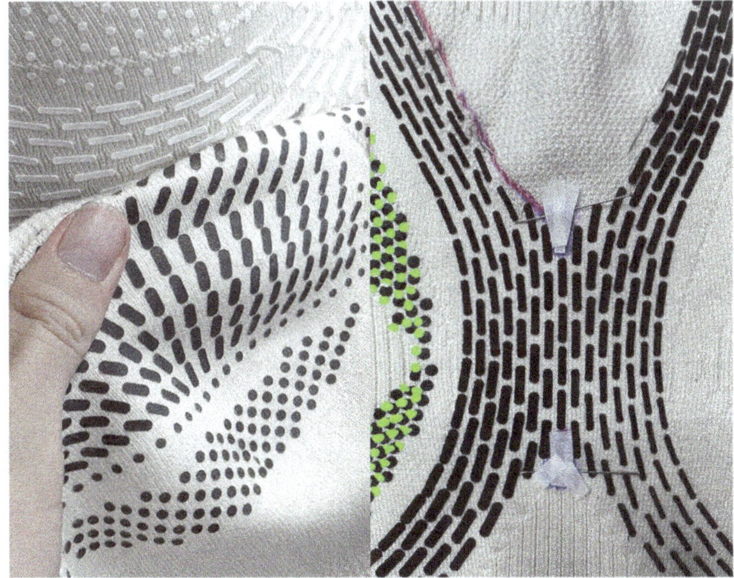

Fig. 7: Details from prototype fittings with traces of communication through the prototype.

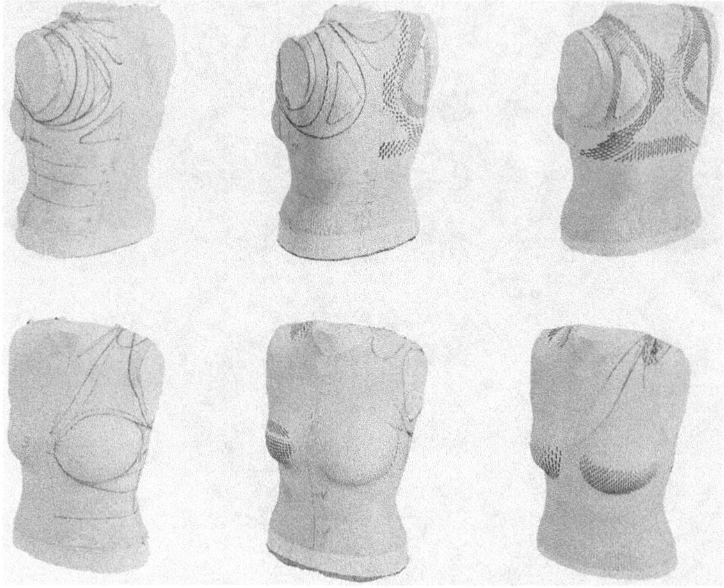

Fig. 8: Three 3D scans of prototypes (the upper row showing the back, the lower row the front).

The Magic Moment

Through providing insight into the design practice, we argue that there is no such thing as a "magic moment" or creative thunder that results in ideas and concepts. Rather, creating moments of inspiration is an iterative, communicative, and performative process—a game that is played with the material or together with the collaborators, aimed at constantly improving the object of interest. The process of tentative prototyping is a dense experience that is volatile and therefore challenging. Designers face this challenge by embracing moments of instability with the intrinsic claim that every iteration is a productive development of the one before. But how to know when to stop this process, when is a prototype finished? In this phase, the role of the designer changes once again. He or she is no longer an actor in the play but steps out of it to filter the variety of grains of material. Decisions on functionality, behavior, or spatial anchoring are given clear contours through filtering. It is, finally, a filtering process that leads to conclusions that manifest in material.

Lorenzo Guiducci and Heidi Jalkh

On Material Grammar: Learning from and Designing with Unstable Behavior in Thin Sheet Materials

Abstract

Structural instabilities in thin materials can both cause failure (e.g., when compressing a soda can) and give rise to beautiful morphogenetic phenomena (e.g., undulations in leaves edges). Here, we present our recent work on the design, mechanics, and performativity of a designed thin sheet undergoing such structural instabilities. We laser cut an inverted honeycomb from a thin sheet of polymeric foam, thereby bestowing it with auxetic properties. This design is then modified in order to introduce a self-locking mechanism, allowing the hexagonal cells to lock in an expanded, open state. The result is a sort of "augmented surface" that can be operated by simple pulling actions on its edges, activated and reset, programmed and reprogrammed. In this process, we realize that this surface has a material grammar, describing how single cells (as letters/symbols) are put together into periodic structures (as sentences) that can transform or create complex behavior (as meaning). For the reader, we hope this article illustrates some of the virtues of such transdisciplinary investigations, and how the resulting physical outcomes are indeed boundary objects, prefiguring a fertile space for research questions and opportunities in the fields of engineering, architecture, design, and performative arts.

Structural Instabilities in Thin Sheet Materials

In engineering, instabilities have been a matter of avoidance and prevention for a long time. Modeling how a technical system works, developing a suitable design, predicting possible sources of misbehavior or failure, these were—and still are—an engineer's primary goals, with instabilities and nonlinearities prefiguring the most concerns. At the same time, quite the opposite perspective can be derived from the biological sciences, where instabilities are usually at the origin of many—and beautiful—morphogenetic processes (such as the unfolding of petals and flowers upon blooming). In the last few decades, this positive connotation toward instabilities has been embraced by engineering sciences, originating the new field of mechanical metamaterials: elastic materials with a controlled geometry undergoing reversible structural instabilities, providing material scientists with a novel, geometric route toward material properties design.[1] One of these "augmenting" properties, particularly relevant for this contribution, is *auxeticity*.[2]

Researchers have been investigating auxetic surfaces by endowing sheets with an ordered pattern of cuts (as in the ancient Japanese art of *kirigami*) for some years.[3] Some of these examples show programmable textures where the cuts imprint a visual pattern on a sheet of material that is being stretched.[4] The pattern is invisible in the original relaxed state of the sheet and appears when the sheet is stretched; it therefore works as a physical data support in which information is read through a pulling action. Auxetic surface applications extend the domain of material science and can also be found in the disciplines of design, art, and architecture, where they are used as physical, responsive interfaces between humans and the built environment, for their unusual and unexpected mechanical behavior.[5] Other notable examples have shown innovative ways to leverage auxetic surface properties to design and fabricate self-supported, free-form surfaces.[6] Chen et al. proposed a specific cut pattern that not only endows a sheet with auxetic behavior, but also makes it bistable: thus, the sheet can switch between a closed, contracted state and an open, expanded one, with the possibility to program a structurally stable, target 3D surface.

In the spirit of these research works, our aim was to explore the deformation behavior of thin sheets bestowed with a pattern of cuts. We were fascinated by the possibility to create physical surfaces with unexpected morphing behavior using the "tools" of geometry and structural instabilities, to formulate new research questions about how this works, or possible uses. The exploration of the design space had to be primarily performed through simple, table-top experiments, allowing us to combine our different skill sets (from design and engineering) in a curiosity driven, transdisciplinary research experience.

---

1 Dennis M. Kochmann and Katia Bertoldi, "Exploiting Microstructural Instabilities in Solids and Structures: From Metamaterials to Structural Transitions," *Applied Mechanics Reviews* 69, no. 5 (2017), https://doi.org/10.1115/1.4037966.
2 A material is called auxetic if, upon stretching, it expands transversally to the stretch direction.
3 Ahmad Rafsanjani and Katia Bertoldi, "Buckling-Induced Kirigami," *Physical Review Letters* 118, no. 8 (2017), https://doi.org/10.1103/PhysRevLett.118.084301.
4 Ning An et al., "Programmable Hierarchical Kirigami," *Advanced Functional Materials* 30, no. 6 (2020), https://doi.org/10.1002/adfm.201906711.
5 Nassia Iglessis, "Urban Imprint," (2019), accessed March 15, 2022. https://www.nassia-inglessis.com/works-recent#/urban-imprint-1/.
6 Tian Chen et al., "Bistable auxetic surface structures," *ACM Transactions on Graphics (TOG)* 40, no. 4 (2021): 1–9, https://doi.org/10.1145/3450626.3459940.

Auxetic Surfaces Based on the Inverted Hexagonal Cell

As the design space of possible auxetic structures is practically boundless, we focused on the well-studied reentrant honeycomb design (in which two opposite internal angles of the base hexagonal cell are larger than 180 degrees) for which an intuitive understanding of the deformation behavior was possible. In addition, digital fabrication methods and craft-making have been used as an experimental research activity of making and thinking. The models for this work were manufactured with laser cutting, using three-millimeter-thick foam (EVA rubber) as the base material, allowing many tests to be carried out in short periods of time. Such a mix of methods, facile prototyping, and different disciplinary backgrounds allowed us to quickly test hypotheses and explore the available design space for cellular structures comprising inverted hexagonal cells.

A useful consequence of auxeticity is the larger shear to bulk modulus ratio: auxetic materials can be more easily stretched or compressed (large direct deformations) than distorted (small shear deformations). In an inverted hexagon, this can be traced back to the kinematics of a geometrically identical six-sided closed linkage. Compared with convex hexagons, inverted hexagons can achieve larger deformations upon stretching along their short axis: much like in the closed linkage, the reentrant sides of the hexagon rotate outward until a convex hexagonal shape is reached (fig. 1). By assembling convex polygons and inverted hexagons into a single cellular structure, the large extension of the inverted hexagon is frustrated by an adjacent convex polygon. In a thick honeycomb material, this incompatibility would cause a localized deformation in the cell walls; but since our prototypes are cut out of a thin sheet, the cell walls buckle, participating into a global out-of-plane deformation. Thus, in a simple structure consisting of four convex cells and four inverted hexagons, a simple pulling action already generates a large flat-to-spatial transition that was used to create a gripping mechanism (fig. 2).[7]

Exploration of 3D Morphing and "Locked" Cell Design

While the out-of-plane deformation resulted from differences in local expansion (i.e., at the cell level), the actual flat-to-spatial morphing also depended on the size of the system (i.e., number of cells in the structure). Therefore, we kept exploring several possible geometric designs, looking for new and unexpected behavior emerging from different combinations of the same cells. Importantly, we worked in a hands-on manner,

---

7 Heidi Jalkh, *Making Matter Active through Form: Fabricating Bio-inspired Behavior with Auxetic Structures*, master's thesis (Humboldt-Universität zu Berlin, 2020).

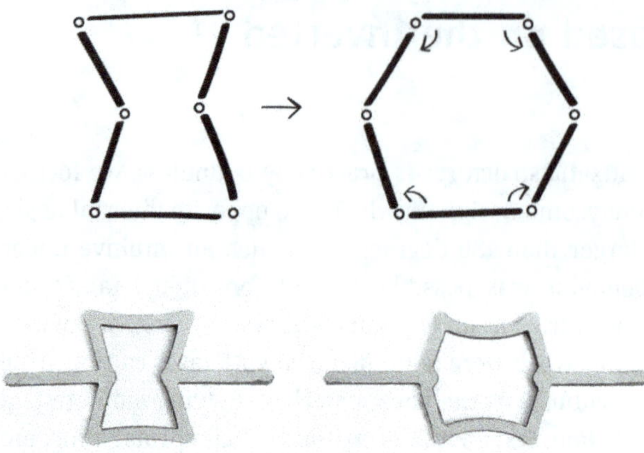

Fig. 1: When an inverted hexagon is stretched, the sides bend to reach a convex shape. The overall shape change is similar to the kinematics of a rigid six-sided closed linkage with floppy joints.

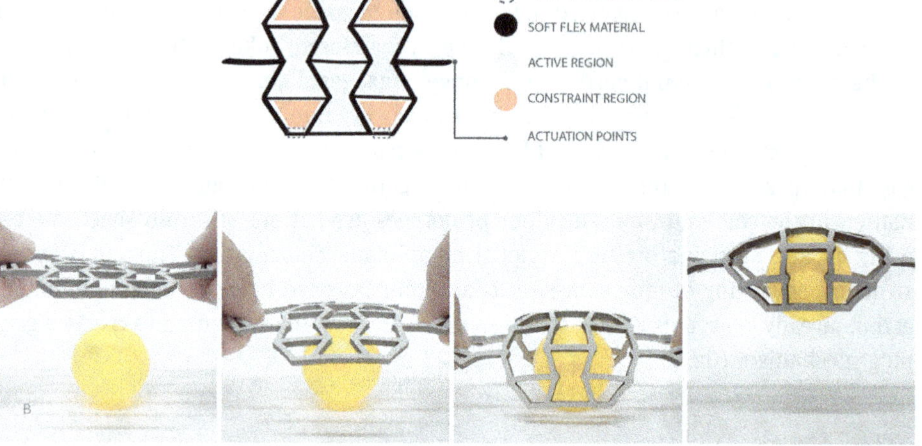

Fig. 2: 2a: 2D geometry and parameters for the design of an active structure that morphs out of the plane upon stretching; 2b: Gripping demonstration of an auxetic soft gripper picking up a 32 mm ping-pong ball.

proposing designs with a certain expected behavior, verifying our hypotheses and assumptions, and then modifying them according to our observations. By comparing, discarding, isolating, and modifying, we developed a certain intuition about what was possible to achieve in terms of spatial morphing, but also realized potentially interesting qualities of these auxetic surfaces, and bumped into some of their limitations. For example, the cellular structures worked as simple springs, that is, they could transform the elastic work from the pulling action (stimulus) into a more complex spatial reconfiguration, but this elastic energy could not be stored after the stimulus ended. While fixing the structure on its outer contour could be a solution (see fig. 3), we aimed at locking the deformation in the structure itself, thus creating a material/surface that would function as an integrated spring and latch system.

To achieve this, we reconsidered the hexagonal cell kinematics upon stretching (fig. 1), and particularly how the inclined beams almost perfectly rotate around their endpoints: thus, we reasoned, by adding appendages on two opposite inclined beams (represented by the pins piercing through the beams in fig. 3), we could create a state of self-contact in the cell while preserving its symmetries. Finally, we shaped the appendages in the form of darts such that they could be set in mechanical engagement, locking the cell in its open state.

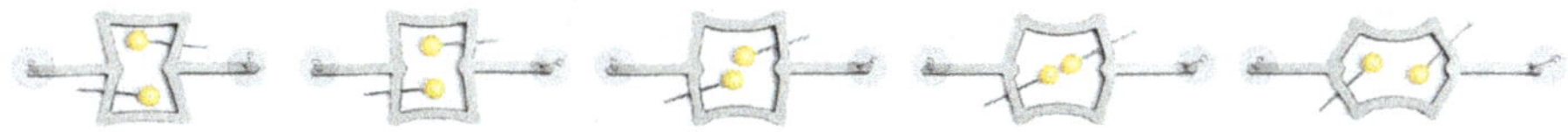

Fig. 3: An inverted hexagonal cell, at increasing stretching, externally fixed to a support (gray circles). The yellow pins showcase the rotation of two opposite inclined walls upon stretching.

Understanding Cell Operation

In more detail, a stretch along the cell's short axis pulls the dart elements apart and makes them rotate (in clockwise direction in fig. 4) until the darts' tips glide past each other. At this point, if the pulling action is removed the dart elements will be in contact and engage mechanically. Once the cell is locked in an open, engaged position, it can resist some compression. A further pulling action will not easily disengage the dart elements because these get in contact with the cells' walls; therefore, a much larger force capable of stretching the walls and making more room for the darts' rotation would be needed. Instead, in order to disengage the cell, it suffices to pull the cell along a different axis (namely, parallel to the engaging surfaces), shearing the cell while exerting a much lower force. More interestingly, if the cell is locked in its open state, it cannot be disengaged easily by pulling it along other directions (see fig. 5), since this would be impeded by the contact between the dart-shaped elements. An interesting aspect is that switching between close and open states is completely reversible, as long as the material used to fabricate the cell is elastic: this allows the programming of the state of the cell, but also the resetting or reprogramming it multiple times. The triangular shape of the dart elements prompts two different orientations for glide and engaging surfaces, and almost automatically suggests how to "operate" the cell to achieve a rest or locked position: and yet, this relationship between shape and behavior was first mediated by a physical experience, and only later extrapolated in geometrical terms.

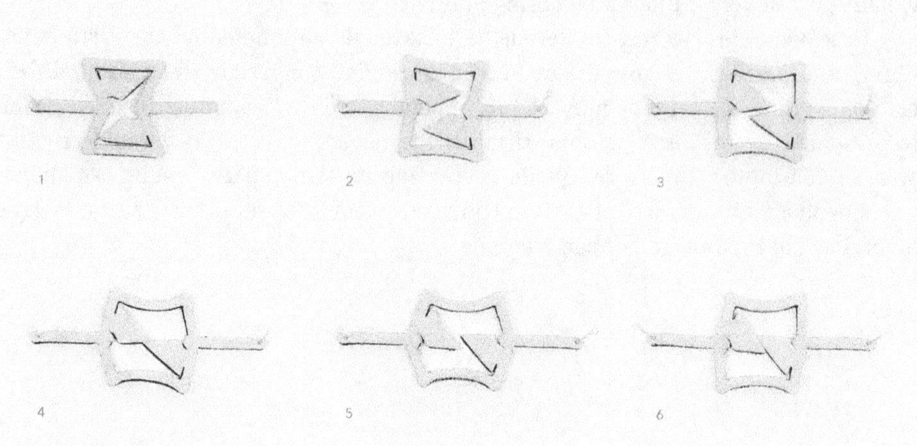

Fig. 4: Image sequence of the cell-with-darts locking mechanism.

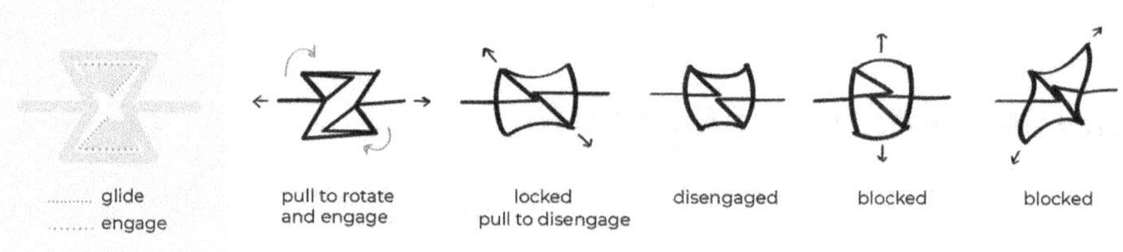

Fig. 5: Possible pulling actions to operate the cell, allowing engagement and locking, disengagement or blocking.

Opportunities in Surface Design

In its open, locked state, the cell has a larger area than in the closed, rest state; at the same time, in the open state, the cell is in a state of self-stress (where the dart elements are in compression while the cell walls are in tension). We took advantage of these two characteristics and fabricated structures composed of many such cells that could be shaped into three-dimensional, stable surfaces (fig. 6). For example, an auxetic sheet composed of mirrored cells grouped into four distinct quadrants morphs into a quadruple dome. Some additional effects can be noticed: the appearance of visual patterning, with continuous bands appearing, which also contribute to stabilizing the domes by increasing the resistance to compression.

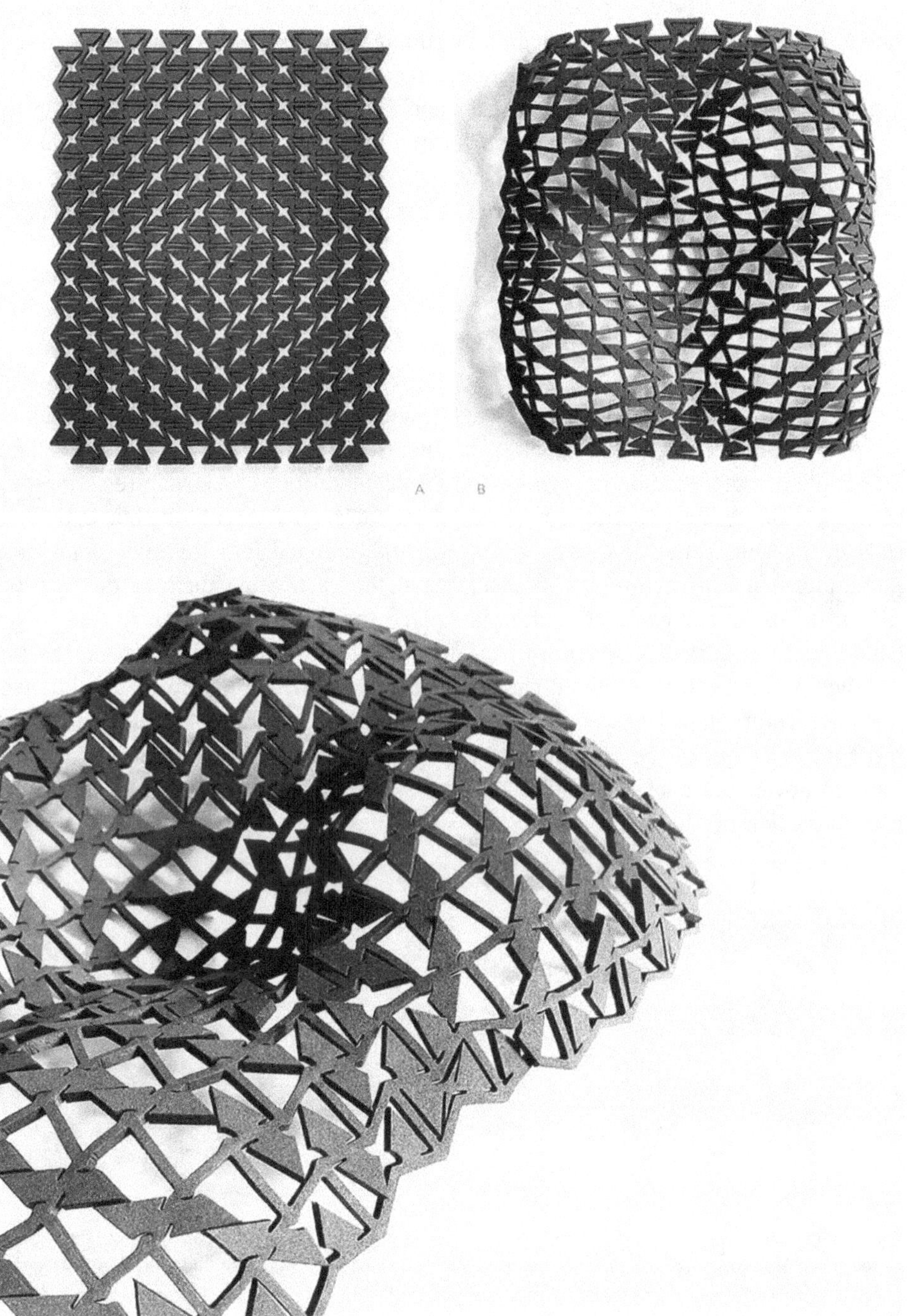

Fig. 6: An auxetic sheet composed of mirrored cells grouped into four distinct quadrants morphing into a quadruple dome.

This opens the way to program—or at least approximate—a target three-dimensional shape into a thin sheet by means of a simple yet quite powerful geometric design at the lower-scale structural units (the inverted cells-with-darts), much like observed in *kirigami* surfaces. What is most interesting about our approach is the possibility to address the stability of the morphed 3D surfaces, a beneficial aspect in architectural scenarios, where the surfaces could be conceived as self-supported structures (bearing their own weight), avoiding external tension or compression elements such as cables or masts.

Edge Activation and Surface Deployment

As the flat-to-spatial morphing in our auxetic sheets needs a local area change taking place during the cell's closed-to-open transition, the individual cells must be stretched. This could be cumbersome or impossible to implement for large structures consisting of many cells. While handling our prototypical surfaces, we realized that individual cells in the sheet interior can be addressed and snapped into the open state by a non-uniform pulling action at the sheet edges, thus activating them at distance (as shown in fig. 7). For example, pulling a square sheet horizontally on its side edges would cause a clockwise rotation of the dart elements. A pulling action on the top and bottom edges instead would cause a counter-clockwise rotation of the dart elements, therefore impeding the cells' transition to the open, locked state. Leaving a segment in the center of the top and bottom edges relaxed allows for the ability to partially remove such a block, resulting in the selective opening of a few cells in the central portion of the sheet.

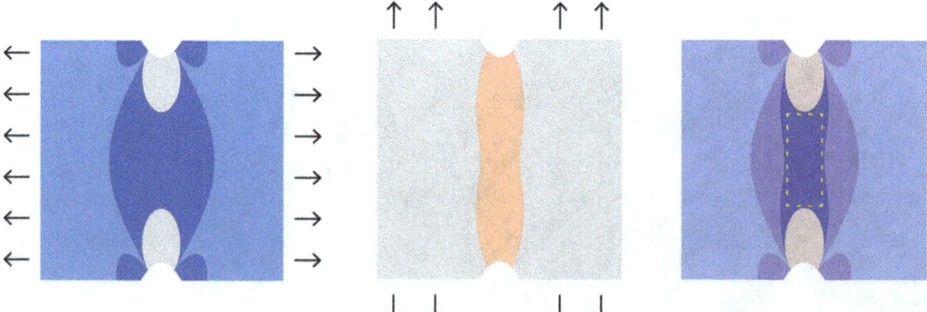

Fig. 7: Edge activation of an auxetic sheet (represented as a continuous square patch) showcasing how to selectively snap open cells at distance by pulling on the sheet's edges. Here, this procedure is targeted at opening only the cells in the sheet's central portion. In the left image, blue areas undergo large horizontal stretching. In the center image, red areas undergo low vertical stretching. In the left and center images, gray color represents areas in which the cells cannot lock in the open state. Therefore, only the cells in the central portion of the sheet—where blue and red areas overlap—will lock in the open state (dashed white rectangle in the right image).

The non-uniform edge activation that we described here could allow the deployment of a free-form architectural surface (e.g. a pavilion) without the need of additional structural support, such as scaffolding. Finally, beyond the evident prospect toward architectural applications already outlined, the specific qualities of the cellular auxetic surfaces we presented can be used to obtain more technical functionalities: for example, creating meta-surfaces with specific transfer of mechanical signals or waves.

An Analogy to Language and Biology

In this hands-on exploration (fig. 8), led by discovery rather than predefinition, we have adopted a trial-and-error strategy, leaning toward geometric features that were of value for us and learning how they influence the structure to fine-tune the in-plane and out-of-plane transformation. This process has led us to design and fabricate auxetic surfaces that can be three-dimensionally programmed, reset, and reprogrammed.

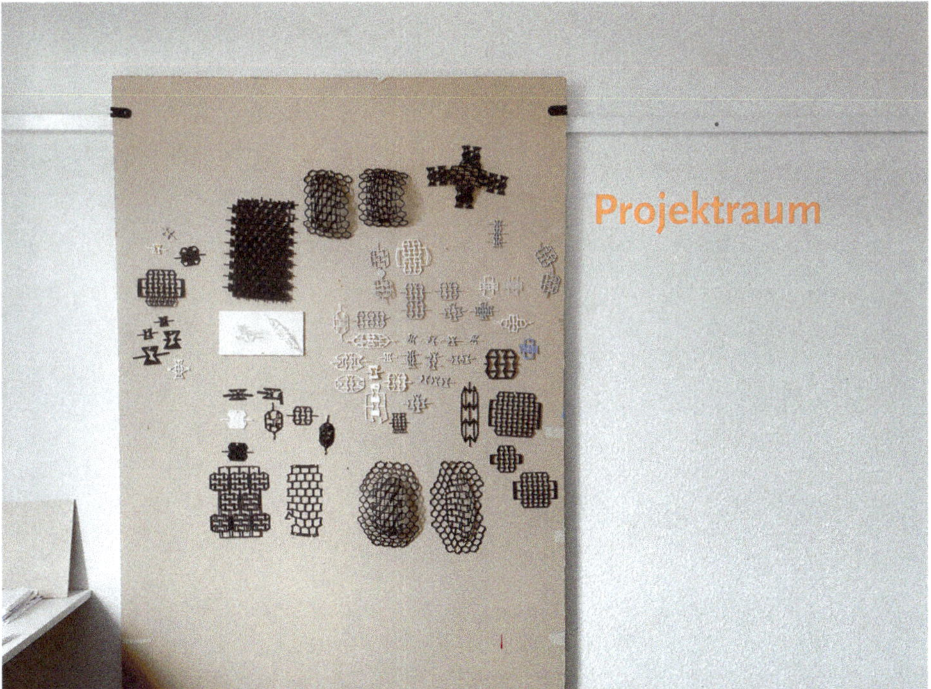

Fig. 8: Physical outcomes of the ongoing research.

Like in biology, certain functions programmed at the cell level can have an effect at a higher, systemic level, creating new, emergent behavior, depending on how cells are assembled.

Tracing a metaphor, the overall process was similar to creating a new language, and then learning how to use it. We first needed to fix the shape of the cells (as letters/symbols) and identify a set of geometric rules to operate them (as a "material grammar"). The single cells are then put together into a spatial arrangement of periodic structures that can transform and create complex morphing, just like letters are assembled into words and create meaningful sentences through grammar. Moreover, some additional visual and sound patterns emerged as bound to this specific material, just like idiomatic expressions are situated in a specific language.

Frank Bauer and Yoonha Kim

Points in Making

Introduction

> Humankind, in order to fully develop human capacities, had first to tear apart these capacities (i. e., engineering, art, philosophy), separate them, isolate them and thereby go for an insane sanity. Renaissance man set out to do so. C. P. Snow formulated the clash, or rather the split. . . . But he went on, in an addendum (1963), to again look at modernity and modernization, the beautiful world of design and the terrifying world of rationalization, and to foreshadow the coming of a Third Culture.[1]

Following these thoughts from an obituary for Romanian pioneer of HCI (Human–Computer Interaction), Mihai Nadin, one may render "points in making" as the making of a scientific method that is all-too-often presented as a given. Written by Frieder Nake, a mathematician on rather extraordinary pathways, his 1960s augmentation of computing from scientific to creative purposes in fact anticipated many questions on transdisciplinary takes on technology we are posing today. At the same time, they show how matters of methodology are inherently conditioned through epistemological turns—in other words, a "new culture of the material world"[2] the current Cluster of Excellence "Matters of Activity" (MoA) is investigating.

The present article highlights and discusses some of its current PhD cohort's "points in making," which is highly interdisciplinary from its conception.[3] Rather than an abstract theory, the intention of this piece is in an explicit account on the actual lived practice of interdisciplinarity. MoA's structured doctoral program hosts researchers from various disciplines including natural and engineering sciences, humanities, and design. Integrating both theoretical analysis and experimental approaches, the program supports intensive interdisciplinary exchange through embedding predoctoral researchers within research projects composed of diverse branches of knowledge. Employing more informal, open, and encouraging meetings within the PhD cohort, we —the two authors of this paper—attempted to give an account of how researchers adapt and adopt different kinds of collaborative practice.

Relating ourselves to such differing disciplinary constellations, we will introduce current methodological frameworks focusing on stabilization versus continuity, and

---

1 Frieder Nake, "Third Culture Man," in *A Mind at Work: We Are Our Questions*, ed. Mercedes Vilanova and Frederic Chordá (Heidelberg: Synchron, 2003), 27.

2 Wolfgang Schäffner, "Matter and Information," in Materials Research: Inspired by Nature—Innovation Potential of Biologically Inspired Materials (acatech DISCUSSION), ed. Peter Fratzl, Karin Jacobs, Martin Möller, Thomas Scheibel, and Katrin Sternberg (Munich: acatech, 2020), 79.

3 In this paper, we employ the term "interdisciplinary" in its broadest sense, involving multiple fields of knowledge. The term "transdisciplinary" is utilized in specific instances where collaborations across different disciplines result in moving beyond discipline-specific approaches.

the discursive space that emerges along with their juxtaposition. On the one hand, we will draw on Andrew Pickering's argument of how the "success of science and engineering (and all sorts of other practices) shows us that there are . . . islands of stability in the flux of becoming."[4] On the other hand, our text will follow philosopher Yuk Hui's recent book *Art and Cosmotechnics* as it challenges epistemic understandings, deriving from Ancient Greek philosophy, of antagonisms as something to be overcome by renewing "contradiction as limit without searching for a reconciliation."[5] Against this discursive background, a qualitative survey among our fellow researchers presents some tentative analytical categories to investigate how collaborative research is situated in complex and organic environments, often torn between indeterminacy and variety—how reciprocal perspectives on our tools confuse and augment the whys and hows of our endeavors, and, through such processes of placing continuity over reconciling contradiction, challenge seemingly static disciplinary boundaries of our research culture (fig. 1).

Disciplinarity: Discarding Stabilization for Continuity?

Obviously, such an argument builds heavily upon existing work on notions of disciplinarity. Our working definition of disciplinary boundaries, and, notably, the manifold ways of gathering and crossing them, locates such actions as a "purpose-oriented" way of working: A) intra- or monodisciplinary, that is, working within a single discipline; B) cross-disciplinary, that is, viewing one discipline from the perspective of another; C) multidisciplinary, after what physicist Basarab Nicolescu refers to as "studying a research topic not in just one discipline but in several at the same time . . . yet limited to the framework of disciplinary research;"[6] D) interdisciplinary, that is, integrating knowledge and methods from different disciplines, using a real synthesis of approaches; and E) transdisciplinary, that is, creating a unity of intellectual frameworks beyond disciplinarity.[7]

The resulting diagram (fig. 2) may be read as a reflection on the productivity of boundaries, how researchers stabilize and destabilize their relation to each other as well as their body of research—an approach that may invoke sociologist Andrew Pickering, whose abstract to a 2014 paper titled "Islands of Stability" argued:

---

4 Andrew Pickering, "The Ontological Turn: Taking Different Worlds Seriously," *Social Analysis* 61, no. 2 (2017): 140, https://doi.org/10.3167/sa.2017.610209.

5 Yuk Hui, *Art and Cosmotechnics* (Minneapolis: University of Minnesota Press, 2020), 45.

6 Basarab Nicolescu, "In Vitro and In Vivo Knowledge: Methodology of Transdisciplinarity," in *Transdisciplinarity: Theory and Practice – Advances in Systems Theory, Complexity, and the Human Sciences*, ed. idem (Cresskill, NJ: Hampton Press, 2008), 1–21.

7 Helga Nowotny, Peter Scott, and Michael Gibbons, *Re-Thinking Science: Knowledge and the Public in an Age of Uncertainty*, Repr (2001; repr., Cambridge: Polity Press, 2011).

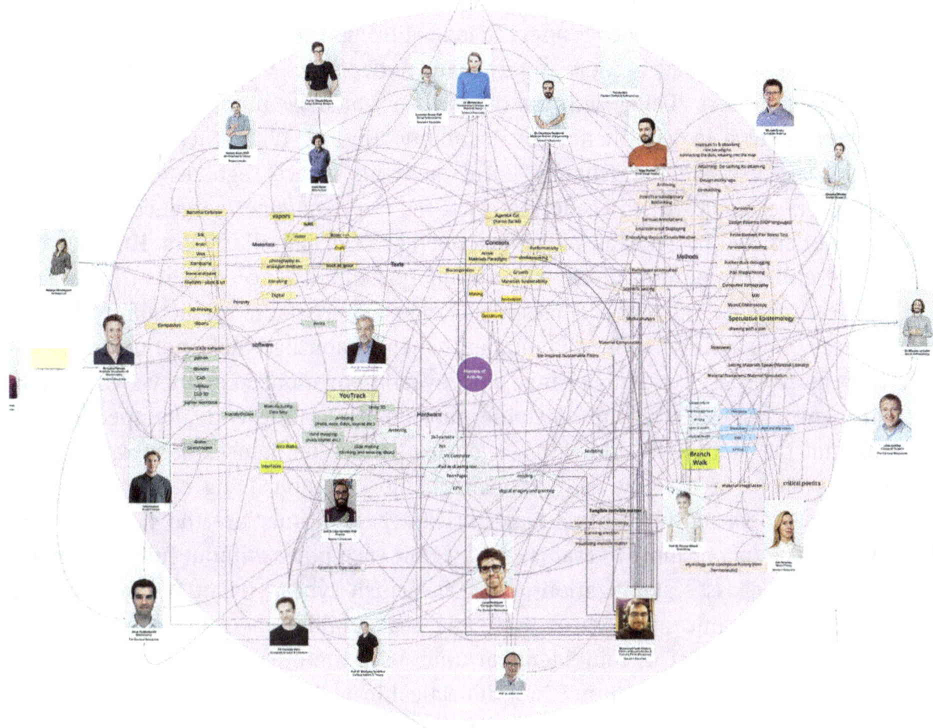

Fig. 1: Collaborative Miro board screenshot of the workshop "Share your Toolbox" during Brown Bag Brunch at the Cluster of Excellence "Matters of Activity," 2021.

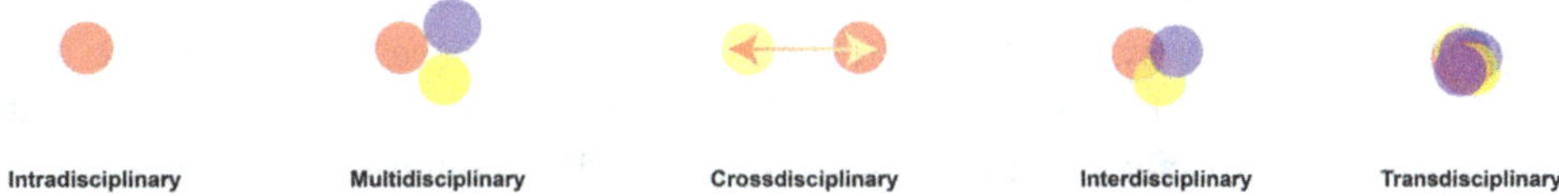

Fig. 2: Frank Bauer and Yoonha Kim, "Disciplinarities," 2022, after Alexander Refsum Jensenius, "Disciplinarities: Intra, Cross, Multi, Inter, Trans," blog post, 2012, https://www.arj.no/2012/03/12/disciplinarities-2/.

Instead of considering "being with" in terms of non-problematic, machine-like places, where reliable entities assemble in stable relationships, STS conjures up a world where the achievement of chancy stabilisations and synchronisations is local. We have to analyse how and where a certain regularity and predictability in the intersection of scientists and their instruments, say, or of human individuals and groups, is produced.[8]

---

8 Andrew Pickering, "Islands of Stability: Engaging Emergence from Cellular Automata to the Occupy Movement," *ZMK* 5, no. 1 (2014): 121–34, here quoted from the abstract, 166, https://doi.org/10.28937/1000106410.

While Pickering declares stabilization of relationships as a way of engaging epistemic emergence, we would invite our readers to look at things the other way around. Adding on our introductory image of MoA as a dynamic network, we would pose the question: What if one considers "being with" as the way in which seemingly rigid disciplinary boundaries dissolve in "confluences of instability" (fig. 3)? While Pickering's "Islands of Stability" focuses on the "chancy" local moments where regularity is produced, this perspective would invite readers to look for the dynamic emergence of vortices A, B, C, to I depicted by Thomas Wiering, a northern German illustrator. His contribution to the enlightenment as another period of epistemic change renders the demand for more contingent dances of agencies that lead anywhere but to predictability. Rather, confluences of instability invoke the moment when established tools, notions, and materialities more or less suddenly change and reveal innovation. Soon, they swirl together and form irregular and unpredictable movements dealing with the specific constraints that tag along with individual agencies. These confluences of instability can be thought about in the line of what anthropologist Anna Tsing calls "productive 'seams' between varied intellectual trajectories, where 'seams' are the visible line in the swirl where two or more currents of water hit each other without fully merging."[9]

Pickering discusses stabilizations, such as British cybernetician W. Ross Ashby's homeostat, a mechanical apparatus that retrieves stability when disturbed. The machine composed of a set of four identical units uses the others as a source of their input and direction of their output in a dialectical logic. With every additional player added to his game, however, his search soon becomes a practical impossibility. Therefore, one may ask whether an alternative way of dealing with the "chancy stabilizations" of knowledge could begin once we embrace the necessary initial "mess" of any interdisciplinary work. This bridges to recent formulations of "slow science," which, as Belgian philosopher Isabelle Stengers puts it, "needs to enable researchers to accept what is messy, not as a defect, but as what we have to learn to live and think in and with."[10] Let us complement this argument with a line of thinking that sees polarity not so much as contradiction—that is, discontinuity—but as continuity. Yuk Hui in his book *Art and Cosmotechnics* remarks how Ancient Greek tragedy and Daoist thinking use oppositions as necessities for unfolding their respective logics (and cosmologies), although oppositions are resolved in different ways:[11] while the ancient Greek hero tries to overcome the clash between freedom and destiny through sheer will, in his Eastern, Daoist example, "[c]ontradiction doesn't mean there is need for reconciliation,"[12] as opposite forces bring together a dynamic system of trans-

---

9 Anna Tsing, "When the Things We Study Respond to Each Other: Tools for Unpacking 'the Material'," in *Anthropos and the Material*, ed. Penny Harvey, Christian Krohn-Hansen, and Knut G. Nustad (Durham, NC: Duke University Press, 2019), 221–43.
10 Isabelle Stengers, "'Another Science Is Possible!': A Plea for Slow Science," in *Demo(s)*, ed. Hugo Letiche, Geoffrey Lightfoot, and Jean-Luc Moriceau (Leiden: Brill, 2016), 66.
11 "Opposition" here meaning a contrast or antithesis.
12 Hui, *Art and Cosmotechnics*, 45.

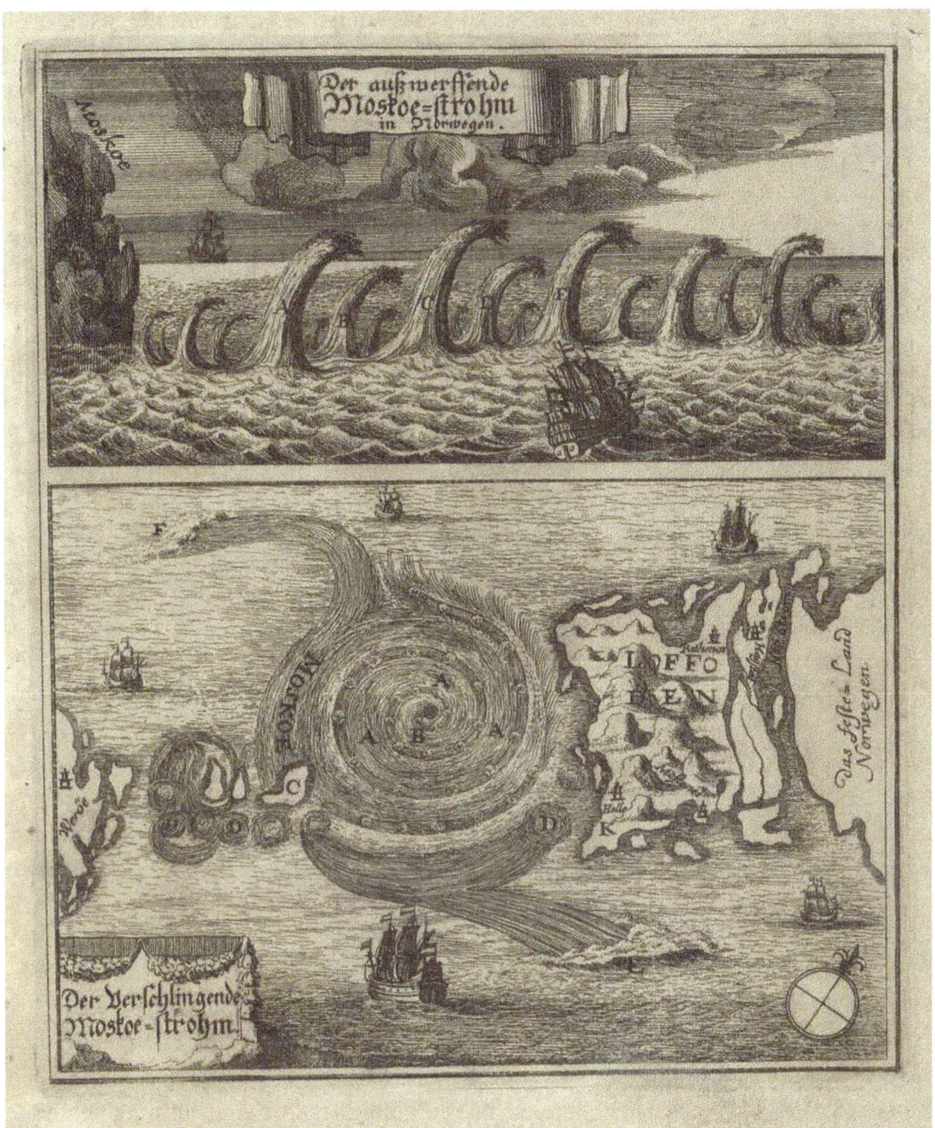

Fig. 3: Thomas Wiering, Der außwerffende Moskow=Strohm in Norwegen, Der verschlingende Moskow=-Strohm, 1683, copperplate print, 15.5 × 13 cm.

formation.[13] In part, this also brings a shift of perspective, turning Pickering's "islands," the moment when and where predictability or regularity is produced, into

---

13 Which is often rendered as "The successive movement of yin and yang constitutes the Way (Dao)"

more contingent confluences of instability. To operationalize such conceptions through methodology, we invited our fellow PhD researchers into the process of developing this paper and asked them to share and discuss inquiries, insights, and incidents that may be useful to qualitatively test its hypothesis. As a result, the following paper will discuss four approaches to interdisciplinary work in the program that we would refer to as "points in making."

Dance of Agency

Doing science in and for a messy world can be described as "dance of agency,"[14] another term borrowed from Pickering. As such, he directs his interest toward knowledge production, rather than scientific cognition and reasoning, toward interactions between people and things, processes that oscillate between phases of activity and passivity—a form of performative experimentation as these varieties meet and turn into an organic movement. Arguably, such performative notions of structure may only materialize in performative methods of research.

> What we need is a space that will allow us to merge these heterogeneous structural fields in order to make room for something new, something yet unknown. ... This is the challenge facing us today: instead of a theory of a new structuralism, we rather need multiple practices of an interdisciplinary laboratory focused on structure.[15]

Resonating with Wolfgang Schäffner's 2016 ambition to reveal a "New Structuralism" as an epistemic vehicle for research, MoA's "Experimental Zone" may be seen as a form of performative experimentation (fig. 4). If a science lab controls the environment, this space allows living organisms and modern devices across diverse fields to flow. The mildew smell of a self-assembling "fabric" from cellulose merges with a hint of chili following an explosion of a jar from a fermentation experiment in another corner. The scented air drifts to the corner of busy hands forming willow structures that are later put on bodies, sparking discussions on wearables as homeostatic membranes. As the scent floats without knowing the borders of disciplines, contingent physical actions and conversations emerge across modes of thinking. Physically locating oneself in such an interdisciplinary setting increases the chance of contingently encountering new ideas and methods. As our fellow Michael Tebbe explains, "[w]hen I'm stuck in

---

(一陰一陽之謂道). See I Ching, "Appended Remarks," pt. 1, ch. 5. Cf. James Legge, "The Yî King," in *Sacred Books of the East* (Oxford: Clarendon Press, 1899), XVI:355–56.

14 See Andrew Pickering, *The Mangle of Practice: Time, Agency, and Science* (Chicago: University of Chicago Press, 1995), 21–24.

15 Wolfgang Schäffner, "New Structuralism: A Field of Human and Materials Science," in *GAM 12: Structural Affairs: Potenziale und Perspektiven der Zusammenarbeit in Planung, Entwurf und Konstruktion / Opportunities and Perspectives for Cooperation in Planning, Design and Construction*, ed. Technische Universität Graz (Berlin: Birkhäuser, 2016), 31.

my work, I don't need to plan and go see an expert. I just ask someone nearby to see how it is in their field. And something new emerges."[16]

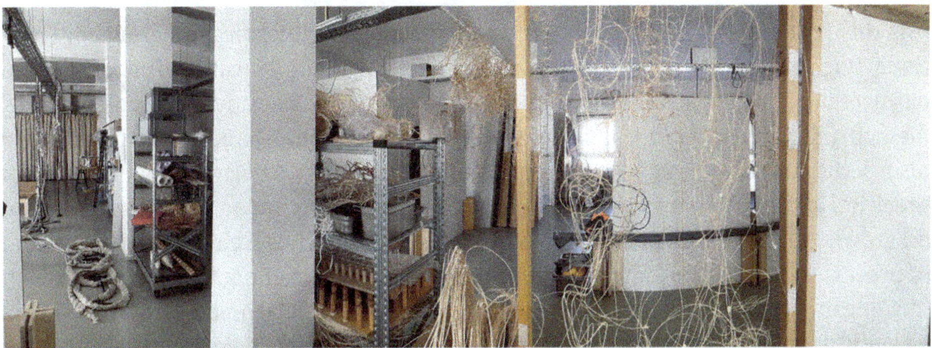

Fig. 4: Experimental laboratory at the Cluster of Excellence "Matters of Activity."

Oppositional Continuity

Bridging rather complex fields of knowledge often begins by asking for simple analogies and differences, whether formal or functional—even though they, too, may eventually turn out to be fiction or failures. Adding to Hui's thoughts on reconciliation, one may frame this as "oppositional continuity," referring to antagonism rather complementing than contradicting each other to invoke a larger unifying moment. As another dimension of our survey, we would see traces of such "oppositional continuity" in Binru Yang's example. Dealing with clean and optimized data sets from forward design[17] and engineering throughout her previous studies, her current work in collaboration with biologists and material scientists finds inspiration in inverse design logic, forcing her to deal with the opposite: a superimposed, unpredictably huge amount of biological data from examples such as a stingray bone, with complex material properties reflecting gradual evolution over hundreds of millions of years. Being trained to look for stabilized settings, she was perplexed to encounter dynamic data in biology where things evolve overtime, turning "messy," to speak with Stengers again:

> A civilized mode of appreciation would imply never identifying what is experimentally well-controlled and precise, with a truth that is valid in emergent circumstances and in a messy world. Laboratory science rarely transcends the boundaries of its assumptions. Under such and such restricted conditions, something may be true; but in the complex reality of life, subjected to unpredictable change and characterized by indeterminate multiple relations, it may be nearly impossi-

---

16 Michael Tebbe, notes on "Points in Making," 2021.

17 Referring to the antagonism of forward and inverse design logic, where the former would go from cause to effect, and inverse ones from effect to cause.

ble to identify any such 'truths'. What is messy from the point of view of fast science is nothing else than the irreducible and always embedded interplay of processes, practices, experiences, ways of knowing and values, which make up our common world.[18]

It is this paradigm of accuracy in engineering, which architectural theoretician Francesca Hughes recently challenged as a rather misfortunate outcome of modern standardization and industrialization. Taking her ideas on alternative and possibly augmented relations of matter and authorship seriously, this would mean to embrace disorderly information as natural complexity, while safely navigating the "redundant precision iceberg"[19] Binru was spotting as well. In response, and in order to collaborate on growing data sets, engineers and biologists continuously develop a filtering system to process outliers in the data reflecting each characteristic of organic matter. Through this ongoing correspondence, the biological data becomes clean enough for engineers to work with it while it still contains valuable meaning for biologists. The threshold would be considered for each case individually, with respect for the characteristics and requirements of both fields. Conventionally, opposing methodologies enter into a productive dialogue, whose recursive logic brings an epistemological surplus for all involved parties' epistemologies—in Hui's words, "to recognize the differences in order to reflect upon the question of diversity as a possibility for moving forward,"[20] in what he then frames as an "overtaking of recursive machines."[21] In Binru's case, it appears that precisely these dichotomies complement each other, constituting continuous reciprocal relations as her team aims to find explanations for the emergence and adaptation of certain patterns to functional or environmental stimuli.

Externality as Catalyst

The introduction of foreign knowledge and/or skills into production processes is a well-known creativity technique. While Horst Bredekamp traces it back to Leibniz's exploration of a notion of active spaces resonating between external impulse and internal activity,[22] Claudia Mareis maintains its significance for the emergence of knowledge

---

18 Isabelle Stengers, "'Another Science Is Possible!': A Plea for Slow Science", in *Demo(s)*, ed. Hugo Letiche, Geoffrey Lightfoot, and Jean-Luc Moriceau (Leiden: Brill, 2016), 66.
19 Francesca Hughes, *The Architecture of Error: Matter, Measure, and the Misadventures of Precision* (Cambridge, MA: MIT Press, 2014), 5.
20 Hui, *Art and Cosmotechnics*, 48.
21 Ibid.
22 Horst Bredekamp, "Leibniz's Concept of Agens in Matter, Space, and Image," in *Active Materials*, ed. Peter Fratzl et al. (Berlin: De Gruyter, 2021).

in all sorts of creative processes.[23] We propose to unravel this concept usually reserved for the realm of art and design in the general scope of the interdisciplinary setting.

Felix Rasehorn, a practice-based researcher in product design, encountered a moment of externality functioning as a catalyst in bringing forth his research project. While attending one of the regular Zoom meetings of the Tessellated Material System group, Lennart Eigen from Nyakatura Lab[24] shared a micro-CT scan of the outer skin layers of a boxfish (*Lactoria cornuta*). An image from Eigen's research displayed the differences in stiffness and material density of the boxfish's scutes (tiles). It induced Rasehorn to translate the same image in terms of its structure. To him, this image demonstrated the constructive forces of nature expressing a geometrical principle that is the duality of patterns. He went on to summarize:

> These specific moments feed the development of my own thesis project. Sometimes I feel the thesis is a place where all those "moments" are being accumulated into one continuous stream of thought . . . . It is less your own work or findings that are essential but rather the accumulation of things that others see in your project.[25]

In an interdisciplinary milieu, Rasehorn's research, underlined by such externalities, potentially becomes a kindling, igniting whole of thoughts and practices for others who cross paths with it. Projects in such settings, arguably, accumulate greater significance when they catalyse thought generation and establish new practices with peers. Consequently, the completion of a confined research project within a single discipline may not be the paramount goal. Rather, the ensuing dialogue—particularly what "others see in your project"—might hold precedence. This dialogue mandates surmounting challenges related to diverse vocabularies and conceptual understandings, employing visualization techniques and analogies as bridging mechanisms.

Reasoning on these moments and their relevance for his working process with Rasehorn, one may partially connect them with anthropologist Annemarie Mol, using a baked French dessert (clafoutis) to discuss *what it is to hang together.* Mol outlines how in the making of clafoutis, diverse elements from the worlds of agriculture, cuisine, nutrition, and sensuousness are not only juxtaposed but rather assembled and recomposed into a new entity.[26] Speaking with her, this would mean to understand one's project as a cumulative embodiment of perspectives from externalities. Moreover, this emerging dialogue could be seen as "making the difference between a random jux-

---

23 Claudia Mareis, "Methodische Imagination—Kreativitätstechniken, Geschichte und künstlerische Forschung," in *Kultur- und Medientheorie*, ed. Martin Tröndle and Julia Warmers (Bielefeld: transcript Verlag, 2011), 203–05.

24 Humboldt-University of Berlin's Research Laboratory for Comparative Zoology.

25 Felix Rasehorn, notes on "Points in Making," 2021.

26 Annemarie Mol, "Clafoutis as a Composite: On Hanging Together Felecitously," in *Modes of Knowing from the Baroque*, ed. John Law and Evelyn Ruppert (Manchester: Mattering Press, 2016), 247–48.

taposition and a felicitous composite"[27] that catalyzes the *hanging together* of heterogeneous parts.

Werkzeug as Denkzeug

> A werkzeug (device for making) is finite and programmed, that is: perfectly suited to serve its purpose. . . . The denkzeug (device for thinking) is very different from this: it requires the permanent and open-ended interaction with its user. . . . A werkzeug, which is used as opposed to its conception, for other purposes and thus gets deconstructed, may at any time become a denkzeug.[28]

The embracing of such external perspectives may at times be complicating our workflows but can work as an inspiring device to use our tools differently. Michael Tebbe shared an example of a transdisciplinary knowledge-making process between disciplinary (here: the developer) and external knowledge (here: the domain expert, i.e., a person with particular knowledge and/or skills in the field), which required a change in perspective from standard procedure. As part of a workshop, developers were showcasing the "Similarity Checker" to domain experts; a module on the software platform Jupyter Notebook to compare sentences for their semantic similarity. Inserting two sets of sentences, one can compare and calculate a score between 0 and 1 on how much the lists mirror each other. In the early stages of the development of the pipeline, this module has been used as a first test of the model, and inputs were given without much concern for the actual content. Such probing practices are common in software development and serve to test the functionality of the written code. This is, however, usually not available as a function in a finished software product, because it does not serve any further purpose. However, as the external user started questioning the content of the probing, the "device for making" turned into a "device for thinking." Domain experts came up with sentences such as, "Where are my glasses?" or, "Can I have a glass of water?" to compare it with the word "empty." Which sentence out of the two is more similar to "empty"? Or what if we compare "empty" with a meaningless sentence such as "thing thing thing thing"? In this case, the content of the probing practices turned out to be very valuable as a way to understand the dynamics of the model and has potentially been made available outside of software development. At the same time, the developer had to change perspective by making the practices during the development process transparent, which Tebbe would consider a crucial result of *Critical Technical Practice.*[29]

---

27 Mol, "Clafoutis as a Composite," 258.

28 Thomas H. Schmitz and Hannah Groninger, "Über projektives Denken und Machen," in *Werkzeug – Denkzeug Manuelle Intelligenz und Transmedialität kreativer Prozesse*, ed. idem (Bielefeld: transcript Verlag, 2012), 19–20. Translation by the authors.

29 Geoffrey C. Bowker, ed., *Social Science, Technical Systems, and Cooperative Work: Beyond the Great Divide* (Mahwah, NJ: Lawrence Erlbaum Associates, 1997).

Conclusion

This paper attempted to embrace *Points in Making* in their challenging of those disciplinary constellations and configurations that are taken as a given. While this given—Nake's distinct scientific methods and conventions—tend to solidify disciplinary boundaries and may therefore become an obstacle in conducting truly transdisciplinary research, there appear to be good reasons to take a more critical and challenging stance —encouraging ourselves to "share our toolbox," or to pose the question, "Where does our knowledge come from?" Through these and similar inquiries addressed in workshops by the authors and their fellows over the course of the past year at MoA, we can start to reveal oppositions and boundaries not only horizontally, but also vertically. This would mean entering into a discourse with positions like Marwa Elshakry's argument for more contingent and diverse knowledge-making practices from nineteenth-century Egypt and China:

> Perhaps the greatest lesson we can learn from revisiting the construction of global histories of science in this way, then, is why some forms and communities of knowledge have come to matter more than others. . . . Showing how international vectors of knowledge production, how missionaries and technocrats created new global histories of science through the construction of novel genealogies and through a process of conceptual syncretism.[30]

Reality is entangled, situated, and ambivalent. Rather than dissolving or constraining such complexity, transdisciplinary methodologies encounter and embrace such multiplicity. And while a simplified ontological plane guided by one type of logic can only reveal a partial perspective, complex realities call for just as complex ways of research.[31] While such disciplinary work will merely focus on specific and therefore partial perspectives, "a new culture of the material" may only emerge through a truly transdisciplinary collaboration of humanities, sciences, and design within a more-than-human world. Rather than tearing apart our capacities into different bits and epistemologies, to speak with Nake, the MoA PhD cohort from diverse disciplines deals with this reality pars pro toto by learning from and actively tipping methods, notions, and subjects of research they are initially assigned to in their respective fields. One could certainly argue how writing this text and selecting and highlighting certain aspects of confluences may also be thought of as generating stabilization. However, as this paper is contested and brought into dialogue again during the process of editing and sharing, it rather continues to be an active and instable piece. As each successive

---

30 Marwa Elshakry, "When Science became Western: Historiographical Reflections," in *Isis* 101, no. 1 (2010): 109, https://doi.org/10.1086/652691.

31 Basarab Nicolescu, 'Gurdjieff's Philosophy of Nature', in *Gurdjieff: Essays and Reflections on the Man and His Teaching*, ed. Jacob Needleman, George Baker, and Mary Stein (New York, NY: Continuum, 2004). Quoted in Linda Neuhauser, 'Practical and Scientific Foundations of Transdisciplinary Research and Action', in *Transdisciplinary Theory, Practice and Education*, ed. Paul Gibbs (Cham: Springer International Publishing, 2018), 31.

PhD cohort proceeds with its very own reflection on the confluences of instabilities, we hope to encourage not only the recognition of an inherent dynamic in interdisciplinary collaboration, but also foster an ecosystem that is vibrant and ever open to transformation.

Acknowledgements:

We would like to thank the PhD cohort of "Matters of Activity" and Franziska Wegener for the invaluable conversations on working across disciplines. We are particularly grateful to Léa Perraudin and Clemens Winkler for their insightful feedback on this text.

____

Growing Form

Regine Hengge and Myfanwy Evans

Buckling, Wrinkling, and Folding: Microstructure, Active Matter Behavior, and Geometric Modeling of Bacterial Biofilms

Biofilms—The Multicellular Form of Bacterial Life

Bacteria are usually seen as single cell organisms that can swim around in the environment or in our bodies. But bacteria have also invented multicellularity—in fact, they even prefer to live in structured communities termed *biofilms*. By definition, biofilms are large aggregates of bacterial cells held together by an extracellular matrix of self-produced biopolymers.[1] However, these communities are not just simple heaps of bacterial cells: rather, they can form complex three-dimensional patterns that become visible to the naked eye.[2] In the laboratory, these biofilms are called macrocolonies. They are usually grown over several days on agar plates until they reach a diameter of two to three centimeters (fig. 1).

Macrocolony biofilms of many bacterial species form intriguing folding patterns that are quite variable. These patterns depend on the genetic makeup and can thus differ for certain mutants, but they are also influenced by the actual growth conditions, for example the chemical composition of the growth medium, the growth temperature, and the humidity of the agar support (fig. 1 shows macrocolonies of several *Escherichia coli* strains obtained at three different temperatures). Three-dimensional patterns arise by buckling and folding of entire areas of the macrocolonies. At first glance, the explanation for these surprising large-scale movements is quite simple. These cells stick together because of their extracellular matrix (fig. 2a). As a consequence, increasing numbers of growing cells cannot just pile up independently, but they behave like a growing tissue—in order to fill additional space, they have to buckle up and fold as a connected consortium.

The dynamic appearance of these folding and wrinkling patterns has been analyzed in time-lapse movies that for *E. coli* macrocolonies on nutrient-rich media usually cover seventy-two hours. To start biofilm growth, a tiny droplet containing *E. coli* bacteria is put onto an agar plate that often also contains the dye Congo red, which turns red upon binding to the extracellular matrix. Therefore, a growing macrocolony turns dark red as soon as the cells start to produce the fibrous matrix components, which

---

1 Hans-Curt Flemming and Stefan Wuertz, "Bacteria and Archaea on Earth and Their Abundance in Biofilms," *Nature Reviews Microbiology* 17 (2019): 247–60, https://doi.org/10.1038/s41579–019–0158–9.
2 Diego O. Serra, Anja M. Richter, and Regine Hengge. "Cellulose as an Architectural Element in Spatially Structured Escherichia coli Biofilms," *Journal of Bacteriology* 195 (2013): 5540–54, https://doi.org/10.1128/JB.00946–13.

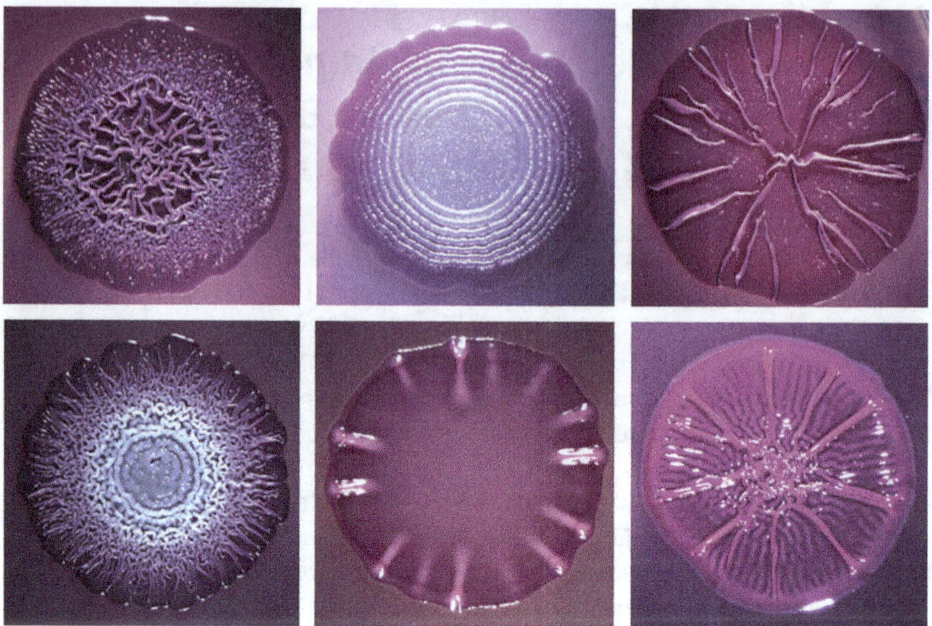

Fig. 1: Morphological diversity of bacterial macrocolony biofilms: biofilms of different strains of the human gut bacterium *Escherichia coli* were grown for three days at different temperatures (between 20° C and 37° C) on agar plates containing a complex mixture of nutrients as well as the dyes Congo red (which binds to the extracellular matrix) and Coomassie Brilliant Blue (which binds to various proteins). The diameter of the colony biofilms is approximately 25 mm.

happens after approximately twelve hours of growth. By twenty-four hours, small regular ripples begin to appear in a ring just behind the outer edges of the round macrocolony. Over time, these ripples become larger buckles that after forty hours further develop into radial ridges that propagate towards the colony center, where they finally intertwine and mix with smaller local wrinkles.[3]

The Interplay of Material, Form, and Function in Macrocolony Biofilms

Why are wrinkling macrocolony biofilms interesting to researchers at the Cluster of Excellence "Matters of Activity"? The answer is that these biofilms are a superb model to study fundamental relationships between material, form, and function.

---

3 Diego O. Serra and Regine Hengge, "Experimental Detection and Visualization of the Extracellular Matrix in Macrocolony Biofilms," in *C-di-GMP Signaling: Methods & Protocols—Methods in Molecular Biology*, ed. Karin Sauer (New York: Humana Press, 2017), 133–45.

Since biofilms are living matter, these relationships depend on an interplay of genes, which means an endogenous source of information that defines the entire repertoire of what is possible, and the environment, which determines which part of this genetic potential is actually realized and when and where this happens.

The biological *function* of biofilm formation becomes clear when we think of a biofilm as a "bacterial fortress." Within this fortress, bacteria are well protected against most physical or chemical stresses and also against other microbes that love to eat bacteria. In addition, biofilm formation provides the potential for homeostasis, which means bacteria can "design" their immediate extracellular space, which is internal to the biofilm, according to their specific requirements. The result is a strong reduction in maintenance energy, which means that multicellularity is thermodynamically favored and therefore can be considered an emerging property of cellular life itself.[4] A biofilm thus represents a microbial example of a space of "extended physiology," as it has been described in detail for the large-scale structures built by social insects such as termites, ants, or bees.[5]

In the following, however, the focus lies on the relationship between *material*—at the molecular and cellular scale—and macroscopic *form*. Thus, we have to take a closer look at the "building materials" of biofilms and how these arrange into a biofilm-internal microarchitecture. We also have to consider the origin of the tissue-like elasticity that allows buckling and folding without breakage. Furthermore, the question arises of what drives the macroscopic activity of the system, that is, the morphogenetic folding movements that generate the complex patterns of ridges and wrinkles. Finally, we address the geometry that emerges during the transition from a flat layer of proliferating cells to an elaborately folded and wrinkled macrocolony.

The Molecular Material Basis of the Extracellular Biofilm Matik

The building materials of a biofilm are bacterial cells and the extracellular matrix, with the latter basically consisting of a network of entangled fibers around the cells (fig. 2a). In the case of the *E. coli* biofilms that are used here as a model, there are just two types of fibers. On the one hand, these are "curli" fibers which consist of proteins in the very stable β-amyloid conformation (which is very similar to the structures

---

4 Regine Hengge, "Linking Bacterial Growth, Survival and Multicellularity—Small Signaling Molecules as Triggers and Drivers," *Current Opinion in Microbiology* 55 (2020): 57–66, https://doi.org/10.1016/j.mib.2020.02.007.

5 See J. Scott Turner, *The Extended Organism: The Physiology of Animal-Built Structures* (Cambridge, MA: Harvard University Press, 2000).

found in Alzheimer plaques).[6] When certain strains of *E. coli* produce only curli fibers as an extracellular matrix, these fibers are tightly packed around the producing cells like little baskets (fig. 2b). Curli fibers alone are non-elastic and break easily.[7] However, many *E. coli* strains also produce another matrix component, a chemically modified form of cellulose,[8] which forms long-range fibrillar connections and sheets (fig. 2c). Such a cellulose network is highly elastic, but it cannot support large structures because it does not hold on to the cells. By contrast, wildtype *E. coli* cells produce both fibers, that is curli fibers and cellulose fibrils, which together form a fibrous composite with superb material properties—it is elastic, it can stand large forces and it fully surrounds and holds cells at their place (fig. 2d).

A Matrix Architecture Much Larger than the Cells that Collectively Produce It

The images in figure 2 show the extracellular matrix arrangement at the surface of a macrocolony biofilm. However, the two matrix components can also be visualized inside a macrocolony biofilm by exploiting their binding of the green fluorescent dye Thioflavin-S (TS) added to the growth medium of the agar plates. After several days of growth, the macrocolonies can be shock-frozen, cut into thin vertical slices, fixed, and analyzed by fluorescence microscopy.[9]

At the rapidly growing outer edge of a macrocolony, which consists of densely packed cells everywhere, a simple two-layer structure can be observed, meaning no matrix at the bottom and the green fluorescent TS-stained matrix at the top. However, when taking a look into the older more central area of the macrocolony, a more complex architecture is revealed (fig. 3). Also here, a matrix is found in the top layer, but this matrix has developed into a complex architecture with several strata showing different patterns.[10] At the upper surface, a "dense brickwork"-like matrix arrangement consists of a composite of curli fibers and cellulose. Somewhat deeper, the matrix

---

6 Margery L. Evans and Matthew R. Chapman, "Curli Biogenesis: Order out of Disorder," *Biochimica et Biophysica Acta (BBA)*—Molecular *Cell Research* 1843, no. 8 (2014): 1551–58, https://doi.org/10.1016/j.bbamcr.2013.09.010.

7 Diego O. Serra, Anja M. Richter, Gisela Klauck, Franziska Mika, and Regine Hengge, "Microanatomy at Cellular Resolution and Spatial Order of Physiological Differentiation in a Bacterial Biofilm," *mBio* 4, no. 2 (2013): e00103–13, https://doi.org/10.1128/mBio.00103–13.

8 Wiriya Thongsomboon, Diego O. Serra, Alexandra Possling, Chris Hadjineophytou, Regine Hengge, and Lynette Cegelski. "Phosphoethanolamine Cellulose: A Naturally Produced Chemically Modified Cellulose," *Science* 359 (2018): 334–38, https://doi.org/10.1126/science.aao4096.

9 Serra and Hengge, "Experimental Detection and Visualization of the Extracellular Matrix in Macrocolony Biofilms."

10 Gisela Klauck, Diego O. Serra, Alexandra Possling, and Regine Hengge, "Spatial Organization of Different Sigma Factor Activities and c-di-GMP Signalling within the Three-Dimensional Landscape of a Bacterial Biofilm," *Open Biology* 8 (2018): 180066, https://doi.org/10.1098/rsob.180066.

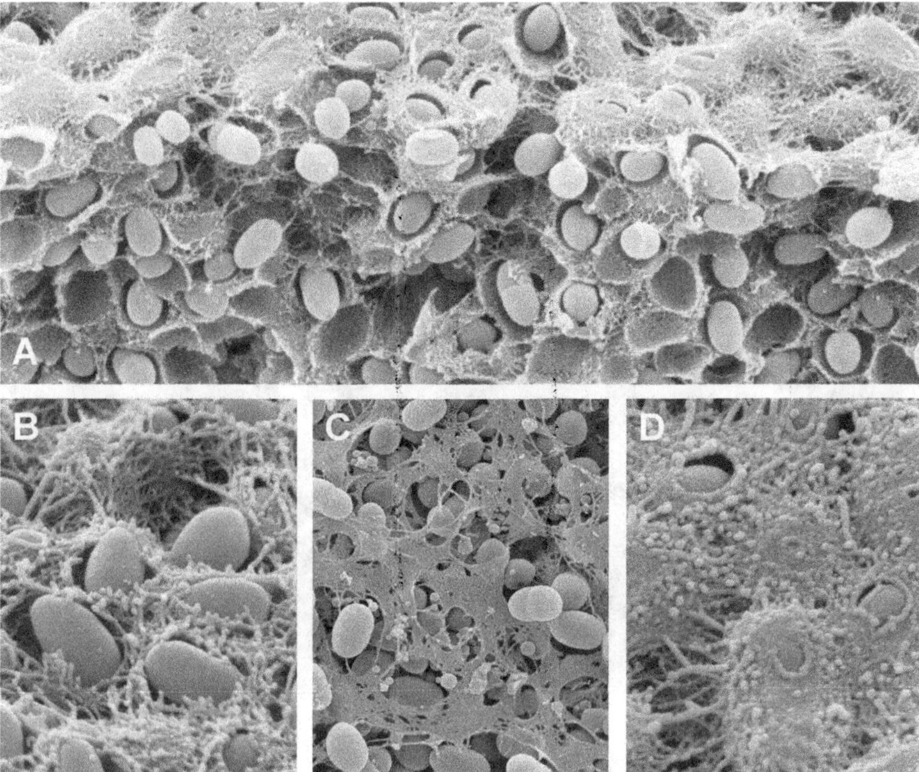

Fig. 2: The fibrous extracellular matrix of *E. coli* macrocolony biofilms as visualized by scanning electron microscopy (SEM). In (2 A) the surface of a macrocolony of an *E. coli* strain, which produces a tight composite of amyloid curli fibers and cellulose, is shown. The macrocolony was broken open in order to show the fully matrix-covered colony surface (also at larger magnification in D) as well as the dense "brickwork-like" matrix arrangement within the biofilm upper layer. In (2B) an *E. coli* strain is shown (cells false colored in blue) that produces curli fibers only, whereas the *E. coli* strain in (2C) synthesizes only cellulose (false colored in green). Scale: the shorter diameter of the ovoid *E. coli* cells corresponds to approximately 1/1000 mm.

forms "vertical pillars," which mostly consist of long cellulose fibrils and sheaths, and a "loose horizontal network" consisting only of curli fibers can be detected further toward the bottom. In contrast to the upper dense brickwork layer, the vertical pillar and horizontal network zones also contain "dark areas" with matrix-free cell clusters right next to the matrix-producing cells, which means matrix production is quite heterogeneous. Finally, at the very bottom of the biofilm, there is a network of entangled flagella, which are a type of extracellular filaments that do not stain with the fluorescent dye TS, but that can be visualized by scanning electron microscopy (fig. 3).

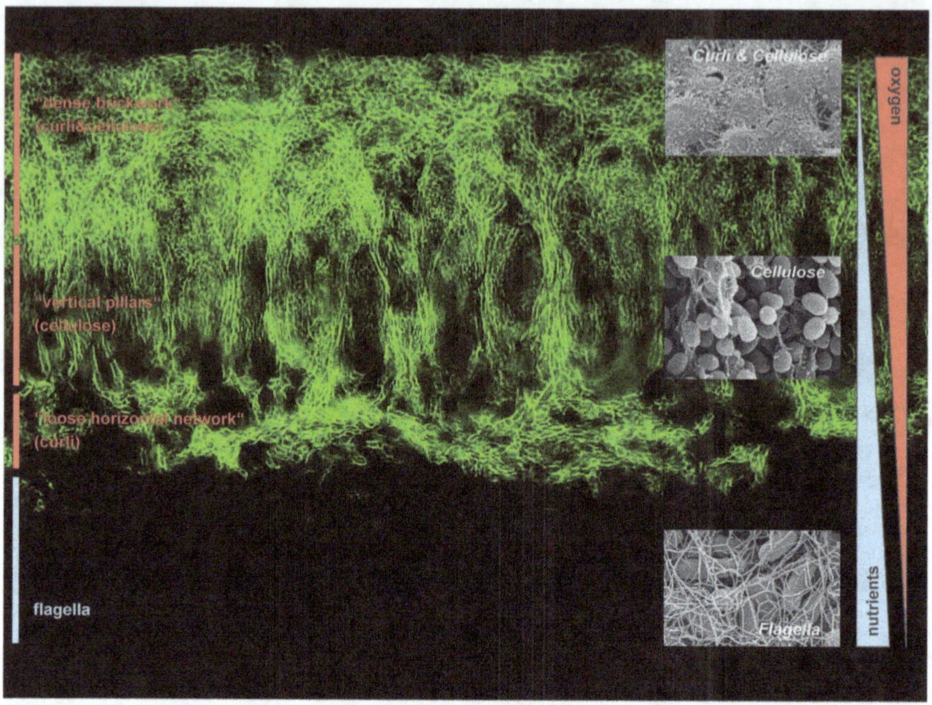

Fig. 3: The extracellular matrix architecture within *E. coli* macrocolony biofilms: in a vertical thin section through an *E. coli* macrocolony biofilm, the extracellular matrix was visualized by staining with the fluorescent dye Thioflavin-S (TS), which was present already during growth of the biofilm. TS tightly binds to both curli fibers and cellulose, thus allowing detection of these matrix fibers by fluorescence microscopy. The entire biofilm is densely packed with bacterial cells, but only matrix-embedded bacterial cells can be seen as small black dots surrounded by the green fluorescent matrix dye. Note that the entire matrix architecture, in which three morphologically distinct strata can be distinguished as indicated, is approximately 50-fold larger (on the vertical axis) than the cells that coordinately produce it. The insets show larger magnification scanning electron microscopic images of cells in the respective zones. The matrix-free cells in the dark bottom layer produce a network of entangled flagella, which do not bind the fluorescent dye, i.e., this entire bottom layer of cells remains dark.

The Elaborate Large Scale Matrix Architecture Matters for Biofilm Elasticity

Over the last twenty-five years, the intricate genetic control network that senses and transduces local environmental signals into the formation of this three-dimensional matrix architecture has been elucidated.[11] Steep chemical gradients, mainly of nutrients (from the agar support below the biofilm) and oxygen (from the air on top of

---

11 Hengge, "Linking Bacterial Growth, Survival and Multicellularity."

the biofilm), have turned out to serve as major signal inputs. Thus, only cells experiencing nutrient-limitation can synthesize matrix components, which explains why the matrix builds up in the upper layer of the macrocolony that is far away from the nutrient-providing agar support. Knowledge about the regulatory network underlying these processes also allows us to introduce mutations that knock out distinct regulatory genes and proteins. The result are changes both in the biofilm-internal matrix architecture and in the wrinkled macrocolony morphology, thus demonstrating that the macroscopic form is genetically controlled.

One of the most interesting mutations eliminated a regulatory protein called PdeR, which serves as an inhibitor of matrix production by sensing and degrading the intracellular biofilm-promoting signaling molecule c-di-GMP. In the absence of this key regulator, heterogeneity of matrix production is lost, whitch means *all* cells in the upper layer are now homogenously producing matrix components. The consequence of this change in matrix architecture is drastic: these biofilms are stiffer, which makes buckling more difficult and therefore a rarer event. Moreover, closer inspection by scanning electron microscopy showed deep breaks all over the surface of the macrocolonies—in other words, there is a large-scale elasticity problem. Therefore, the tissue-like elasticity does not only depend on the material properties of the matrix fibers, in particular on the highly elastic cellulose fibrils, but also on the much larger matrix architecture, in particular on the presence of the matrix-free cell clusters within the matrix layer.[12]

Spatial Self-organization into Two Distinct Cell Populations Generates the Matrix Architecture and Triggers Morphogenetic Movement

So, what is the exact role of these matrix-free cell clusters in biofilm buckling and elasticity? Introducing a green genetic marker—designed to light up in cells that grow and divide rapidly[13]—showed that these globular matrix-free cell clusters represent the fastest growing cellular subpopulation in the older central area of a macrocolony. By contrast, the massively matrix-producing cells in the same biofilm zone grow only slowly into the thin cellulose-sheathed vertical matrix pillars. So, there is a division of labor between two cell populations in different physiological states: one that uses the available resources for rapid growth and expansive proliferation, and another one that invests these resources in building the matrix architecture.

---

12 Diego O. Serra and Regine Hengge, "A C-di-GMP-Based Switch Controls Local Heterogeneity of Extracellular Matrix Synthesis Which Is Crucial for Integrity and Morphogenesis of Escherichia coli Macrocolony Biofilms," *Journal of Molecular Biology* 431 (2019): 4775–93, https://doi.org/10.1016/j.jmb.2019.04.001.
13 A reporter fusion of the gene for green fluorescent protein (Gfp) linked to the ribosomal *rrnB* gene (Klauck et al., "Spatial Organization of Different Sigma Factor Activities").

Most intriguingly, the rapidly proliferating matrix-free cell clusters were found to change their position while the biofilm buckles up and folds into the high ridges, that is, when the initially flat area goes through a transient bending event, which is associated with strong compression and stretching of distinct microzones of the biofilm. This indicates that by just growing rapidly, these proliferating cell clusters first *build up* local compression forces and therefore local instability, which triggers local buckling. In addition, during the actual bending event, these cells also contribute to *dissipating* these local forces by getting squeezed through the matrix network—this explains how these non-matrix-fixed cell clusters, which can be flexibly pushed around, prevent breakage of the biofilm at large during bending.

In summary, morphogenetic activity of macrocolony biofilms is driven by a division of labor between physiologically different subpopulations of cells, which self-organize in space to produce the large-scale elastic matrix architecture and to first generate and then dissipate the local forces that trigger buckling and folding during biofilm growth.[14]

Local Exponential Growth Drives the Buckling and Folding of Biofilms

When we look at a macrocolony from above, that is, quasi in 2D (as in fig. 1), its morphogenetic movements are not randomly distributed but seem to develop into a distinct geometry. At the outer biofilm edges, rapidly growing cells drive the radial expansion of the colony. The first buckling always starts somewhat behind these outer edges, which is where cells begin to produce matrix in the upper layer of the macrocolony. It turns out that this initial buckling can be described as a simple consequence of geometry. When a round bacterial colony is growing rapidly at its outer edge—which for real macrocolonies is an oversimplification as somewhat slower growth also occurs in specific zones inside the macrocolony as described above—, the macrocolony radius (r) increases over time (t). During this process, the colony circumference grows in a linear manner ($2\pi r$), while its area grows by the square (πr^2). However, the cells that drive this expansion, grow exponentially ($N=N_0 e^{kt}$), which means, in absolute terms, cell numbers initially increase only slowly, then accelerate in growth and finally "explode." As a practical consequence, the growing area of a macrocolony biofilm can no longer be accommodated in a flat circular plane—there is no other possibility than buckling out into the third dimension. In geometrical terms, a surface increasing by exponential growth becomes hyperbolic, meaning it is necessarily "negatively" curved. In fact, nature makes ample use of hyperbolic surfaces—we find them for instance in marine flatworms, cabbage leaves, or fungi. The mechanisms behind generating many of these biological hyperbolic forms are the same, that is, local proliferation of cells—

---

14 Kim Nguyen and Regine Hengge, unpublished results.

with multiple cell divisions representing an exponential growth pattern—forces initially planar tissues to curve out of the flat plane.

Curvature and Hyperbolic Surfaces

We have used the terms *hyperbolic* and *curved* here to describe the geometry of the biofilm surface as it buckles as a result of the exponential proliferation of cells. Understanding these ideas in a more precise way enables us to consider the biofilm geometry in a more complete way. To talk about hyperbolic surfaces and the hyperbolic plane, we first need to discuss what it means for a surface to be *curved.*

There are numerous strategies to define curvature from a mathematical perspective. It is easiest to start with a curved one-dimensional line in the plane, before moving onto the more complicated case of a two-dimensional surface embedded in our three-dimensional world. There are two distinct perspectives that we can take here, looking at how the line curves over its whole length—the *total curvature*—and how the line curves at a particular point—the *pointwise curvature.* The total curvature of a line can be considered by constructing a series of lines parallel and with increasing distance to the original line. As the distance (t) from the original line increases, the length of the line changes linearly with a weighting that is the curvature: a straight line will have parallel lines of the same length, and a curved line will have parallel lines that increase in length with increasing t. This is a very intuitive way to capture the total curvature of a line. From the pointwise perspective, for each infinitesimally small segment of the line, we can imagine it wrapping around a circle of a given radius. This is known as the *osculating circle* of a line at a point. When the line is curving gently and the circle is big, the curve is said to have a large radius of curvature (R, the radius of the circle), but a small curvature given by 1/R. When the line curves more strongly and the circle is small, the curve has a small radius of curvature and hence a larger curvature. These definitions capture our intuitive notion of the curved line.

On a two-dimensional surface, things are a bit more complicated, as there are fundamentally different ways that a surface can curve. We can consider the same idea of total curvature of a surface as we did for a line, only now we look at parallel surface and their change in area as t increases. It turns out that this area changes quadratically with t, with a curvature component that changes linearly with t, and a curvature component that changes quadratically with t^2. The linear component is known as the *mean curvature,* and the quadratic component is the *Gaussian curvature.* These quantities can also be computed at a particular point on the surface: the curvature at this point is encapsulated by considering the directions where the surface curves the most, and where it curves the least (which may also involve curving in the opposite direction). More precisely, you could make a slice of your surface through the point (where the slice plane contains the surface at the point) and look at the curvature of the line along that slice. So, at a point on the surface, we can now consider the curvature of a line radiating from the point in each direction on the surface. If we consider all of these curved lines radiating from a point, we can find the directions where the curvature is maximal and minimal (fig. 4). The curvature in these directions are the *principal curvatures* at a given point on the surface, and they are always perpendicular to each other.

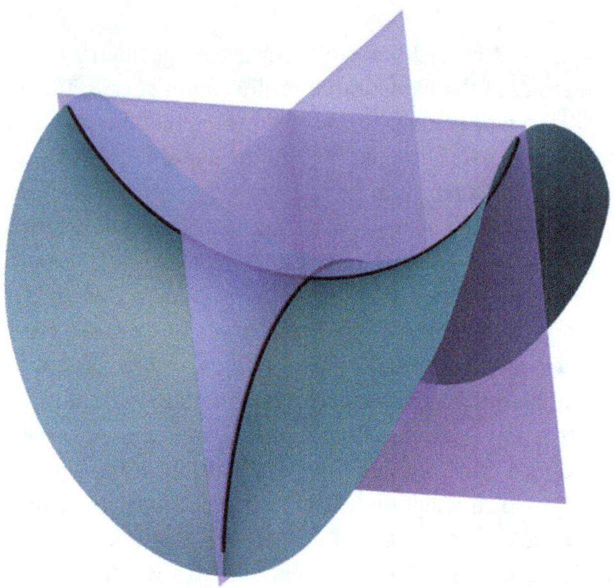

Fig. 4: Principal curvature lines on a hyperbolic saddle surface. A hyperbolic saddle surface is shown in green, along with the planes (purple) intersecting the surface along the principal curvature directions (shown in black) at a particular point on the surface. The two principal curvatures curve in opposite directions, and thus have the opposite sign, which means that their product—the Gaussian curvature—will be negative. All negative Gaussian curvature surfaces are termed hyperbolic surfaces.

The concepts of mean and Gaussian curvature at a point on the surface can be described using these two principal curvatures. The mean curvature is the average of these two principal curvatures, and refers to an extrinsic notion of curvature, related to how the surface is embedded in space. The Gaussian curvature is the product of the two principal curvatures and is an intrinsic property of the surface. In general, the sign of the Gaussian curvature tells us what kind of surface we have: when it is positive, we have something spherical, when it is zero, we have something flat, and when it is negative, we have a hyperbolic surface. The surface in figure 4 has the principal curvatures curving in opposite directions, so one is positive and one is negative, which is the characteristic shape of a hyperbolic surface. We now have a definition of a hyperbolic surface with which to think about the biofilms, namely that at each point on the surface, it has negative *Gaussian curvature.*

A Wrinkled Macrocolony Biofilm Can Be Described as a Finite Disc That Is Intrinsically Hyperbolic

One very natural way to explore such a hyperbolic surface is through a crochet model (fig. 5). The idea comes from the mathematician Daina Taimina, who first introduced the model in 1997.[15] The addition of stitches in a particular proportion with each round of the crochet dictates a pattern of local growth in the crocheted surface. Local growth can be described by what is known as the *metric* of the surface, which characterizes the distances between points. This metric is precisely what defines the *intrinsic* properties of the surface, so it is related to the Gaussian curvature of the surface. In this case, the exponential growth implies that the surface has negative Gaussian curvature, making the crocheted surface hyperbolic. The textile model was a breakthrough for people trying to understand hyperbolic geometry because it was suddenly possible to touch and experience a hyperbolic surface. The construction of a hyperbolic surface using the exponential growth of the stitch numbers mirrors the emergence of hyperbolic surfaces through cell proliferation in the biofilm.

Given a finite disc that is intrinsically hyperbolic, we can consider how this disc arranges itself extrinsically, in what shape it embeds itself in space, which is captured by the *mean curvature* of the surface. For a hyperbolic disc, there are many ways to create an extrinsic embedding, where the disc can take on a variety of undulations, folds and oscillations while satisfying the intrinsic curvature conditions. The different presentations of the hyperbolic crochet in figure 5 show some of the possibilities. In an unconstrained elastic disk, such as the crochet or various other physical systems, this extrinsic embedding will be found as an equilibrium between bending and stretching energies, where homogeneous local curvature can lead to embeddings with large scale folds and small-scale oscillations.[16]

In the case of the biofilm growth, an important geometric consideration is that the hyperbolic nature of the biofilm is fundamentally incompatible with the flat substrate on which it is growing. Thus, when the tissue-like biofilm grows, we actually have a growth process together with adhesion to a substrate. This now becomes a balance of growth within the film and the hyperbolic geometry of the disc, the elasticity of the material, and the adhesion properties of the film to the substrate.[17] The collective solutions to these conditions, as well as the idea of inhomogeneous growth patterns

---

15 Daina Taimina, *Crocheting Adventures with Hyperbolic Planes: Tactile Mathematics, Art and Craft for All to Explore* (Boca Raton: CRC Press, Taylor & Francis, 2018).
16 Yael Klein, Efi Efrati, and Eran Sharon, "Shaping of Elastic Sheets by Prescription of Non-Euclidean Metrics," *Science* 315, no. 5815 (2007): 1116 – 20, https://doi.org/10.1126/science.1135994.
17 Martine Ben Amar and Min Wu, "Patterns in Biofilms: From Contour Undulations to Fold Focussing." *EPL* 108 (2014): 38003, https://doi.org/10.1209/0295-5075/108/38003; Julien Dervaux and Martine Ben Amar, "Morphogenesis of Growing Soft Tissues," *Physical Review Letters* 101 (2008): 068101, https://doi.org/10.1103/PhysRevLett.101.068101.

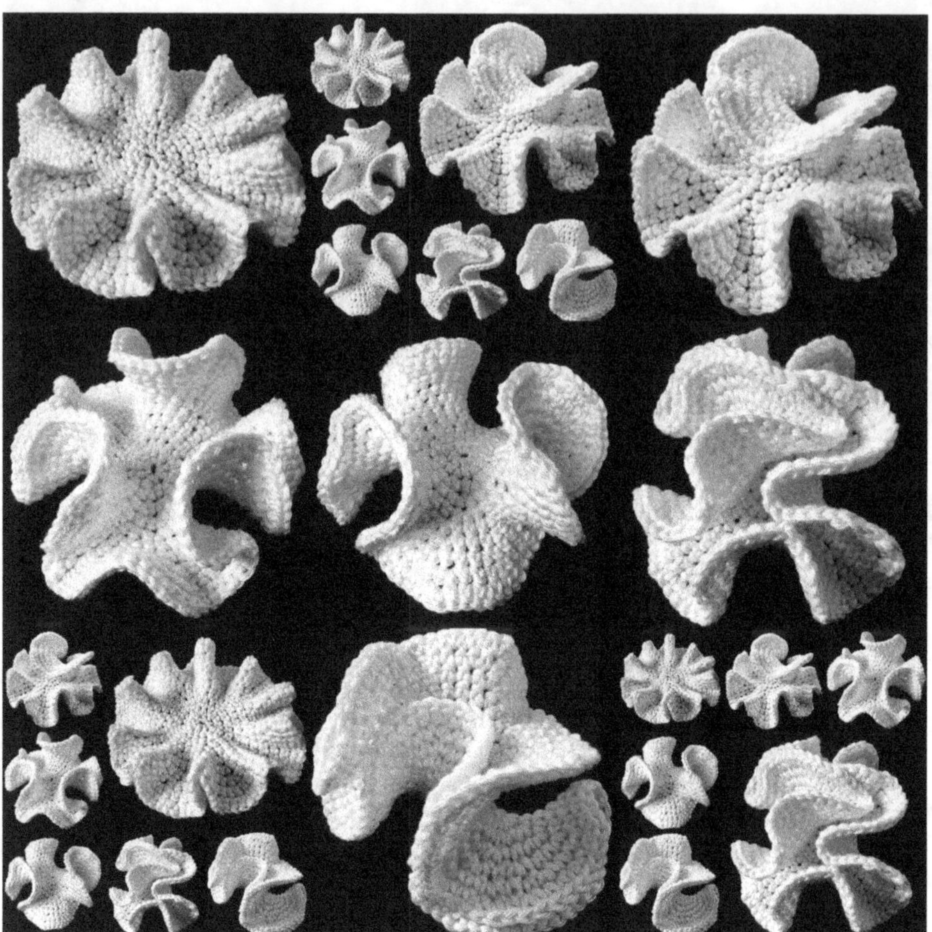

Fig. 5: A crochet model of a hyperbolic plane. This particular model started from a small circle of ten stitches, with stitches added in a regular manner in each new round along the circular outer edge (four stitches over three stitches in the previous round). This represents an exponential growth of the number of stitches that leads to buckling in the outer area, which resembles the initial buckling of macrocolony biofilms. This similarity (in particular to the colony shown in the middle of the lower row in fig. 1) is best seen in the photograph in the upper left corner of the panel shown here. Note that all images show a single object, with the different forms illustrating the potential of the hyperbolic disc to take on a variety of undulations, folds and oscillations compatible with the intrinsic curvature conditions.

over the surface, has the potential to explain the rich variety of biofilm folding, buckling, and wrinkling patterns observed experimentally.

In conclusion, if we want to fully understand how macrocolony biofilms grow into their intriguing three-dimensional form, we should explore not only the spatial control of matrix gene expression in response to environmental and cellular signals and the material properties of the extracellular matrix components and large-scale architecture, but also the rules and possibilities of hyperbolic geometry.

Bastian Beyer and Iva Rešetar

Entangled Scales: Structures and Environments of Cellulose Biofilms

The collective imagination of how scale connects the living and nonliving has rarely been so evocatively depicted as in Charles and Ray Eames's short film *Powers of Ten* (1977).[1] This virtual journey through successive scales of matter takes us from the familiar scale of the human body to urban space (10^2), via satellite images of weather phenomena (10^6), and on to the planetary scale (10^7). Returning to the human body, we enter the fibrous structure of tissues (10^{-4}) and finally the finest atomic scale of vibrating particles (10^{-15}). With demonstrative continuity, the film leaves us with the impression that human life, together with the structures and objects surrounding us, occupies an intermediate dimension—neither too large nor too small. The human and architectural scales seem interwoven at the center of these scalar transitions and, at the levels above and below them, matter is less tangible and belongs to disciplinary fields other than architecture.

But what is meant by "scale"? The understanding we are most familiar with is relational—the size or extent of something in comparison to something else. The etymology of the term is more diverse, referring to parts of animal shells, measuring instruments, but also to acts such as scanning, examining, and looking closely.[2] In architecture, as in Eames's film, scale is often associated with human proportions, either in a "cosmological" sense, in diagrams where the human figure stands as a symbolic measure of the world,[3] or in a normative context, where it defines a blueprint required to plan or make something.

This perception of the human scale as the focal point for design—a measure of things existing in the middle of continuous metric space—is a topic we aim to examine critically in this text. We argue that, whether in processes of design, making, or growth, one can speak of *scalar entanglements*, in which structures are influenced by others at different scales that are neither immediately apparent nor clearly ordered. As designers, we usually deal with materials at a scale convenient for human use, though they still depend on relationships and behaviors that go far beyond their own scale or materiality. Microorganisms, for example, are almost imperceptible to the human senses, even though they live and form complex communities in all the spaces we inhabit, from our bodies to the outer environment. Being indiscernible in large-scale building

---

1 *Powers of Ten and the Relative Size of Things in the Universe*, directed by Charles and Ray Eames (Eames Office, 1977). The film was based on the earlier version from 1968, which juxtaposed human and earthly space and time scales, https://www.eamesoffice.com/the-work/powers-of-ten-a-rough-sketch/, https://www.eamesoffice.com/the-work/powers-of-ten/.
2 Eric Partridge, *Origins: An Etymological Dictionary of Modern English* (London: Routledge, 2006).
3 Adrian Lahoud, *The Problem of Scale: The City, the Territory, the Planetary*, PhD diss. (University of Technology Sydney, 2013).

practices, however, means that they are unlikely to be seen as active participants in shaping these spaces and their livability.

Based on our experiments with cellulose-producing bacteria, conducted in collaboration with microbiology and materials science,[4] we would like to turn our attention "inward," to the living and growing processes at the microscopic scale, and ask: How does architectural design engage with them? When a design process revolves around the microorganism, how does this differ from dealing with the conventional materials? And, what becomes entangled in the fibrous structure of bacterial cellulose that can be considered across the scales—microbial, material, environmental, and infrastructural?

Engaging with different scales creates uncertainties about their transitions and continuities, as they tend to filter what belongs to a particular frame of reference and what falls outside of it. Architectural theorist Luis Fernández-Galiano notes that, on small scales, the usual distinction between matter and energy no longer applies, one being routinely described in terms of the other. Indifferent to scalar conventions, energies travel across scales and materials, linking them to processes of life, transformation, degradation, and renewal.[5] For architect Adrian Lahoud, the problem of scale is a disciplinary "trap."[6] He draws attention to the unexpected juxtaposition of the small and the planetary, the near and the far, by showing, for example, how the trajectories of aerosol particles are linked to climate changes. Moreover, scalar categories, he argues, have historically emerged around explanations and representations of specific problems, and in turn have partitioned knowledge in accordance with inherited conventions.

Similar ambiguities form part of our interdisciplinary explorations with bacterial cellulose (fig. 1, 2). Starting from cellulosic materials as sites of disciplinary and scalar entanglements, we reflect on novel frameworks required to account for bacterial activity in architecture. In exploring the intrinsic structure and properties of cellulose, we disclose its complex relationships with the environment, and the history of processing of this hierarchical biomaterial, in which bacterial cellulose has only recently gained attention. Some of the practices discussed, such as fermentation and co-weaving textiles with bacteria, go beyond conventional design methodologies and ultimately offer insights on how collaboration with microbiology and materials science can expand the current architectural field of vision beyond the anthropocentric.

---

4 Experiments with bacterial cellulose are carried out at the intersection between microbiology, materials science, and architectural design, within the research projects Weaving and Material Form Function, and in collaboration with the Hengge Group at the Department of Microbiology of the Humboldt-Universität zu Berlin, Adaptive Fibrous Materials and Biofilm-based Materials groups at the Max-Planck-Institut für Kolloid- und Grenzflächenforschung in Potsdam.

5 Luis Fernández-Galiano, *Fire and Memory: On Architecture and Energy*, trans. Gina Cariño (Cambridge, MA: MIT Press, 2000), 2.

6 Adrian Lahoud, "Scale as a Problem, Architecture as a Trap," in *Climates: Architecture and the Planetary Imaginary*, ed. James Graham et al. (New York: Columbia University Press, 2016), 111–19.

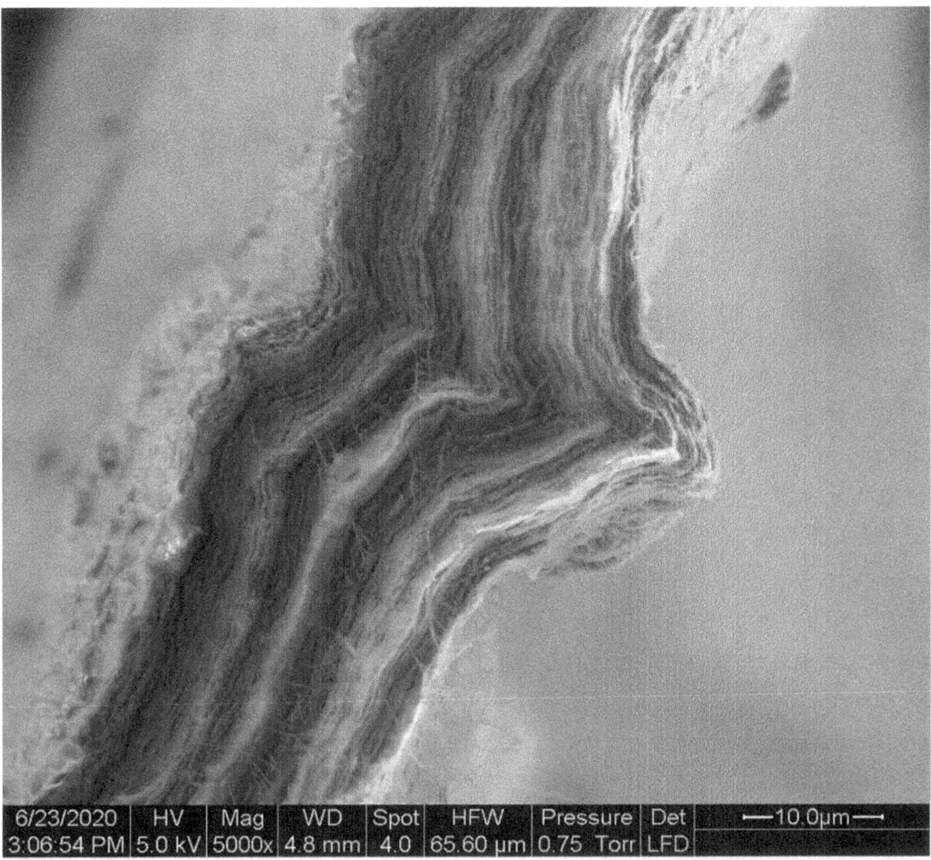

Fig. 1: Microscopic image of the fibrous structure of bacterial cellulose visible in the 10 µm range.

Cellulose: Structures and Processes

Cellulose is the most abundant organic compound occurring in nature—and found in biological materials across their scales. As the skeletal component of the cell walls of all plants, cellulose possesses a unique fiber morphology consisting of elementary fibrils (1.5–3.5 nm), microfibrils (10–30 nm), and micro fibrillar bands with a lateral dimension of around 100 nm.[7] In plants, in combination with hemicellulose, lignin, and pectin, cellulose is arranged in hierarchical organizations that give these biological materials a range of outstanding properties, such as great yet variable tensile strength. In

---

7 Dieter Klemm, Brigitte Heublein, Hans-Peter Fink, and Andreas Bohn, "Cellulose: Fascinating Biopolymer and Sustainable Raw Material," *Angewandte Chemie International Edition* 44, no. 22 (2005): 3358–93, https://doi.org/10.1002/anie.200460587.

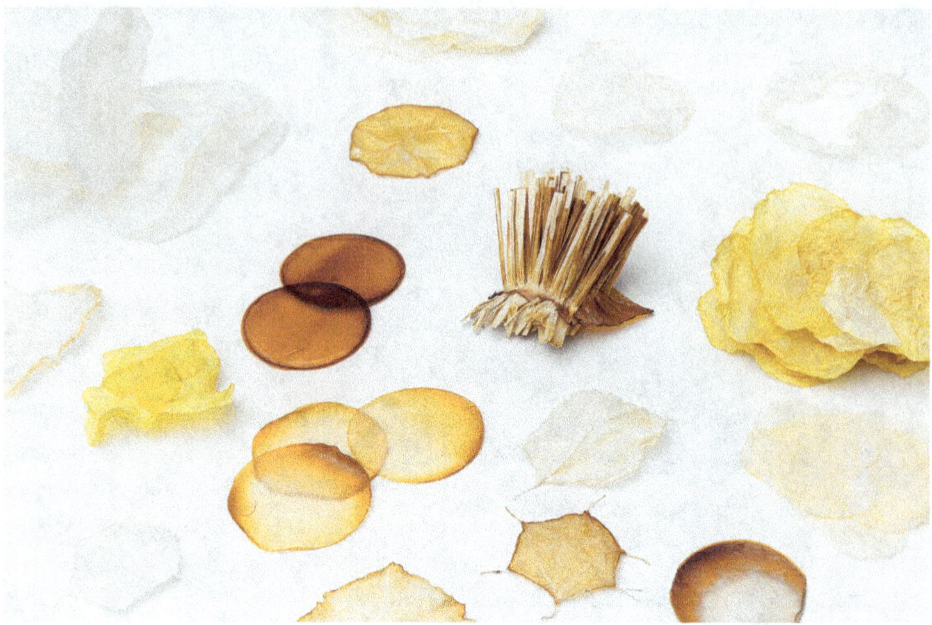

Fig. 2: Open-ended design explorations with bacterial cellulose.

addition to plants, certain bacteria and algae are also capable of synthesizing cellulose. These biological composites differ from the homogenous, monolithic structures: with a few constitutive elements and a multitude of possible configurations, they grow and adapt to changing mechanical and environmental conditions, which are reflected in their structure.[8]

Human culture is deeply intertwined with cellulosic materials. Throughout history, we have built, clothed ourselves, consumed, transported, and communicated using various forms of cellulose, such as wood, straw, cotton, flax, and hemp. In architecture these materials have proved equally important for construction and combustion—both for making structures and for supplying them with energetic flows.[9] Their use has been particularly widespread in societies that rely on plant-tending and on the construction of lightweight and portable, rather than permanent and massive structures.[10]

The knowledge and skills required to manipulate, construct and cooperate with plant matter have developed tacitly through interaction and negotiation with the ma-

---

8 Peter Fratzl and Richard Weinkamer, "Nature's Hierarchical Materials," *Progress in Materials Science* 52, no. 8 (2007): 1263–1334, https://doi.org/10.1016/j.pmatsci.2007.06.001; Michaela Eder, Shahrouz Amini, and Peter Fratzl, "Biological Composites—Complex Structures for Functional Diversity," *Science* 362, no. 6414 (2018): 543–47, https://doi.org/10.1126/science.aat8297.

9 Fernández-Galiano, *Fire and Memory: On Architecture and Energy*, 7.

10 Mark Jarzombek, *Architecture of First Societies: A Global Perspective* (Hoboken, NJ: Wiley, 2013).

terial.[11] Here, understanding the material on different scales—its internal structure and fiber orientation—was key to working with it successfully.[12] Practitioners, such as carpenters or instrument makers, have perfected this art and developed methods to fine-tune and amplify certain properties of materials. In the example of archery, bows were made of thin strips of wood laminated according to a certain predefined pattern in order to improve their strength and resistance. This marked a profound change in the way plant materials were treated: instead of using structures as they occurred, recomposition and restructuring was introduced to activate their inherent properties.

Industrialization gave rise to a new economy and conception of plant matter as a resource and a raw material involving the manipulation of materials on a molecular level. Wood is no longer perceived and utilized as a complex, grown structure, but broken down into a series of ingredients—lignin, hemicellulose, cellulose—through the development of processes of fractionation (fig. 3). These semi-finished products are then becoming the source of numerous substances (for example, cellulose acetate, cellulose nitrate, methyl cellulose), which in the "formless" state can easily be shaped into a variety of products.

Vast industrial landscapes have evolved around cellulose and its extraction that not only involve energy-intensive and chemically laden manufacturing, but also demand deforestation or long-term land use for monocultures. From source materials for extraction to semi-finished products and derivatives, cellulose now occupies an intermediate position—processed from other substances it serves as a raw material for further processing.

To rethink the role of cellulosic materials in the context of architectural design, these dependencies and production methods need to be called into question. Bacterial cellulose has the potential to bypass the logic of extraction: because of its small scale and the fact that it is created by a living organism, this material can remain close to the environments and processes of growth. A new collaborative setting—between species and between disciplines—provides an opportunity to observe and reveal its potential for design.

Microorganism and Biofilm Growth

By familiarizing ourselves with the growth of cellulose at the microscale, we can begin to understand how such a process differs from industrial production. Microbial cellulose does not involve extraction from other materials, as the bacteria produce chemically pure cellulose. Few resources and little energy are needed in the process, which

---

11 Tim Ingold, "The Textility of Making," *Cambridge Journal of Economics* 34, no. 1 (2010): 91–102, https://doi.org/10.1093/cje/bep042.

12 Michaela Eder, Wolfgang Schäffner, Ingo Burgert, and Peter Fratzl, "Wood and the Activity of Dead Tissue," *Advanced Materials* 33, no. 28 (2021): 2001412, https://doi.org/10.1002/adma.202001412.

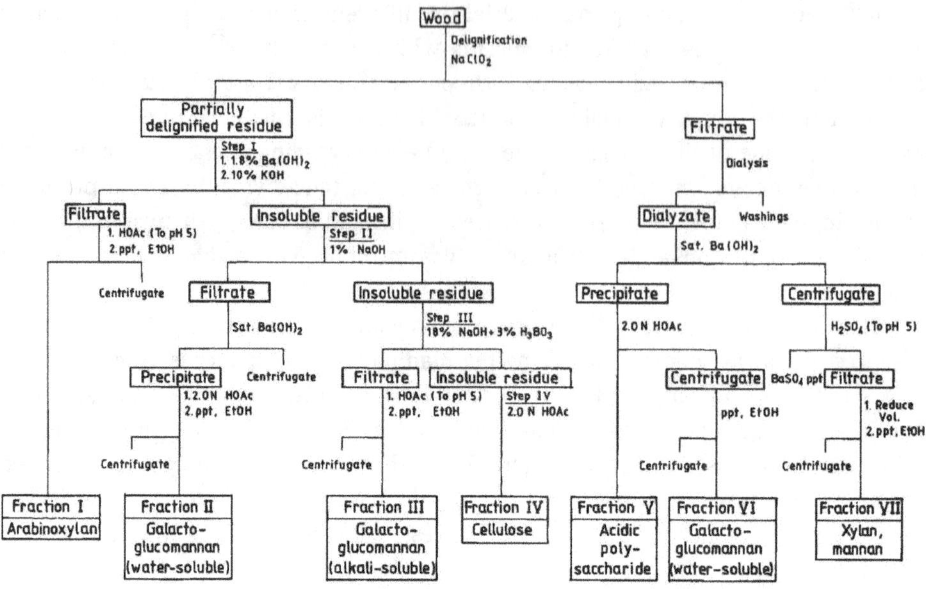

Fig. 3: Pathways of industrial material processing: fractionation of wood.

can be undertaken in domestic or citizen science contexts as well as in the laboratory. Recently, a research landscape in design and architecture has emerged with interest in taking the cellulose growth in various interdisciplinary directions, raising critical questions about care, symbiosis, and cohabitation.[13] Simultaneously addressing bacterial, human, and urban scales, its projects include proposals to use organic waste as a nutrient source for bacterial cellulose, deploying energy infrastructure on an urban scale, imagining growth through community knowledge and within a public space,[14] and actively engaging with the cellulose livingness in making everyday objects and artefacts.

Most of these projects do not obtain cellulose from plants or through industrial processes, but rather grow the material in the laboratory. But how is cellulose grown in a microbiology laboratory? What species, conditions, and scales are entan-

---

13 Elvin Karana, Bahareh Barati, and Elisa Giaccardi, "Living Artefacts: Conceptualizing Livingness as a Material Quality in Everyday Artefacts," *International Journal of Design*, 14, no. 3 (2020): 37–53, https://www.ijdesign.org/index.php/IJDesign/article/view/3957/923.

14 Numerous design research projects have been undertaken with bacterial cellulose, see for example the following: "Vibrant Tissue," IAAC, accessed August 31, 2021, http://www.iaacblog.com/programs/vibrant-tissue_-augmented-microbial-cellulose/; "Bio-Fabric," IAAC, accessed August 31, 2021, http://www.iaacblog.com/programs/bio-fabric-microbial-cellulose/; "GrowPak – A step towards closing the loop," accessed August 31, 2021, https://julianajschneider.com/growpak; *Metabolizing Urban Waste into Layered Morphologies*, YouTube video, 4:15 min., uploaded by ecoLogicStudio eLS, January 24, 2018, accessed August 31, 2021, https://www.youtube.com/watch?v=Ds9qk3oFIRI.

gled in the structure of biofilms and in experiments at the intersection between micro-biology and architectural design?

Bacteria produce cellulose to form biofilms—fibrous assemblages that gather colonies of individual microorganisms. As symbiotic communities of bacteria, biofilms are sites of complex interactions where diverse multispecies act as a collective entity in cooperation and conflict.[15] Bacteria build, maintain, and live in this gelatinous matrix of a fibrous microstructure (fig. 4). The material created during biofilm growth is also termed nano-cellulose, since cellulose particles possess at least one dimension in the nanometer range (1 nm to 100 nm). We have been experimenting with the particular bacterial strain of the genus *Komagataeibacter* called *K. xylinus, which* belongs to the group of acetic acid bacteria known for its extensive cellulose production. Unlike that of photosynthetic plants, its growth does not require light, but a warm environment and nutrients. Although the fibrils form on the nanoscale, the biofilms woven by *K. xylinus* are visible on the human scale, allowing for the direct interaction with its growth process and hence for speculation on how it might become part of more symbiotic, bacterial-human design processes.

Biofilm growth follows several stages of materiality. When the bacteria are introduced into the culture medium, the film is formed at the interface between two environments: a nutrient-rich liquid and the air. Formation begins with pellicles floating on the surface of the medium and, after a few days, when the hydrogel membrane covers the entire surface, bacterial activity stops or slows down. Once the cellulose is taken out of the medium, it dries and turns into a skin-like, translucent, lightweight structure (fig. 5).

In this process, the biofilm as a living entity undergoes transformation constantly: it changes from a liquid and gelatinous to a dry and solid material. The life within it also fluctuates, since, after the nutrients and oxygen have been depleted, the microorganisms enter a dormant phase. This gradient of nutrients and livingness is not only temporal; it is also spatial and linked to the internal structure of the biofilm and to its fluid boundary with its surroundings.[16] The most active part lies in its contact with the air where the newly-made fibers assemble above already formed layers. Oscillating between wet and dry, living and inert, in many ways the biofilm evolves at the interface of different environments, which in turn become part of its microstructure.

---

15 Alexander May et al., "Kombucha: A Novel Model System for Cooperation and Conflict in a Complex Multi-Species Microbial Ecosystem," *PeerJ* 7 (2019): e7565, https://doi.org/10.7717/peerj.7565.
16 Diego O. Serra, Anja M. Richter, and Regine Hengge, "Cellulose as an Architectural Element in Spatially Structured Escherichia Coli Biofilms," *Journal of Bacteriology* 195, no. 24 (2013): 5540–54, https://doi.org/10.1128/jb.00946-13.

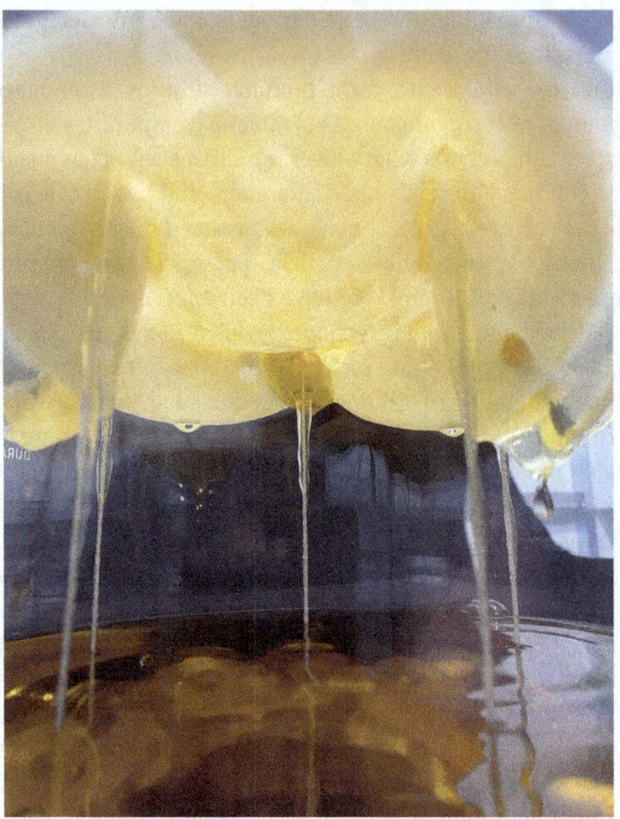

Fig. 4: Gelatinous stage of cellulose biofilm.

Microbial Environments and Fermentation Practices

One might rightfully ask: how do microbial life and practices of fermentation relate to the field of architecture? How can processes on such a scale, imperceptible to the human eye, influence the spaces we inhabit? In the field of architecture, microbial activity is generally connoted with negative effects such as rot or mold. By conditioning spaces, by dehumidifying or sanitizing surfaces, we try to prevent, delay, or mitigate these processes. In spite of our efforts, we are still surrounded by a plethora of microbial life—be it in the food we consume or the diverse spaces and environments we move through and inhabit.

If we take these different scales into account, we can see that architecture is in constant microbial exchange with its environment and inhabitants. Every space and building hosts a unique microbiome in constant flux due to its use, as well as with

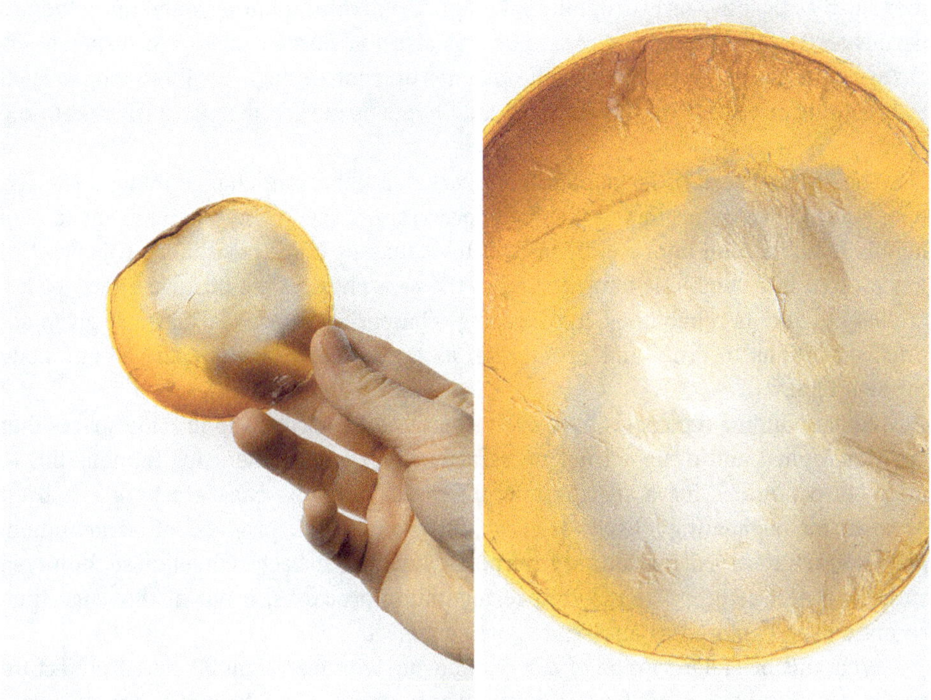

Fig. 5: Dry cellulose biofilm: a heterogenous, skin-like structure with a gradient of opacity.

the indoor and outdoor climate.[17] Historically this has been mainly studied in the context of pathogens; more recent research, however, suggests that commensal and benign bacteria also exist in large numbers. In addition to building materials, architectural spaces and surfaces can, in this context, be understood as interfaces between their inhabitants and various microbiota.[18] Recent studies suggest that being exposed to a diverse microbiome can have wide-ranging benefits, from mental health to allergy resistance.[19] The study of these interdependencies and complex relationships between outdoor microbiota, the built environment microbiome and the human microbiome, transgresses scales and disciplines. Architectural materials—the interfaces—play an active role in this regard as they mediate colonization processes through surface qual-

---

17 Simon Lax et al., "Longitudinal Analysis of Microbial Interaction between Humans and the Indoor Environment," *Gautam Dantas* 15, no. 6200 (2014): 1048–52, https://doi.org/10.1126/science.1254529; Jack A. Gilbert and Brent Stephens, "Microbiology of the Built Environment," *Nature Reviews Microbiology* 16, no. 11 (2018): 661–70, https://doi.org/10.1038/s41579-018-0065-5.

18 Simon Lax et al., "Our Interface with the Built Environment: Immunity and the Indoor Microbiota," *Trends in Immunology* 36, no. 3 (March 2015): 121–3, https://doi.org/10.1016/j.it.2015.01.001.

19 Andrew J. Hoisington et al., "The Microbiome of the Built Environment and Mental Health," *Microbiome* 3, no. 1 (2015), https://doi.org/10.1186/s40168-015-0127-0.

ities, such as porosity or hydrophilic behavior.[20] This continuum between microbiomes permeates not only built environments but also the human body. We disperse, exchange, and absorb microbiota from our environment through bodily functions such as breathing or touch,[21] in the exchange that happens mostly unconsciously and unnoticed.

The consumption of fermented food marks an important change though. The fermentation of food relies on the metabolic processes of specific microorganisms to help modify nutrients and improve the taste of food, making it less perishable. This practice has evolved in a symbiotic manner and in the awareness of the positive effects of microbial life and its capabilities. It is a caring relationship where, bacteria are given optimal conditions to grow and proliferate, as they become a vital part of our body through food.[22]

Whole building typologies have evolved around this process to provide spaces that are conditioned and designed not necessarily as an environment for humans but to maintain conditions for a specific microbiome to develop—cheese cellars, wineries, or barns for fermenting tobacco or cacao, for example. The products of fermentation, however, are reserved for culinary purposes. Looking at bacterial cellulose, however, suggests a new setting. The result of fermentation processes, it has at the same time recently been explored as an architectural material.

What this brief description of our relationship with microbial life and architecture suggests is that we are confronted with an interdependent fabric of n environment, material, and scale, as well as with various life forms which are in constant change and interaction. Our research interest lies precisely in this relationship between nurturing environments, architecture, and microbiology. We have chosen not only to focus on bacterial cellulose as an outcome of fermentation processes, however, but on the changing roles of designer and organism within these shared environments and processes of making.

By seeking to understand bacterial cellulose as simultaneously a material and a living organism, the notion of design and manufacturing changes fundamentally, prompting us to acknowledge its intrinsic activity and aliveness. Instead of relying—as in the case of synthetic polymers—on a predetermined and precisely timed process, designing with fermentation disrupts this procedure and questions our role as designers. This is a shift from simply envisioning a final product and its materialization, toward designing boundary conditions in which an organism and its structure evolve (fig. 6). It differs starkly from the common conception of materials as passive and "obedient" entities within the design process. Bio-design calls for a bottom-up approach that guides materials during growth throughout different scales, rather than imposing

---

20 Marcos Cruz and Richard Beckett, "Bioreceptive Design: A Novel Approach to Bio-Digital Materiality," *arq: Architectural Quarterly* 20, no. 1 (2016): 51–64, https://doi.org/10.1017/S1359135516000130.
21 Lax, "Our Interface with the Built Environment: Immunity and the Indoor Microbiota."
22 Sandor Ellix Katz, *The Art of Fermentation: An in-Depth Exploration of Essential Concepts and Processes from around the World* (White River Junction, VT: Chelsea Green Pub, 2012).

shapes onto raw materials. In other words, the primary task is to design an environment rather than forms and structures.

The settings for these environments of growth are not limited to the enclosure of a container or an incubator. For instance, observing the development of a biofilm in a Petri dish, one would assume that it is a contained process, a closed system with defined and rigid boundary conditions: the glass sides of the Petri dish, the medium with its nutrient composition, and a lid that should seal everything off. But this is hardly the case: even though the Petri dish may appear encapsulated, it is in fact embedded in an environment and permeated by energy flows that affect its temperature and humidity. Air circulates through minute gaps between the dish and its lid while radiation and light permeate the system. Variable levels of nutrient, temperature gradients, and microstructural changes in the growth medium all influence the growth of biofilm to a certain degree. Instead of viewing these boundary conditions as geometrically defined and enclosed, they can be understood as fluid zones of exchange, where different scales intersect and guide the growth process.

Donna Haraway describes these relations as "sympoietic,"[23] or more specifically as a "holoent" condition, so as not to privilege "the living but to encompass the biotic and abiotic in dynamic sympoietic patterning."[24] Processes of growth are intrinsically interdependent and their environments fluctuate. As designers we are also part of that environment. Compared to working with conventional materials, the degree of control and influence we have over such conditions is often limited. Instead of imposing forms onto materials we start to think about designing environments of their growth. Instead of approaching additive assembly using screws and bolts and subtractive processes with drilling and milling, the toolset shifts to preparing a nutritious broth on which microorganisms can feed and multiply. It is the way of guiding growth and structure through nutrients, viscosities and temperature. This interaction with a living organism and the environment resembles more closely the cheese- or wine-making practices we mentioned earlier: it is more a task of care than of control.

Textilic Milieus and Co-Weaving

We understand the biofilm as a textile entity, composed of grown nanocellulose entangled in a fibrous network. For us, therefore, "weaving" and the "textilic" start on the microscale through microbiological activity. From this microbial textile, a microenvironment emerges that actively supports the development of the bacterial community.[25]

---

23 Donna Haraway, *Staying with the Trouble* (Durham, NC: Duke University Press, 2016), 25, https://doi.org/10.2307/j.ctv11cw25q.

24 Haraway, *Staying with the Trouble*, 26.

25 Karin Krauthausen and Regine Hengge, "Das Ereignis der Faser / The Event of a Fibre," *Gropius Bau Journal* (article accompanying the exhibition *Kosmos Weben* by Hella Jongerius), 2021, 1–8.

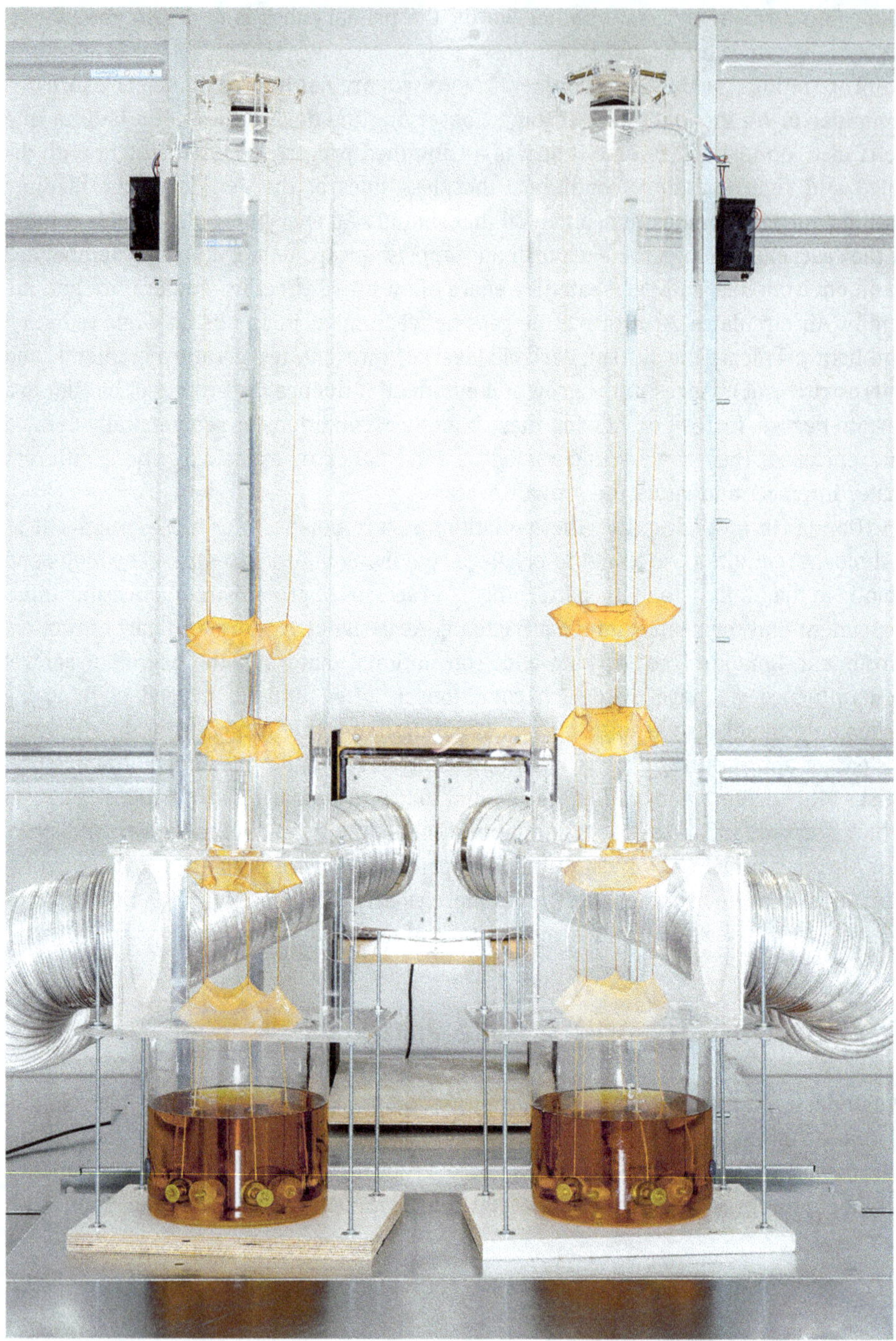

Fig. 6: Environment of growth: designing with boundary conditions in the laboratory.

It provides a buffer zone, and its permeability allows for nutrients and bacteria to travel within what is a *textilic milieu.*

For anthropologist Tim Ingold, the notion of "textilic" encompasses a negotiation process with the material during the process of making.[26] Ingold emphasizes that making and textility are not a question "of imposing preconceived forms on inert matter but of intervening in the fields of force and currents of material wherein forms are generated."[27] Both the designer and the bacteria play a crucial role in the fields of force and currents in the material: we set boundary conditions in which the microbiological "weaving" can take place.

In our experiments, a scaffold—another textile entity on a different scale—is introduced to this setting that not only structurally mediates between the nano-cellulose and a macroscopic fiber structure, but similarly allows for new microenvironments to emerge during growth (fig. 7). In contrast to the biofilm, this secondary textile scaffold follows a designed pattern and a geometry on a macro scale to provide a support for the biofilm to develop. The two textile systems, the biofilm and the thread, intersect in certain areas or fully integrate with one another. During growth, a gradient between the fuzzy boundary of the threads and the biofilm emerges, where the fibers of the scaffold intertwine with the nanocellulose to form a structural bond. Rather than being glued to each other, these structures grow together, intertwining the human-made and the bacterial: in this web-like milieu, we observe the reciprocal and fluid relationship between the nano-, micro-, and mesoscale in a nurturing and active environment.

Even after growth, we notice that the exchange with the environment continues to contribute to the "fields of force" as a generative impulse for transforming material structures and configuring their relationship to their surroundings. In studies of the morphologies of drying, for example, both the shape and internal structure of the biofilm are reconfigured through several wetting and drying cycles (fig. 8). Through rehydration, the cellulose returns to its original shape, but the fiber alignment remains in the "memory" of the material. All these changes take place on the level of the fiber: swelling causes the biofilm to temporarily become gelatinous before returning to the dry state, where the fibrous structure reappears. In the drying process, both elastic and thermal energy pass through the material and are exchanged with the environment: elastic energy is stored and released during the shape-change, while heat escapes by evaporation—a phase transition of water from liquid to gas. In these processes, environment and structure intermix and become part of each other's composition, so that the environment acts not as a passive, immaterial domain surrounding the structure, but is incorporated into the material (figs. 9, 10).

---

26 Tim Ingold, "The Textility of Making," *Cambridge Journal of Economics* 34, no. 1 (2009): 93, http://doi.org/10.1093/cje/bep042.
27 Ingold, "The Textility of Making," 2.

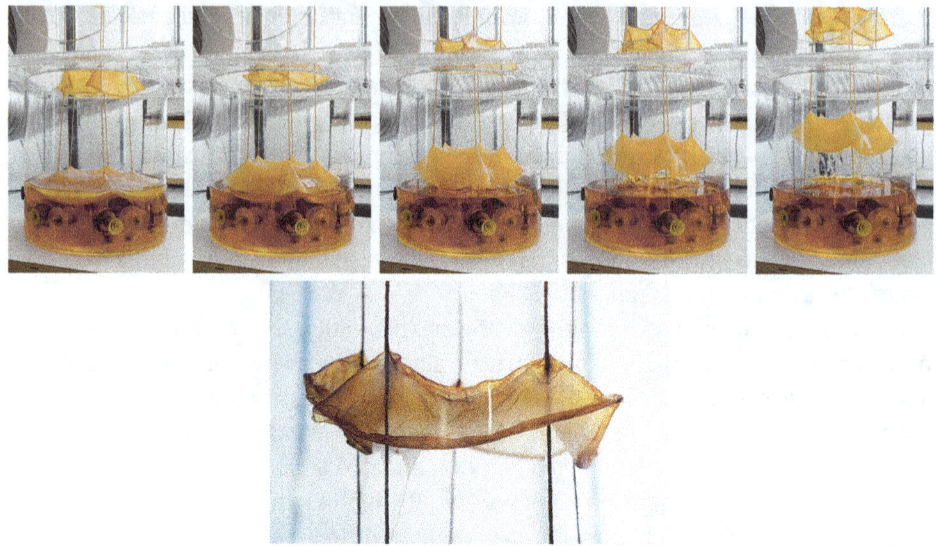

Fig. 7: Experimental setup with a perpendicular thread.

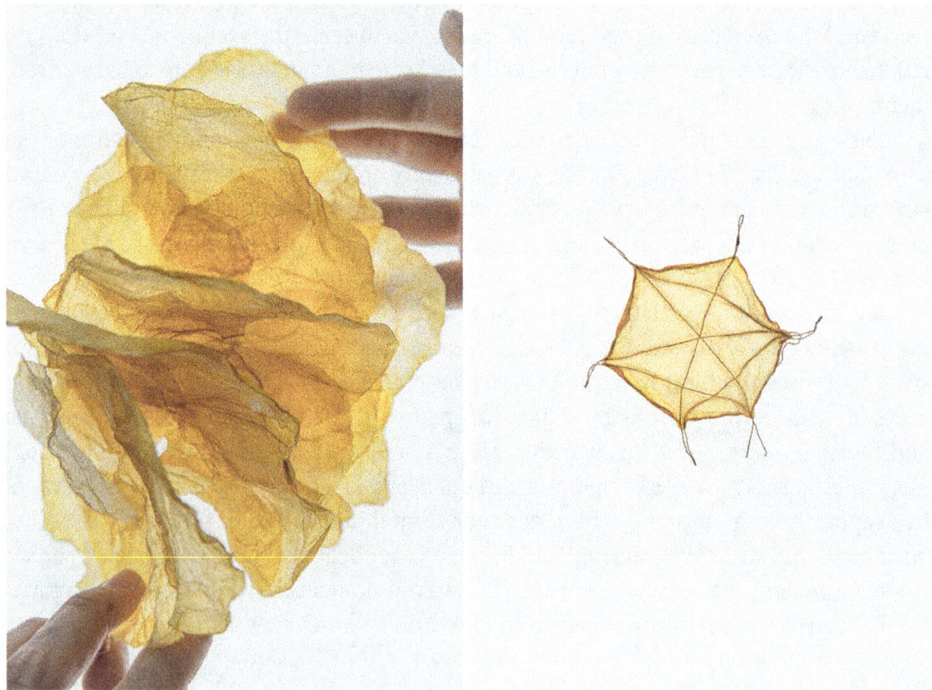

Fig. 8: Morphologies of drying.

Again, if we consider how Ingold differentiates between the "fields of force" through which artifacts are shaped and "morphogenetic fields" that give rise to biological

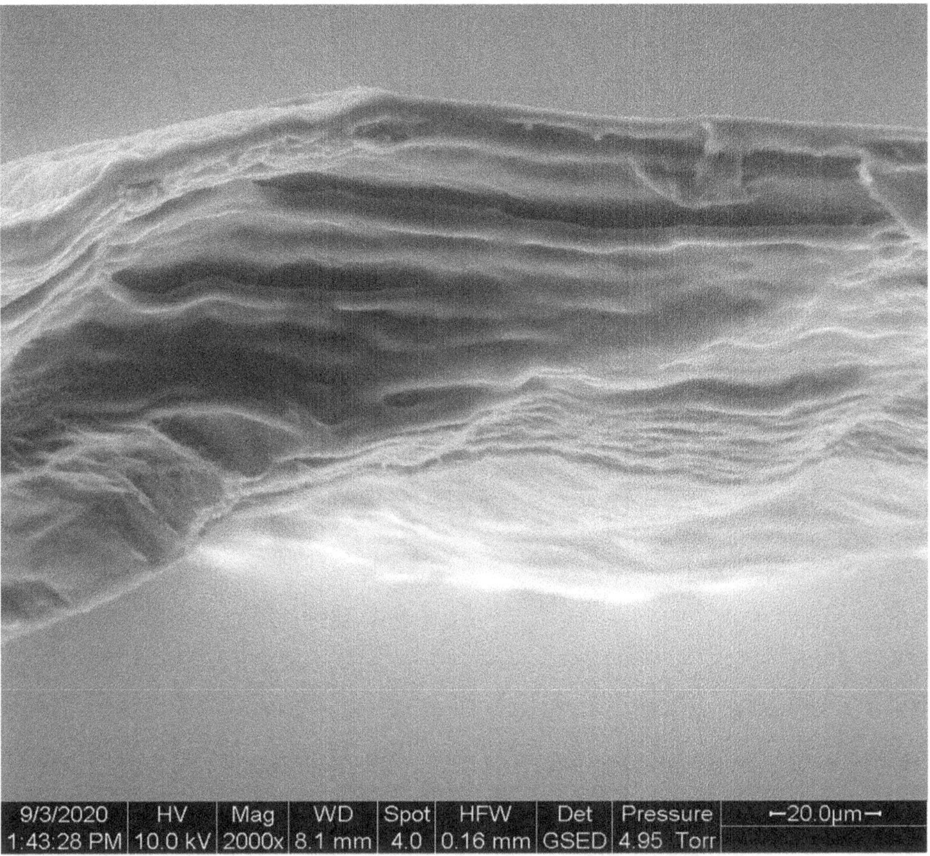

Fig. 9: Wet, gelatinous internal structure of the biofilm exposed to moisture.

form, we need to take a closer look at this duality as it affects our experiments. Ingold states that "[b]oth kinds of field cut across the developing interface between the object (organism or artefact) and an environment which, in the case of the artefact, critically includes its 'maker'."[28] While he distinguishes between "organism or artifact," in our experiments, these dualities start to blur, since they form a continuum. In fact, we can observe an overlap between what Ingold describes as "fields of force" and "morphogenetic fields," where the growth of the biofilm is guided by the textile structure.

In his essay "The Textility of Making," Ingold discusses the craft of woodworking, describing how the "blade enters the grain [of wood] and follows a line already incorporated into the timber through its previous history of growth."[29] The act of making is

---

28 Tim Ingold, *The Perception of the Environment: Essays in Livelihood, Dwelling and Skill, Mind, Culture, and Activity*, vol. 9, 2002, 345.
29 Ingold, "The Textility of Making," 92.

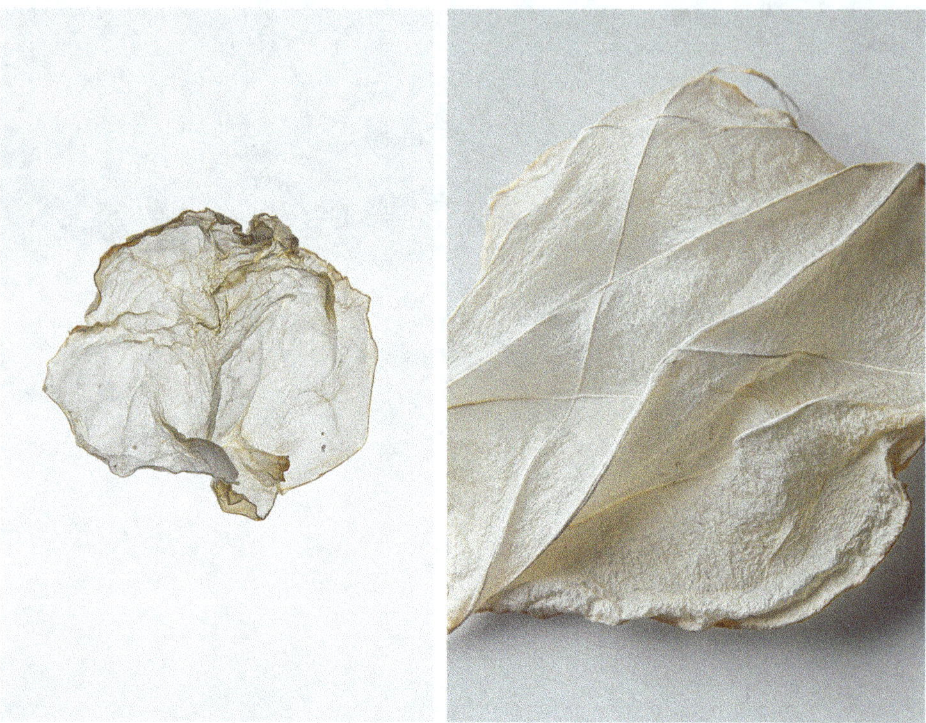

Fig. 10: During drying, spontaneous folding occurs along the textile threads and the entire biofilm structure.

described as temporarily disconnected from the process of growth. Through direct interaction with bacteria during the processes of growth, a simultaneity between growth and making takes form and it suddenly becomes possible to co-weave structures together with the organism and to guide what Ingold refers to as "lines" in their becoming (fig. 10).

Weaving Scales

In researching and seeking to understand microbial structures, our disciplinary lenses instinctively tend to separate out the material from the biological activity, observing each at their respective scales. In materials science laboratories, we investigate and measure fibril diameters, tensile strength on the micro or macro scale, and material behavior upon wetting or drying (fig. 11). Meanwhile, in the microbiology laboratory, the focus is on the bacteria, on their genomes and metabolism. Even though the data and insights we gain from these distinct disciplinary methodologies are important in understanding parts of the complex workings of the biofilm and its materiality, the individual pieces can scarcely provide a sense of the intertwined ecology of a biofilm.

The relationship between microorganism, structure, and environment is complex yet fascinating and can hardly be described through discrete scientific viewpoints and scales.

The interplay of these viewpoints is far from linear. Charles and Ray Eames's visual journey through the successive scales of matter offered a systematic and ordered picture of the hierarchy of scale, each clearly framed within its own microcosm. But this separation—a "trap" of scale—together with independent disciplinary scientific views, does not capture the simultaneity and fuzziness of the scalar encounters in the design process, nor the tactility of their weaving.

Stepping away from the field of architecture, where the default measure is the human, our attention has shifted towards the seemingly imperceptible processes of textile-making on the microscale. This change of perception leads us to recognize the multitude of lives and structures beyond the human, which, even though not visible, shape and influence our bodies as much as the spaces surrounding us. Our attempt to leave the convenience of a solely anthropocentric view of materials and processes has presented us with new opportunities for designing through growth, but likewise with challenges and limitations. Whenever we encounter contaminated cellulose in Petri dishes in the laboratory, our understanding of control and collaboration is challenged: our design intentions are confronted with the unpredictable reality of the culturing practice that defies isolation and containment. Working with biofilms has shown the fluidity and simultaneity of biological processes, where fibers on different scales emerge and intersect in dynamic environments of growth. To co-design these environments means to acknowledge the dynamic relationships of bacterial cellulose, its structure, and life, together with their entanglement across scales.

Rather than seeking control over these intricate relationships, which Ingold refers to as "fields of force," our approach guided us toward crafting environments that foster and nurture growth. In addition to bacteria, these environments required careful consideration of factors such as nutrients, temperature, and airflow. Our role as designers shifted from thinking about finished objects to creating settings in which bacteria might act independently. This shift calls for a different approach to design, that moves away from imposing predefined shapes onto passive materials and towards embracing roles of care and responsibility that more closely align with practices of fermentation.

As we engaged with bacteria, a novel collaborative environment unfolded for us, fostering not only interdisciplinary connections but also forging collaborations between different species. While our documentation shows the outcomes of this attempt at "collaborative weaving", it is merely a static representation of the dynamic process that takes place between bacteria and its textilic milieu, where making unfolds in a constantly changing environment of entangled scales.

Acknowledgments:
Special thanks go to our colleagues at the Cluster of Excellence "Matters of Activity": Skander Hathroubi, whose support and guidance as part of the Hengge group at the

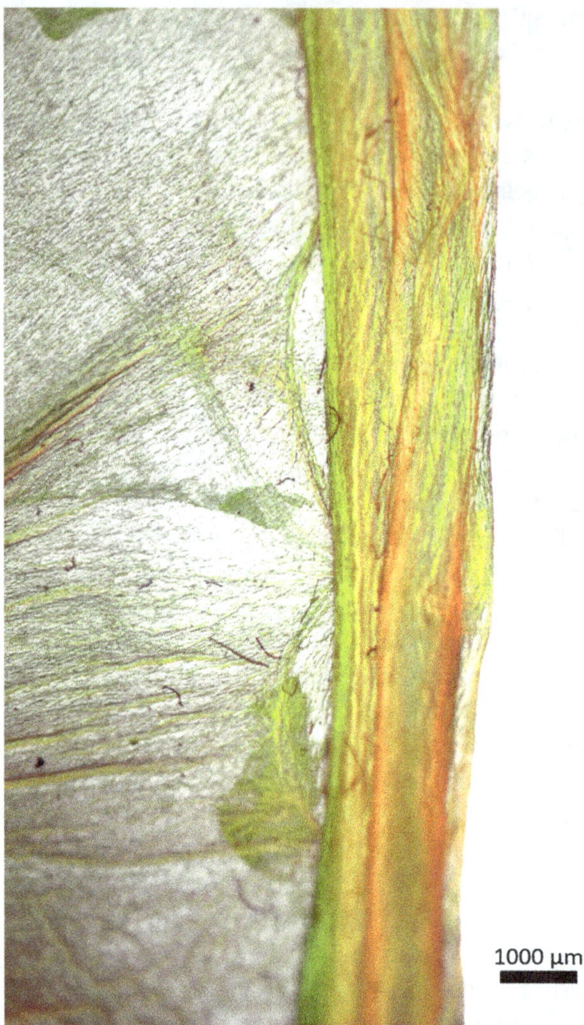

Fig. 11: Microscopic image showing the "guided growth": the alignment of cellulose fibers to the textile scaffold and to the edge of the glass enclosure.

Department of Microbiology at the Humboldt-Universität zu Berlin made this project possible, and to Karin Krauthausen for her invaluable suggestions and feedback on this text.

Khashayar Razghandi

Rethinking Growth: A Bio-Inspired Take on Creative Processes

Introduction

Problem. The concept of *growth*, stemming from our understanding of the natural world, has played a crucial role in the perception and shaping of our material, social, and economic realities. As dystopian futures are unfolding in front of us, there is a growing concern about growth, which urges us to rethink some of our basic concepts around the matter to stand a chance in facing the current and future crises.

Objective. The presented work is a transdisciplinary effort to reflect on the concept of growth from a bio-inspired perspective, rethink it within the current discourse of New Active Materialism, and put it forward as an analytical tool as well as a creation strategy to address the urgencies of our time.

Approach. The approach of the work is to take the knowledge of activity of matter and material systems from materials science and biology; borrow the questions and concepts from various realms within humanities disciplines that are relevant to the *Active Materialism* discourse (cultural studies, feminist new materialism, information philosophy, and so on); weave the reflections on the interdisciplinary creative experiences into the mesh; and rethink the phenomena in light of these new transdisciplinary understandings, perspectives, and concepts, with the ultimate goal of bringing the new insights back into the various relevant realms of analysis and synthesis (e.g., interdisciplinary science and design research and education, sustainability discourse, and so on).

Outline. The most relevant place to start such a rethinking process is the realm of biology. By going through an example from biological material activity, the first chapter, "Active Materials Paradigm," opens up the current discourse around matter and material activity, paints a more performative picture of matter beyond the conventional paradigms, and looks at activity of matter through various *stances.* The second part, "Rethinking Growth as Sympoiesis," highlights some of the core principles in biological material systems and compares them to our industrial *fabrication paradigm,* and, based on those reflections, proposes a different take on growth that is rooted in and reflects the *creative nature of entangled dynamisms and performative structures.* The third chapter, "Rethinking Growth as Creation Strategy," summarizes the proposed model of growth and hints at how such a paradigm shift can serve a *creation methodology.* Chapter four concludes the argumentation and puts forth some *future* perspectives.

Active Materials Paradigm—Activity through Various Stances

Case of an Active Material System

Consider the example of ice plant's hydro-actuated seed dispersal. Ice plants grow in arid areas and have evolved a sophisticated material strategy to ensure the seed dispersal in the right environmental conditions, namely upon rain. The underlying material architecture that enables the hydro-responsive actuation of the dead tissue is traced at various length scales in figure 1. The five seed-containing compartments are closed in a dry state (fig. 1A, top), and unfold and release the seeds only upon wetting (fig. 1A, bottom). The five hygroscopic muscles (keels) are folded inward and keep the valves closed in the dry state, and flex outward only upon hydration (fig. 1B, 1–2 and fig. 1, top right: schematic of flexing of the keels). The keels are made up of a network of hexagonal eye-shaped cells, filled with a highly swellable cellulosic inner layer, which would swell/shrink upon hydration/drying cycles, resulting in the reversible opening/closing of the cells (fig. 1C–E, 4–5) and the expansion/contraction of the honeycomb structure (fig. 1C, 3). Presence of an inert backing tissue (fig. 1B, 2) hinders and resists the hydro-actuated deformation of the honeycomb structure, and, to compensate, the linear deformation translates into a bending one (fig. 1B–C, 1–3), which results in flexing of the keels and opening of the seed capsule (fig. 1A, B, 1).[1]

To Narrate a Material's Activity

The common narrative of *Material → Property → Function* leans on the *Subject-Object-Property* paradigm: an inherently hylomorphic and patriarchal paradigm that takes matter as individuated and passive (fig. 2A).

The Conventional Materials Paradigm from material science and engineering visualizes the matters of material in a tetrahedron (fig. 2B) as an integrated way of addressing different perspectives of understanding and control of various aspects of materials.[2] One can take the perspectives of each of the interrelated four apices of the diagram: particular organization/ arrangement of materials (compositions) into structures over range of length scale; the synthesis and processing of those compositions and arrangements; the properties resulting from the compositions and their arrangements; and, finally, the performance of the material as a measure of its utility in a

---

1 Lorenzo Guiducci et al. 2016, "Honeycomb actuators inspired by the unfolding of ice plant seed capsules," *PLoS One* 11 (2016): e0163506, https://doi.org/10.1371/journal.pone.0163506.
2 National Research Council, *Materials Science and Engineering for the 1990s: Maintaining Competitiveness in the Age of Materials* (LOCATION: 1989).

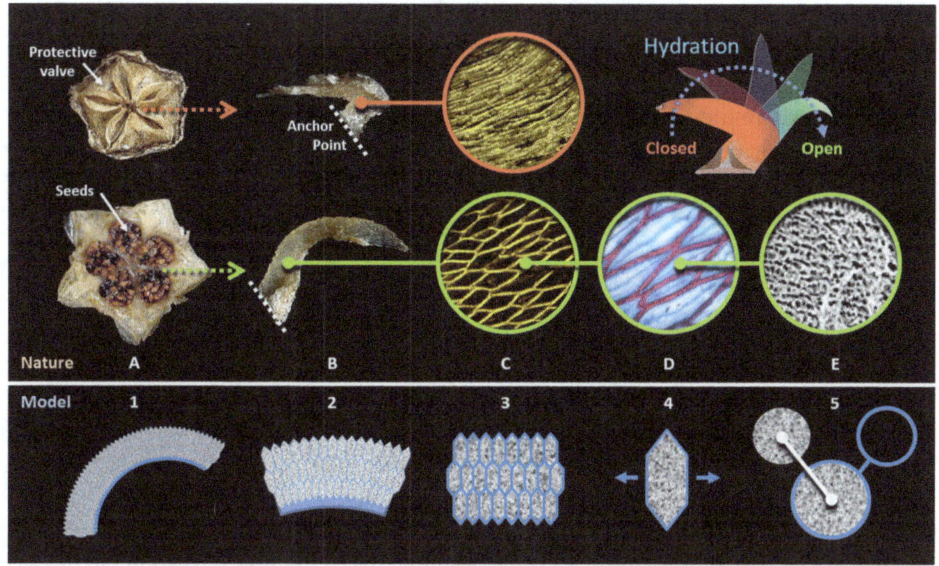

Fig. 1: Ice plant hydro-actuated seed dispersal: an active material system.

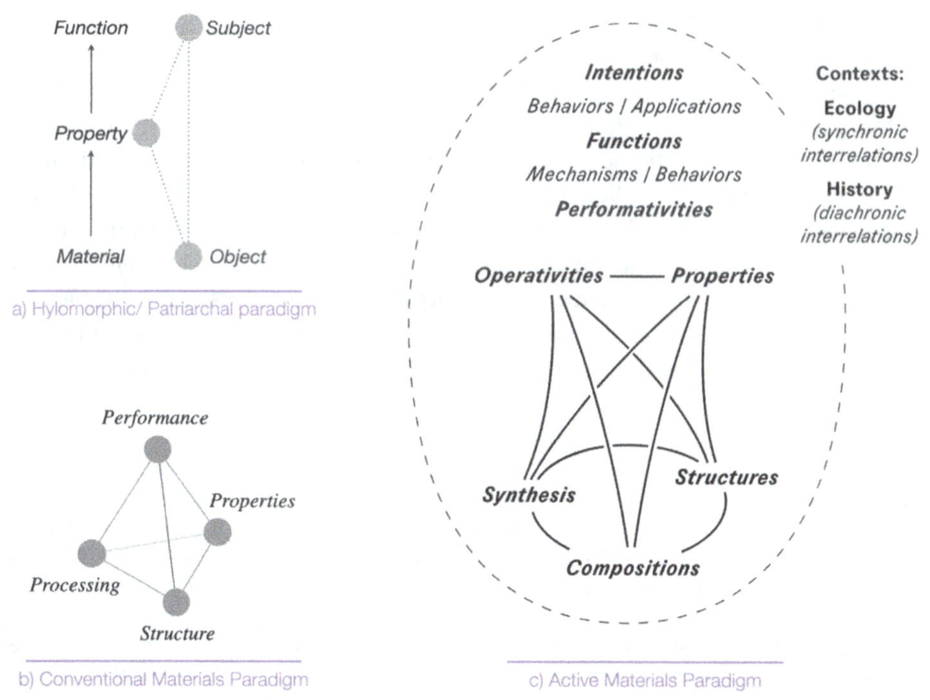

Fig. 2: Toward an Active Materials Paradigm: (2A) author's conception of hylomorphic paradigm; (2B) conventional materials science tetrahedron (National Research Council 1989); (2C) Active Materials Paradigm by Khashayar Razghandi, 2022.

given context. Nonetheless, the hylomorphic view of matter, as a passive building block to be tamed and shaped into forms and structures with specific functions, is also dominant in this conventional materials paradigm.

Going back to the ice plant hydro-responsive actuation, one can try to narrate the activity from various perspectives or stances as different yet entangled levels of description of the phenomena: various sugar molecules as composition; the process of how they are synthesized and put together; the structure and multiscale architecture they are synthesized into; and the role which this material architecture plays in defining the observed properties and functions. In this way we can construct some structure-function relationships. The whole system is made up of few basic components (cellulose, hemicellulose macromolecules, etc.), and the activity of the system (hydro-responsive actuation) is mainly due to the architecture of the material system at various length-scales. Ice plant and similar plant actuation systems are not a rare case; in biology, we encounter various forms of this structure-function or structure-activity relationship. Biology is based on only a few basic elements and building blocks, and the wide range of properties and functionalities we observe in nature are mainly achieved through diverse structures and material architectures.[3] In this narration of the ice plant actuation system, sensing, processing, and response to stimuli (rain) are integrated in one material system: we have "sense-action."[4] Material structure/organization is the key to relate matter, information, and energy.[5]

To continue with the narration, we can place composition, structure, and synthesis as the triad base of the material paradigm giving rise to various properties at various length scales of the system. However, one can take the same system, and narrate the system not only through the lens of composition, structure, and synthesis defining a specific property, but by following the chains of operations: "Water absorbs on sugar molecules the cellulosic inner gel swells, inflating the cells . . . pushing the cell walls from inside and opening the cells . . .," and so on. Looking at the system through the chains of operations can shed a different light on the matter, taking matter as operative. Property and operativity are two interrelated stances one can take to look at what emerges from the triad base of composition, structure, and synthesis.

---

3 See Peter Fratzl et al., *Active Materials* (Berlin: De Gruyter, 2021); Alexander H. King, "Our Elemental Footprint," *Nature Materials* 18, no. 5 (2019): 408 – 09, https://doi.org/10.1038/s41563-019-0334-3; Michaela Eder, Shahrouz Amini, and Peter Fratzl, "Biological Composites—Complex Structures for Functional Diversity," *Science* 362, no. 6414 (2018): 543 – 47, https://doi.org/10.1126/science.aat8297; Yuri Estrin et al., *Architectured Materials in Nature and Engineering* (Heidelberg: Springer, 2019); and Peter Fratzl, Christiane Sauer, and Khashayar Razghandi, "Editorial for the Special Issue: Bioinspired Architectural and Architected Materials," *Bioinspiration & Biomimetics* 17, no. 4 (2022): 1 – 4, https://doi.org/10.1088/1748-3190/ac6646.

4 Mohammad Fardin Gholami et al., 2021. "Rethinking Active Matter: Current Developments in Active Materials," in *Active Materials* (Boston: De Gruyter, 2021), 193 – 222, here 212.

5 Gholami et al., "Rethinking Active Matter"; Fratzl et al., *Active Materials.*

Thus, we end up with a double-horned pyramid as the foundation of a new *Active Materials Paradigm* (fig. 2C).[6] This places *activity* at the core of the materials paradigm.

The first thing this paradigm shift does, is that the boundaries and interfaces within and in between material systems become blurry. If you follow the chain of operations, you will find yourself facing with porous, blurry, and fluid interfaces and event-based boundaries signified by the intra-activities at various scales.[7]

Following this path, *function* turns out to be not just a *telos* at the macro scale, but can be taken as a stance situated within various interrelations in the architecture of the system: a sugar molecule can be a functional unit of the swelling gel inflating the cell, or single cells can be taken as the functional unit of the honeycomb structure, and so on. Here, we have structural and functional interrelations not only throughout the material architecture (between larger or smaller length scales), but also with the surrounding environment.[8]

Ice plants open in response to rain and have evolved to do so: material systems are situated within ecologies and histories.[9] This brings the performativity stance into the materials paradigm (fig. 2C).[10] Biology is the realm of entangled dynamism and performative structures. *Things* are always situated in an entangled interwoven meshwork of things (ecologies), and things are dynamic: they have histories and futures. In this sense, *ecology context* highlights the synchronic sense of interrelations, and *history context* highlights the diachronic interrelations. Looking from the performativity stance, matter can be seen as performative in the meshwork of ontological relations and relevance. We are not dealing with subjects, objects, and their properties; rather, we are dealing with co-beings and co-becomings.

We are dealing with entangled dynamisms and performative structures.

---

6 Khashayar Razghandi, "Rethinking Materials Paradigm: Towards an Active Understanding of Gestalt," in *Design, Gestaltung, Formatività*, ed. Patricia Ribault (Berlin: Birkhäuser, 2022).

7 Khashayar Razghandi and Emad Yaghmaei, "Rethinking Filter: An Interdisciplinary Inquiry into Typology and Concept of Filter, Towards an Active Filter Model," *Sustainability* 12, no. 18 (2020): 7284, https://doi.org/10.3390/su12187284.

8 Gholami et al., "Rethinking Active Matter."

9 Razghandi, "Rethinking Materials Paradigm."

10 Karen Barad, "Posthumanist Performativity: Toward an Understanding of How Matter Comes to Matter," *Signs: Journal of Women in Culture and Society* 28, no. 3 (2003): 801–31; Karen Barad, *Meeting the Universe Halfway: Quantum Physics and the Entanglement of Matter and Meaning* (Durham, NC: Duke University Press, 2007); Jens Hauser and Lucie Strecker. "On Microperformativity," *Performance Research: A Journal of the Performing Arts* 25 (2020): 1–7; Razghandi, "Rethinking Materials Paradigm."

Rethinking Growth as Sympoiesis—Bio-inspired Analysis of Creative Processes

Makings of Biology vs. Fabrication Paradigm

The whys and ways of biology making things are paradigmatically different from the whys and ways we design and manufacture things. This has to do with the difference in the boundary conditions or constraints that shape the *making*.

In biology, things are mainly made through growth processes. Making in biology looks more like following a recipe: it is approximate. Our manufacturing paradigm is more like a blueprint: identical, exact materials, parts, and processes (think of the production lines of early cars or today's smartphones). Situated within specific ecologies (thermodynamical, physiological, evolutionary, developmental, and so on), biology is "poor." It has to make do with available resources: ambient temperature, relatively low forces, and so on. It also has to make do with whatever is available as building blocks to make. All life (on earth) uses only few basic elements (fewer than thirty) from the whole periodic table. Our smartphones use around seventy of the 118 known elements.[11] We are "rich" with "abundant resources"!

Life makes a variety of properties and functionalities possible by playing with material structure. It makes use of a variety of biologically or thermodynamically controlled processes to shape those structures. These processes of making are fundamentally different from the energy- and resource-heavy fabrications that shape the making paradigm of our time. "Heat beat treat" is not just a faint memory of the blacksmiths, or a foggy factory of an old industrial area. Even our most advanced 3D-printing technologies follow the same hylomorphic logic: extract, assemble, and shape the passive matter into the desired forms and functions. We are "rich" with "abundant resources"!

Following these *boundary conditions* and constraints of growth, biology ends up with diverse material architecture and multi-functionalities. Multi-scale structure of material systems gives rise to various properties and operativities and addresses multiple functionalities simultaneously.[12] The core of the fabrication paradigm is about picking of the "right" material (material selection), design and fabrication of different parts, and assembly of those parts into systems and devices.[13]

Thinking along Borgmann's distinction of "things" and "devices," biology is more "thingy," more open and entangled, while our fabrication more often seeks and results

---

11 King, "Our Elemental Footprint."
12 Eder, Amini, and Fratzl, "Biological Composites"; Estrin et al., *Architectured Materials*; Peter Fratzl, "Biomimetic Materials Research: What Can We Really Learn from Nature's Structural Materials?," *Journal of the Royal Society Interface* 4, no. 15 (2007): 637–42, https://doi.org/10.1098/rsif.2007.0218; Fratzl, Sauer, and Razghandi, "Bioinspired Architectural and Architected Materials."
13 Fratzl, "Biomimetic Materials Research."

<table>
<tr><td align="center">

Growth

few basic elements

Recipe   *diverse processes*

diverse structures

multi-scale Structures

multi-functional

"things"

Open & Entangled

entangled Activity

Intra-active

Adaptive design

</td><td align="center">

Fabrication

"abundant" "resources"

Blueprint

"Hear, Beat, Treat"

Materials, Parts, Devices, Systems

specific function

"devices"

Closed & Protected

Imported Activity

Hylomorphic

Secure & fix design

</td></tr>
</table>

Fig. 3: Makings of biology vs. fabrication paradigm, inspired by Peter Fratzl, "Biomimetic Materials Research: What Can We Really Learn from Nature's Structural Materials?," *Journal of the Royal Society Interface* 4 (2007): 637–42, doi: 10.1098/rsif.2007.0218.

in "devices" as closed and protected systems.[14] *Things* in biology are open and entangled material systems, with fluid interfaces and event-based boundaries, where matter, energy, and information flow within the entangled dynamisms and performative structures.

Biology is the realm of entangled intra-activities,[15] within meshworks[16] of entangled dynamisms and performative structures, with interwoven yarn-balls of ontological cares and significances which we call things. Fabrication relies mainly on imported activity. It is about shaping passive matter into forms, parts, devices, and systems, and then bringing the activity into those systems (e.g., cables of energy and information) to serve a certain function for a certain amount of time. Fabrication follows a paradigm of closed and protected, secure and fixed design with imported activity.

The boundary conditions and constraints are the key to emergence of organizations structures. Biology and information philosophy point to the potentiality and contingency of the constraints.[17] The potentiality of not just what happens but also what

---

14 Albert Borgmann, *Technology and the Character of Contemporary Life: A Philosophical Inquiry* (Chicago: University of Chicago Press, 1987); James Auger and Julian Hanna, *Reconstrained Design* (Madeira: M-ITI, 2019).

15 Karen Barad, "Posthumanist Performativity" (see note 10); Karen Barad, *Meeting the Universe Halfway.*

16 Tim Ingold, "Toward an Ecology of Materials," *Annual Review of Anthropology* 41 (2012): 427–42, https://doi.org/10.1146/annurev-anthro-081309-145920.

17 Terrence W. Deacon, *Incomplete Nature: How Mind Emerged from Matter* (New York: W. W. Norton & Company, 2011).

could happen and could have happened, is defined through the boundary conditions and constraints. The entanglements and coupling of the different constraints of different material systems are, on the one hand, maintaining the organizations and structures that matter, and, on the other hand, the loci of the emergence of new structures, organizations, and creations.[18]

This paints a meshwork model of entangled co-beings and co-becoming. The structures are in becoming and the emergence and sustenance of structures/organizations are to be traced in the entangled dynamisms and constraints. These can be imposed by physical, thermodynamic, evolutionary, and developmental processes, among others, and interactions across various scales and stances. Biologically or thermodynamically controlled processes that realize the makings in biology are shaped by the boundary conditions of entangled material systems at various scales and stances. The coupling of the constraints and boundary conditions of different materials' organization results in dynamic fluxes of matter, energy, and information into and in between such entangled structures. Structure is the key to biological material solutions. Structures have open, fluid, and event-based boundaries, entangled with various other structures at smaller and larger length scales of the system, and expanding out into other structures of other dynamisms.

Growth is situated within various entangled histories and ecologies with specific boundary conditions and constraints imposed on it at various scales and stances. Situatedness of growth in biology urges adaptation, not only at the level of evolution, but also at the systems and materials level. The growth of biology, entangled with the historical and ecological contexts, begets adaptations of materials, structures, and functions to the changing entangled dynamisms and conditions. Adaptations is biology's way of dealing with uncertainty and openness.

In the light of this analysis of the whys and ways of making in biology, and through the lens of the active materials paradigm, I propose to reimagine growth along the background of the creative nature of entangled dynamisms and performative structures (fig. 4).

Rethinking Growth as Creation Strategy—A Bio-Inspired Synthesis of Creative Processes

Growth as a Creative Methodology

Complex problems are in essence systemic and dynamic. *Systemic* nature entails that there are multiple actors involved that are in an entangled interrelation with each other: dealing with an ecology rather than mere causal relations of subjects, objects,

---

18 Deacon, *Incomplete Nature.*

and properties. *Dynamic* aspect points out to the fact that any structure is essentially a structure in becoming: the relations defining the synchronic aspects are, in fact, a snapshot of an ever-changing relation in a diachronic dimension. In dealing with such problems, which are multifaceted and ever-changing, a suitable creative approach is one that is simultaneously integrative and adaptive. The proposed bio-inspired model of growth can serve as a methodology that can address these various aspects both in analytical and synthetical practices.

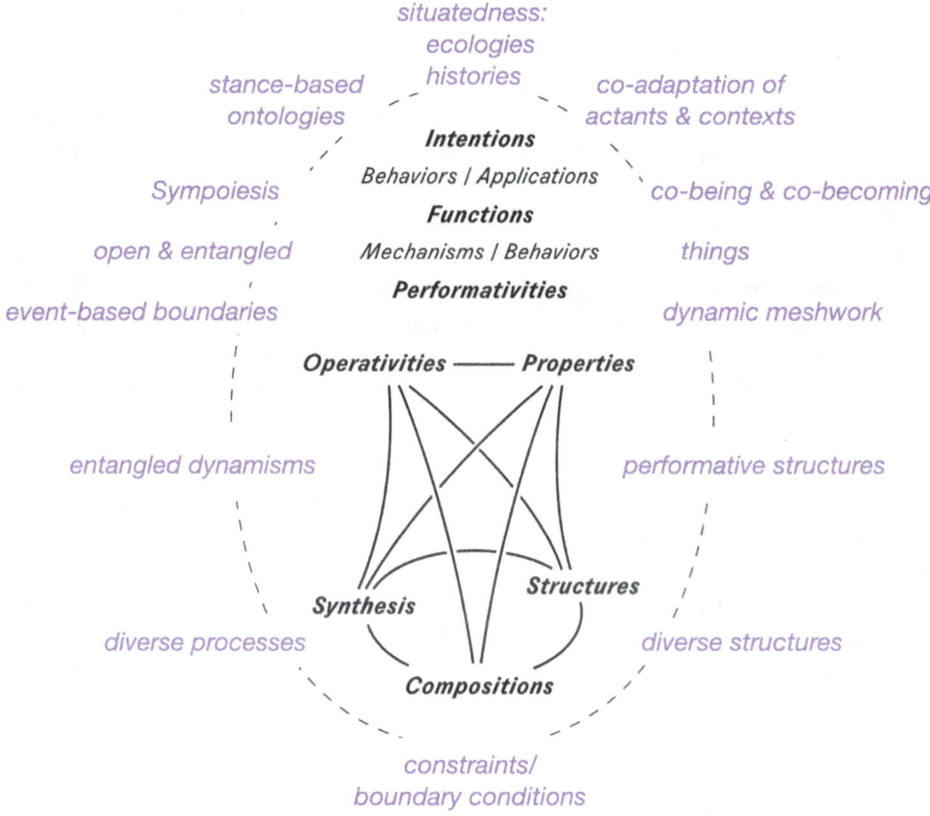

Fig. 4: Rethinking growth: creative nature of entangled dynamisms and performative structures.

Figure 4 highlights the core concepts around the proposed model of *growth as creation strategy* which can be summarized as follows.

The proposed *active materials paradigm* (figs. 2, 4) invites thinkers and makers to address and engage with matter through the lens of a diverse range of stances of *composition, structure, synthesis, property, operativity, performativity, function, intention.* The active materials paradigm diagram can be used as a tool to cultivate a more systemic and dynamic understanding of making:

"Acknowledging and paying attention to the proposed conceptions, as well as following a dynamic shift of perspective between these stances, can serve as a tool for engaging with these various levels of descriptions of a phenomenon, the relevant realms of ontologies and boundary conditions associated with these stances and the interrelation between them. The hope is that such an engagement helps with our thinking, questioning, explaining, or making and nurtures a more comprehensive picture of Gestaltung throughout analysis and synthesis."[19]

Things are systemic and dynamic. Things are situated within ecologies and histories (fig. 4). In various realms of making (e. g., chemistry, biology, engineering, design, etc.) we usually deal with things rather than isolated subjects or objects. This serves as a reminder that things are situated in open and entangled dynamic meshworks of interrelations and intra-actions.

Taking the operativity stance—and following chains of operations—invites us to take a different perspective of borders, interfaces, and boundaries of things. The assumed solid boundaries suddenly seem much more porous, more fluid, blurrier, and we find event-based boundaries assumed, signified, or made through the intra-activities at various scales.

It is important to notice that various stances—as various interrelated levels of description of things—are situated within different historical and ecological contexts (fig. 4). The emphasis lies in engaging with various stances; one deals with—entangled yet different—constraints and boundary conditions. For instance, in engaging with the material composition of a system, one deals with histories and ecologies, which are (entangled yet different) from the ones that one has to take into account as an *intentional stance*. Consequently, paying attention to—and shifting in between—the different constraints and boundary conditions that are acting at each stance becomes a crucial part of the creation process (fig. 4).

Another critical point is that in engaging with different stances one faces entangled yet different ontologies. Figuring out the things that things care for depends on the level of description of the system. The graph of ontological interrelations of a material system can be very different depending on wether the loci of attention is on the *synthesis stance* or the *function stance*. They involve different actants, constraints, ontologies, ontological interrelations, and so on. Thinking of things through these *stance-based ontologies* helps to have a clearer understanding of various actants and boundary conditions through the creation process (fig. 4). Keeping aware of stance-based ontologies, event-based boundaries, and so on, helps to question and rethink some conceptions such as unit, borders, inside/outside, passive/active, and so on.[20]

Things are and happen within entangled dynamisms and performative structures (fig. 4). The performativity stance invites to take a dynamic meshwork understanding of *things*, which allows us to practice seeing things through the lens of *co-beings* and *co-becomings*. In this light, growth—reimagined as the creative nature of such entangled

---

19 Razghandi, "Rethinking Materials Paradigm."
20 Gholami et al., "Rethinking Active Matter"; Razghandi and Yaghmaei, "Rethinking Filter."

dynamisms and performative structures—can be thought of and implemented as a methodology for co-creative processes among human and nonhuman actants.

In such an entangled and dynamic view of things, adaptation—at various stances —becomes an essential feature of the creation strategy (fig. 4): adaptive materials, structures, properties, designs, etc., as well as adaptive processes, collaboration modes, dynamics, and so on.[21] The evolutionary tree (coral) of the creation process can be traced as a tool to: a) follow the growth of the ideas, creations, and the projects as a whole; b) help to signify the adaptations and the constraints and rationales behind them; c) highlight the failures, dead ends, etc., and the constraints and rationales behind them; d) notice complementary ideas across branches; and so on.[22] As such, adaptation becomes one of the key features of growth as creative methodology, and prepares one to notice and address the co-adaptation/co-evolution of various actants and contexts within the creation process. Such *sympoietic* understanding of things serves as a reminder-tool to distance oneself from seeing creations as individuated, closed, and passive objects.

I propose that such entangled and intra-active conception of growth (fig. 4) can serve as a co-creation methodology in various realms of making.

Conclusion and Outlook

Rethinking Growth: Interdisciplinary Co-Creation Methodology

Some realms where the proposed growth model can particularly serve as a co-creation methodology are the realms of science-design interdisciplinary research, education, and innovation practices.[23] Taking growth as creation methodology can help to cultivate such interdisciplinary in-between spaces where diverse range of knowledges can encounter and weave into open, complex, and adaptive co-creation processes. In such interdisciplinary settings, various stages of exposure, inquiry, exploration, creation, adaptation, etc. can be facilitated through the proposed growth methodology (fig. 4). Such open and adaptive co-explorations can provide a fertile ground for nucleation of various new questions and research or design projects within natural sciences,

---

21 Or Ettlinger and Khashyar Razghandi, "Speculative Inquiry and Growth: A Methodological Approach for Creative Knowledge Generation and Interdisciplinary Problem-Solving," *Creativity Research Journal* (forthcoming).

22 Facundo Gutierrez and Khashayar Razghandi, "MotorSkins—A Bio-Inspired Design Approach towards an Interactive Soft-Robotic Exosuit," *Bioinspiration & Biomimetics* 16, no. 6 (2021): 066013, https://doi.org/10.1088/1748-3190/ac2785.

23 Ettlinger and Razghandi, "Speculative Inquiry and Growth"; Khashayar Razghandi et al., *Scaling-Nature – A Reflection on an Interdisciplinary Design-Science Studio* (Bielefeld: transcript Verlag, forthcoming); Gutierrez and Razghandi, "MotorSkins."

humanities, and design disciplines: a playground for realization of *research through education* and *education through research.*[24]

Rethinking Growth: Uncertainties, Urgencies and Open Futures

As dystopian futures unfold in front of us, the powerful yet blurry image of "future as a horizon that is approaching us" holds various uncertainties and urgencies. *Rethinking growth* hopes to offer a different understanding and a way to deal with this *openness.*

Rethinking growth emphasizes a paradigm shift from individual subjects and objects with properties and autopoiesis, to performativity and sympoiesis within active and entangled meshworks. Future-makings are rooted in these co-beings and co-becomings.

The image of the entangled dynamisms and performative structures embraces the diversity and activity of various actants and agencies, and growth can be re-understood within this image. The categorization and dichotomies of the patriarchal dominators—the subject/object, subject/property, organisms/environment, active/passive, and so on—are failing the reality of the entangled dynamisms and performative structures.

Futures cannot follow this prediction-control paradigm of domination and exploitation. The normative discourses of atomization, alienation, universalism, and dominance are the prevailing conceptions and logics of the current passive conception of matter, buttressing the foundations of patriarchy and capitalism. A less hierarchical, more interrelational, more diverse, a more intra-active and performative conception of growth seems to be the way to reflect and face the current ecological and socioeconomic crisis.

Open futures need co-creation paradigms: to pay attention, to care for, and to co-work. We need to acknowledge the diversities—of agencies, of ontological relevance and concerns, of constraints, and of encounters, the entangled dynamisms and performative structures, with multitudes of agencies and relations, colliding and intertwining, bonding and so on.

Rethinking growth hopes to highlight the entangled and performative nature of creative processes, and put forward a new conception of growth to inspire new interdisciplinary and intersectional approaches toward a more comprehensive creation paradigm: growth as co-creative future-makings.

---

24 Ettlinger and Razghandi, "Speculative Inquiry and Growth"; Razghandi et al., *Scaling-Nature.*

—

Indeterminacy, Ambiguity and Liveliness

Mareike Stoll

The In-Between as a Place of *Poiësis:*
Photography of Ambivalence in a Photobook

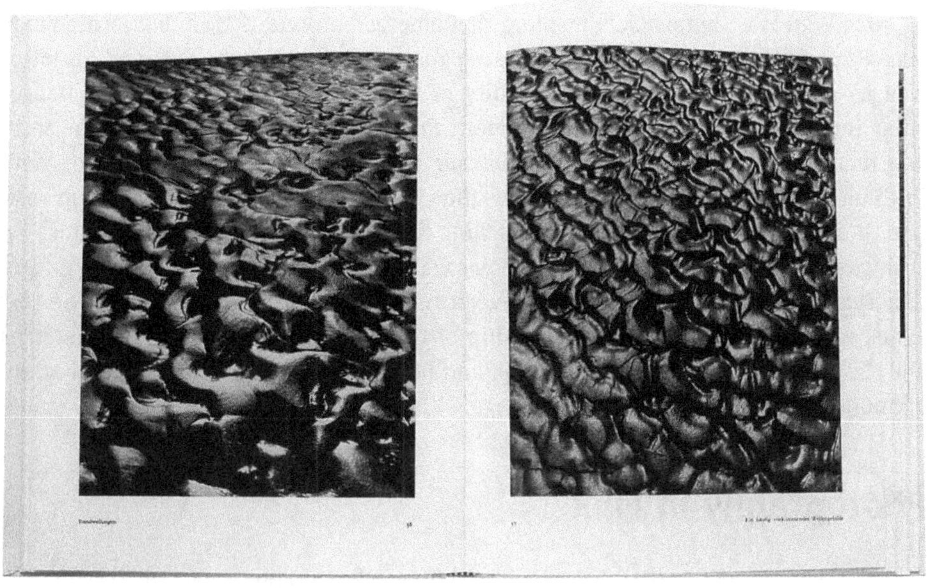

Fig. 1: Alfred Ehrhardt, *Das Watt* (Hamburg: Heinrich Ellermann, 1937), plates 36 and 37.

Sand, sculpted by currents of water and wind, presented in stark contrasts, unusual frames, forming poetic close-ups. Two photographs are positioned next to each other on the double-spread of a book; on the right frame a slither of another photograph shines through from behind. The series continues. The images are accentuated in their verticality by the portrait format but also by the direction of the camera that does not include a horizon line for orientation. On the page, the photos are stabilized visually by captions on the lower right- and left-hand corners respectively, and by the plate numbers ("36" and "37") below the inner edges. The tonality of the photographs' black-and-white nonetheless allows for many shades in between, turning the surface of sand portrayed here into almost tactile landscapes. The close-up studies of sandy surfaces are recognizable as both representational photography *and* abstractions, a fact that gives them their interpretative and poetic richness. Rewarding our close reading with a more nuanced understanding of the morphology of sand formations, they give us an idea of what granular material looks like in photography. They are, above all else, surfaces that provoke tactility, printed photographs in a book. Published in

1937, *Das Watt* (in English this translates to "mudflat") collects ninety-six images that resist any quick interpretation with their opaque photographic, aesthetic surfaces. Celebrating this simultaneity of surface *and* landscape, insinuated relief *and* actual flatness of the page, they are *Kippfiguren*, symbolic material of the region that they capture: the mudflats are both land and sea, constantly in flux between ebb and flow. The landscape evoked by the title is broken into fragments of a landscape, or unstable grounds, by the hands of the beholder who is activating the image sequence of the book.

Das Watt is a photobook, a medium that emerged around 1924 in the German context of avant-garde bookmaking, the newspaper industry, and the growing visual as well as political importance and ubiquity of photography.[1] In a photobook, images enter into an epistemological inquiry about the potential of photography as medium and material, and as a way to construct our perception of any given point of view. In the following, I will briefly sketch out the characteristics of the photobook in relation to ambiguity, and spend time unpacking the notion of the in-between, before focusing on the page of the book that becomes wall-like in the case of the peculiar images that *Das Watt* offers, which hover between representation and abstraction. The text closes with reflections on the relationship of the book and the body of the beholder and the space in between that opens up, and how it might relate to agency and an invitation to step inside the book, to engage with images.

Photography in Flux

A photobook contains very little text and combines photographic images in a sequence, with a particular focus on the in-between. By "in-between" I refer to the space between the photographs within the sequence, the space between all the (visual, tactile, invisible) elements of which the book is composed, between two photos juxtaposed on a double-page, but also to the correspondence and visually rich murmur of connotations, associations, and meanings that comes into existence in this space.[2] Understanding the

---

1 Some of the most iconic German photobooks were published over the course of three years: between 1928 and 1930. August Sander's *Antlitz der Zeit*, Karl Blossfeldt's *Urformen der Kunst*, Albert Renger-Patzsch's *Die Welt ist schön* and Germaine Krull's *Métal* were all published in 1928 and 1929; as was the groundbreaking publication accompanying the *Film und Foto* exhibition in Stuttgart (the catalogue was called *Foto-Auge*). László Moholy-Nagy's *60 Fotos* and Aenne Biermann's publication of the same series both appeared in 1930.

2 At the end of the Weimar Republic, photography as a medium and as material was debated as simultaneously holding the key to a better future of human-kind (the hygiene of the optical, as László Moholy-Nagy put it) as well as being an instrument in manipulation, corruptibility and propaganda (this position was argued by Walter Benjamin, Siegfried Kracauer, and Aby Warburg, among others). In this discourse of what photography as a form of knowledge can offer, the medium of the photobook was recognized as an important instrument to knowledge acquisition. In my previous research I have called attention to the pedagogical program of visual literacy that many publications ascribed to. See Mareike

page of the book as site and stage where words and images form dynamic constellations of epistemological inquiry, I wish to explore questions relating to the human body, imagination, creativity, agency, and resonance. Photobooks are tangible things that need to be held; only by leafing through them does the sequence of pictures unfold.[3] The material support immediately starts interacting with hand and body, eyes and mind of the beholder when the book is picked up. This also means that any photobook is a collaborative medium. They are not necessarily published as such—most often we find one author and a title on the cover; sometimes there is a mention of the editor in the imprint and not everyone involved in the process is named, but often we do not know who was responsible for the layout and graphic design of the books published in the 1920s—but they were conceived between various collaborative instances, as any book is, still to this day: they are part of a dialogue of inspiration and creation, of a back-and-forth between agents, materials, and everything in between.[4] Much of the work, experience, and design knowledge going into the photobook is invisible to the untrained eye. That is why I want to locate this knowledge (and indeed activity) in the space of the in-between. There, we have space to breathe, and listen, truly listen, to the material, the things at hand: this is where we co-create in an open-ended conversation.[5]

This space in between is unstable and messy, perhaps, but an urgent depository of activity that is contagious in the best of ways.[6] The hands of the beholder are essential, as they connect the practice of making (the knowledge of the designer and photographer) that is often silent, taciturn, to the knowledge of letters and the pictorial shape that the book as a medium is tied to; together they create a way of knowing. Investigating the relation of words and images by way of resistances, frictions, as much as associations and correspondences, I wish to place the act of engaging with this productive activity in the hands of the beholders.

Das Watt by photographer Alfred Ehrhardt (1901–1984) presents photographs taken on many extended walks on the coast of northern Germany, close to the city of Cuxhaven, in the area that is defined by the tide, where ebb and flow dominate. Ehrhardt had been teaching abstract painting in the 1920s and the beginning of the 1930s.

Stoll, *Schools for Seeing: German Photobooks between 1924 and 1937 as Perception Primers and Sites of Knowledge*, PhD diss. (Princeton University, 2015).

3 By "unfolding" I am referring to an idea formulated by Walter Benjamin in his essay on Franz Kafka's writing, describing that his texts unfold similar to the bud of rose, not like a piece of paper, but rather expanding, "bringing forth" complexity and layers of meaning: this is what *poiësis* is to me.

4 The photobook is designed to become a machine that helps slow-read photography; it underscores the relevance of hands in the process of understanding the material by grasping it (*begreifen*).

5 Instead of the hylomorphic model and object–subject binary, I use concepts of agency and things, inspired by Tim Ingold's writings. See Tim Ingold, "Textility of Making," *Cambridge Journal of Economics* 34 (2010): 91–102.

6 See Georges Didi-Huberman, *Atlas: How to Carry the World on One's Back?* (Karlsruhe: ZKM, 2010); see Horst Bredekamp, "A Neglected Tradition? Art History as Bildwissenschaft," *Critical Inquiry* 29, no. 3 (2003): 418–28.

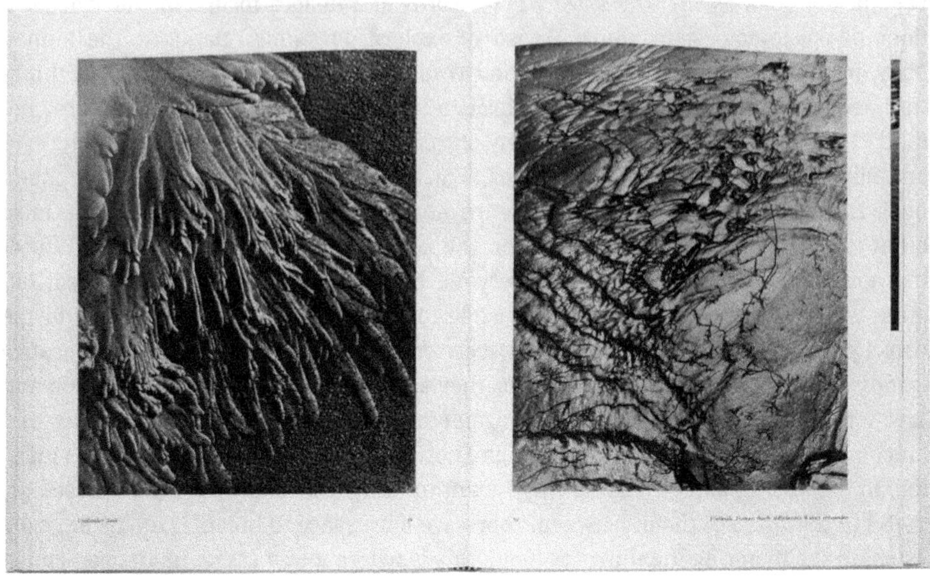

Fig. 2: Ehrhardt, *Das Watt*, plates 58 and 59.

Forced by the Nazis to give up teaching in 1933, Ehrhardt also completely abandoned painting and picked up the camera instead, which for him offered new ways to combine abstraction with representation.[7] In my reading, the photographs examine the sandy ground they aesthetically portray as shifting: the book is about orientation as much as it is about context. No grounds were neutral at the time, not least the term or concept "ground" itself—*Boden* of course was part of Nazi propaganda.[8] In 1937, the year of publication of *Das Watt*, the Nazis organized the exhibition *Entartete Kunst* (Degenerate Art) in Munich, which attacked abstract modernist painting and art as "degenerate."[9]

Already in the 1920s, the medium of the photobook had been tied to contemporary forms of perception and, broadly speaking, to the discourse of modernity and urbanization in relation to subjectivity and a discussion of "nature versus culture," which are

---

7 For more on Ehrhardt, see Christiane Stahl, *Alfred Ehrhardt: Naturphilosoph mit der Kamera: Fotografien 1933–1947* (Berlin: Dietrich Reimer Verlag, 2007).

8 Tied to this notion of space as "Lebensraum" with the connotations of home or "Heimat," the photographic portrait in particular is a genre that in Nazi Germany becomes more and more politicized. See mostly the work of photographers Erna Lendvai-Dircksen, Leni Riefenstahl, and Dr. Paul Wolff, for portraits depicting the phantasm of the "Aryan race" as connected to space, landscape, and environment.

9 See Franz Roh, *"Entartete" Kunst, Kunstbarbarei im Dritten Reich* (Hannover: Fackelträger Verlag, 1962). More recently Anson Rabinbach and Sander L. Gilman, ed., *The Third Reich Sourcebook* (Berkeley: University of California Press, 2013), 483524, and Olaf Peters, ed., *Degenerate Art: The Attack on Modern Art in Nazi Germany, 1937*, exh. cat. Neue Galerie New York (Munich: Prestel, 2014).

also present in the book at hand.[10] German photobooks in 1937 are, however, embedded in a radically different context. The concept of "landscape" in relation to photography needs to be read against this backdrop, paying close attention to the surface depicted as simultaneously fluid and fixed. By celebrating ambiguity visually, *Das Watt* makes the beholder an accomplice in ambivalence.

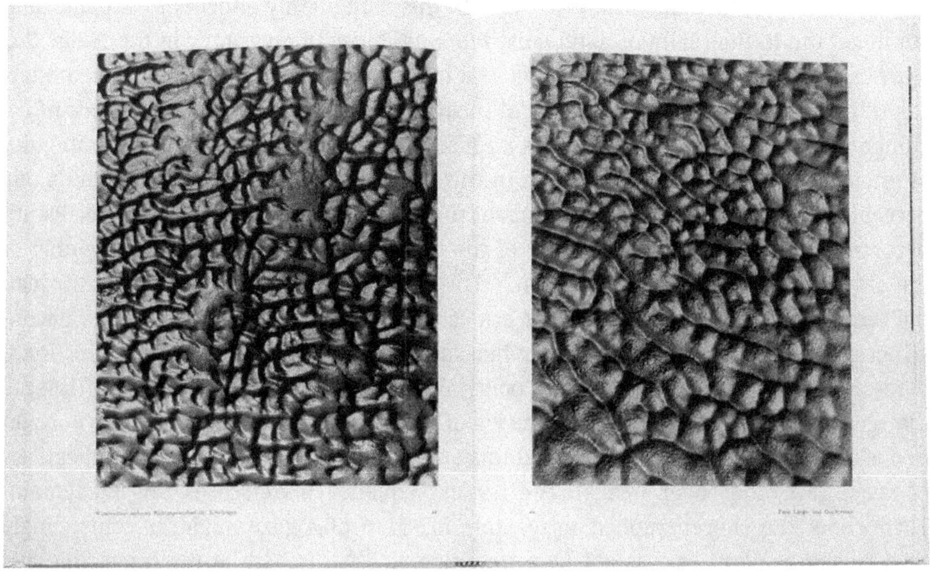

Fig. 3: Ehrhardt, *Das Watt*, plates 28 and 29.

Ehrhardt pointed the camera to his feet, so to speak, to photograph details of the ground. Landscapes are usually defined by format and by inclusion of a visual marker that we find in the horizon. If this is missing, if the familiar becomes unfamiliar by omitting the horizon, how can we make sense of what we see? The photographs thus shift our point of view—with regards to landscape and to abstraction but also to orientation. Where do we stand, metaphorically speaking? Because photobooks always establish a bodily connection as a material thing, I argue that they become a thinking device that is tied to the grasping of photography's intriguing complexity by the hand. The book deliberately employs photography in relation to landscape and abstraction to destabilize notions of regional belonging; it thus transfers the question of agency and taking a stand, the question of orientation and point of view, to us as the beholders. Re-examining *Das Watt* in the context of the Cluster of Excellence "Matters of Activity" allows us to understand this destabilization as de-passivation, as indeed the activity is within the "material" of photography and in the medium of the photobook.

---

10 For more on this see Daniel H. Magilow, *The Photography of Crisis: The Photo Essays of the Weimar Republic* (University Park, PA: Pennsylvania State University Press, 2012), 63–64.

With regard to photographer Albert Renger-Patzsch, already in 1931 Walter Benjamin had called the failure to acknowledge the connective tissue of human involvement ("menschliche Zusammenhänge") in photography the very "corruptibility" of the medium.[11] Photography, in other words, allows us to forget that we are always-already part of what we look at: we are implicated by reading it, making sense of it by having to reconstruct what is invisible, what lies beyond the photograph and forms the connective tissue to political dimensions as well. In this context, my effort is to expand Benjamin's claim to the realm of nationalist *Blut-und-Boden* photography in the 1930s. *Das Watt* calls attention to the construction of a landscape that is always also to be understood in terms of claims of belonging, of inclusion and exclusion on the grounds of arguments that are visually enforced. As a consequence, the corruptibility and ambiguity of photography both come to the fore in the book, literally changing the grounds our perception is based on. With the means of the photobook horizontality and verticality become mobile, making the beholder of the book an active part in the reassembly of the arguments the ground is based on.[12] This is achieved most pronouncedly through the sequence of the photographs that are carefully choreographed in order to deconstruct any stable reading. At a time when any depiction and representation of landscape in Germany is always-already political, this book offers a sequence of images that goes back and forth between images of abstraction (fig. 4): images of a horizon line, sky and ground, and perspectival images without a horizon but clear indications of space (e.g., the cover image). The book's sequence underscores this movement. Throughout the choreographed series, the direction of a gaze might as convincingly be a "looking away," as it might be a "focusing on."[13] It is both at the same time, yet neither completely. The book is also in this sense ambivalent, every photograph in this book is a *Kippfigur*.[14]

---

11 See also Mareike Stoll, "Menschenleer: Der Tatort in Benjamins Schriften zur Photographie," in *Walter Benjamin und die Anthropologie*, ed. Carolin Duttlinger, Ben Morgan, and Anthony Phelan (Freiburg, Berlin, and Vienna: Rombach, 2012), 343–62; Walter Benjamin, "Kleine Geschichte der Photographie [1931]," in *Gesammelte Schriften I.1*, ed. Hermann Schweppenhäuser and Rolf Tiedemann (Frankfurt: Suhrkamp, 1991).

12 The photographic portrait in particular is a genre that becomes more and more politicized. See mostly: Erna Lendvai-Dircksen, Leni Riefenstahl, and Dr. Paul Wolff for portraits depicting the phantasm of the "Aryan race."

13 For the political implications of a "looking away," see Rei Terada, *Looking Away: Phenomenality and Dissatisfaction: Kant to Adorno* (Cambridge, MA: Harvard University Press, 2009).

14 At a time when the depiction of a coastal landscape usually meant a horizon line, Ehrhardt has to be compared to those photographers who depicted seascapes like that. See most prominently Paul Wolff's summer-vacation book from around the same time that depicts this landscape in a very different way. Dr. Paul Wolff, *Sonne über See und Strand: Ferienfahrten mit der Leica: Text und 112 Tiefdruckbilder: Mit einem Schlussbeitrag von H. Windisch* (Frankfurt: Bechhold Verlagsbuchhandlung, 1936). Norderney, where the photos are taken, was advertised in the book as tourist destination, and the local office of tourism was proud to proclaim that it was "German" (which meant banned for Jews, or "judenfrei"). See Lisa Andryszak and Christine Bramkamp, ed., *Jüdisches Leben auf Norderney: Präsenz, Vielfalt und Ausgrenzung* (Berlin: Lit Verlag, 2014) 173–74.

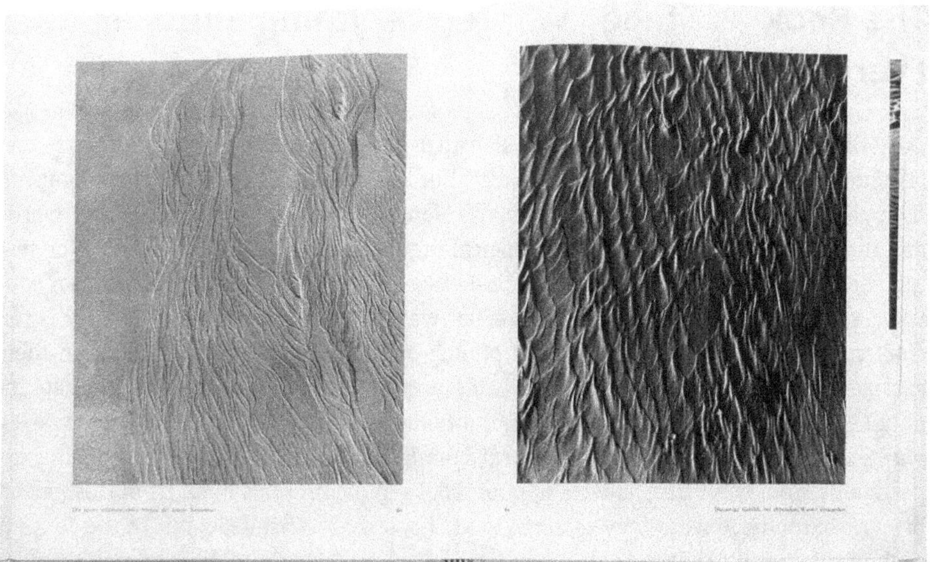

Fig. 4: Ehrhardt, *Das Watt*, plates 60 and 61.

The most abstract double-pages in the book indeed tilt the ground to be represented in such a way that they become wall-like. Photographed by pointing the camera to the ground (and not, like you would expect, oriented toward the horizon), this spread resembles an exhibition wall, showcasing the tilting of the ground upward to an absolute verticality, and to abstraction. Ehrhardt effectively uses the mobilizing elements of the photobook in order to produce the shock of the impenetrable photographic surface, exhibiting abstraction, no less. Ehrhardt's abstract photographs can be related to the internalization and representation of the wall in terms of their verticality.[15] Folded into this destabilization is the potential for a crisis and thus for the material to enlighten, to move us to think critically. When we step out of the passive consumption into an "engaging with," we are acknowledging our agency in the processing of information and in knowledge creation.[16] What is missing from what we see, especially with the photography of sand on a coastline in Germany in 1937? This is the question of context, and of societal, even political connective tissue. Only when we take into account the implied and the invisible does the "bigger picture" come into view with all its messy, urgent, and complicated dimensions.

---

15 For an intriguingly complex, profound, and philosophical investigation of ground with regards to drawing see artist Riet Eeckhout (*1975), especially her *Uncommon Ground.*

16 Roland Barthes, *Wie zusammen leben*, ed. Éric Marty, trans. Horst Brühmann (Frankfurt: Suhrkamp, 2007), 213, where he speaks of the etymological connection of critique and crisis: the purpose of (literary) critic/critique lies in provoking a crisis.

The Book as Stage "Where Everything Shifts at Every Step"

Das Watt was distributed by Heinrich Ellermann, a publisher based in Hamburg, specializing in photography books and poetry, but also children's books.[17] This might not be a coincidence, because children's books, too, are complex constellations of words and images, offering sites of epistemological inquiry, that is, as well as schools for seeing and literacy using the space of the in-between. *Das Watt* works with such an activated space between the photographs. What was at stake and still is to this day is the challenge to learn to fully understand photography and image constellations in their complexity. Because much of it is invisible and happens in the "in-between," it can only be understood through analysis and a taking-apart, only by considering *con*texts, *para*texts, *inter*texts, and wider cultural connotations. Avant-garde book-making experiments from the 1920s, as well as photobooks and children's book from this period have revolutionized what they referred to as "book space."[18] Indeed, it is in the concept of the "in-between" that the simultaneity of page and "stage" unfolds as "architecture" that invites the beholder to step inside. This essay is thus posited at the threshold of two research projects, both dealing with in-between as place for *poeisis*, saturated with activity.

It was Walter Benjamin who in 1926 wrote about an observation that is most fitting in this context. He noted that the child who is focused on reading and looking at a picture book is often so fully immersed in the story unfolding before them, that they enter the book as if it were a stage. The book is not a thing anymore to the child, but the child understands it as its environment and establishes a dialogical relationship with it where both the child and the book (and indeed all the elements in between) act and interact to co-create the experience of reading, the unfolding of the story in the book space.

---

17 Roland Jaeger, "Ennoch und Ellermann: Zwei deutsche Fotobuchverlage der zwanziger und dreißiger Jahre," *Photonews* (Thema Photobuch, 05/2008): 4–5, where Roland Jaeger makes mention of the publisher of Gebrüder Enoch Verlag, Kurt Enoch, and his publishing house that he was forced to sell to "Aryan" owners. Enoch emigrated from Germany to Paris and in 1940 to the United States. Until 1936, the young publishing house Ellermann Verlag had mostly published poetry that was not National Socialist. Trying their hand at the new medium of the photobook, publisher Heinrich Ellermann and his team printed the project with exceptional attention to the range of gray in the photographic reproductions. To this day, Ellermann Verlag publishes children's picture books.

18 El Lissitzky argued that the book-space is made of content as much as form, layout design, and script-image, as much as the arguments and ideas. In July 1923, eight theses on New Typography ("Topographie der Typographie") by Lissitzky were published in *Merz*, no. 4. (cited in Andrea Nelson, "László Moholy-Nagy and Painting Photography Film: A Guide to Narrative Montage," *History of Photography* 30, no. 3 (2006): 258–69, 260).

> Nicht die Dinge treten dem bildernden Kind aus den Seiten heraus—im Schauen dringt es selber als Gewölk, das mit dem Farbenglanz der Bilder sich sättigt, in sie hinein. [Das Kind] meistert die Trugwand der Fläche und zwischen farbigen Geweben, bunten Verschlägen betritt es eine Bühne, wo das Märchen lebt … In solch farbenbehängte, undichte Welt, wo bei jedem Schritt sich alles verschiebt, wird das Kind als Mitspieler aufgenommen.[19]

Benjamin surprisingly turns the tables on us, because we might be expecting the characters to become alive and step out of the book to interact with us. Instead, the child as the beholder enters the book as if it were a stage and there, it finds the content of the images as transformative nourishment (the child feeds on the colors as it is saturated by them—*sättigen* means both). It is a stage that, through play and interaction, facilitates confidence-building, even empowerment. The child is allowed to interact and play as an equal in this world of the book, to act responsibly, "where everything shifts at every step," as Benjamin writes, speaking as if looking at Ehrhardt's photographs.

Das Watt pays attention to the act of construction of perception, while the ground depicted begins to shift from horizontal to vertical and back again. The images engulf us like walls, or allow us to tread on fluid sand (figs. 3 and 4, respectively). Akin to children's picture books or ABC books used in school, photobooks were meant to alphabetize their readers in the language of photography. But they also appeal, like any art worthy of this name, to a greater endeavor: to the orientation within the complex web of meanings that images and photography as medium and material, as thinking devices and practices, and as ubiquitous elements that our environment always encompasses. I want to propose that—in these series of crises that we are facing—we need a space of the in-between, of the ambivalent that at the same time encourages us to become agents in responsible interactions. "The child is accepted/taken in as a fellow actor," Benjamin writes (*wird als Mitspieler aufgenommen*), and the German word *aufnehmen* indeed also means to grasp and absorb, but also to responsibly interact with, and to admit, in the sense of respectfully allowing to "become part of." This activation, becoming, transformation is to be taken literally, not only through nourishment, as Benjamin suggests, but by way of associations, correspondences, and frictions of words and images in our imagination when we become part of the material. This notion of inclusion into the material acknowledges the in-between as a place where knowledge acquisition —and, hence, a transformation—occurs, because we handle the books and immerse ourselves in them, let ourselves be engulfed by them. Beholder, material, and the in-between are part of this transformation toward a change in perspective, and, possibly, also in behavior, understood as handling and acting, *Handeln*, in a world that simulta-

---

19 Walter Benjamin, „Aussicht ins Kinderbuch [1926]," *Gesammelte Schriften IV.2*, ed. Tillman Rexroth (Frankfurt: Suhrkamp, 1991), 609. „It is not the things that emerge from the pages for the child who creates the pictures (image-ing)—in looking, they themselves penetrate these pages as a cloud that saturates/ nourishes itself with the glow of the colors of the pictures. [The child] masters the illusory wall of the surface and between colorful fabrics, colorful hutches they enter a stage where the fairy tale lives. . . . In such colorful, leaky world, where everything shifts at every step, the child is taken in as a fellow actor/ player." Translation by the author.

neously demands resilience, empathy, *and* a compass for what matters when it comes to acting responsibly and humane.

The legacy of avant-garde photobooks, and Ehrhardt's book especially, is exemplary in this regard. It lies in its posed questions of agency and how the material and the beholder, how what we see and our construction of it in our minds are related through our hands, how our bodies are implied in our grasping and holding of photographic images. *Das Watt* offers a training manual in orientation in ambivalent constellations, also, and this is important, because it becomes a thinking device, metaphorically speaking. The in-between in this book is one that solemnly embraces ambiguity. As Benjamin reminds us, the imagination of the child allows for everything to be in flux, and it is decidedly a joyful place, because it has agency and is allowed to play as an equal. Perhaps the takeaway is this: precisely in this capacity to tolerate ambiguity and to enjoy the grounds of interpretation as shifting and in flux, we can more confidently approach this complex world that also lies beyond the book, and accept our responsibility in co-creating instead of being overwhelmed and confused. Researchers in psychology and neuroscience have referred to this capacity to tolerate nuanced shades of gray between black and white as "integrative complexity."[20] It denotes the capacity to simultaneously accept two seemingly contradictory truths as possible and legitimate from different standpoints while still holding on to one's agency and a clear compass for right and wrong. This is the prerequisite for both empathy and diplomacy, so desperately needed in times of crisis and polarized societies. By entering into the book, by acknowledging that in the in-between there is agency, we can perhaps "master the illusory wall of the surface" that Benjamin speaks of, and go on to find out what lies beyond.

Notes:

This text is based in part on arguments developed in my *ABC der Photographie: Photobücher der Weimarer Republik als Schulen des Sehens* (ABC of Photography: Photobooks of the Weimar Republic as Schools for Seeing) (Cologne: Walther König, 2018). I thank Horst Bredekamp for invaluable comments on an earlier version presented at the Annual Conference "Tipping Points: Plastic, Contingent and Unstable Matters" and for allowing me to discover El Lissitzky's designs of children's picture books along the way.

---

20 Vera Békés and Peter Suedfeld, "Integrative Complexity," in *Encyclopedia of Personality and Individual Differences*, ed. Virgil Zeigler-Hill and Todd K. Shackelford (Heidelberg: Springer, 2019); Peter Suedfeld and Philip Tetlock, "Integrative Complexity of Communications in International Crises," *Journal of Conflict Resolution* 21 (1977): 169–84.

Kolja Thurner

Transcending Form: On the Potential of Ambiguous Imagery

Ambiguity and (In-)Stability

In an October issue of 1892, the German humorous weekly *Fliegende Blätter* printed a joke that has since become a staple of visual ambiguity, featured in different variants throughout publications of psychology and works on art and perception.[1] The small original engraving, since dubbed a "duckrabbit," presents a bi-stable image of a rabbit and a duck that plays with the ambiguous co-existence of its two alternative readings, conflated and congruent in one single form (fig. 1). Once recognized, neither figure can be erased again as present. Instead, the forms will become temporally unstable, irrevocably oscillating back and forth at some frequency between either representing a rabbit or a duck. At the same time, it is next to impossible to securely behold the image both as a rabbit's and as a duck's head at once.

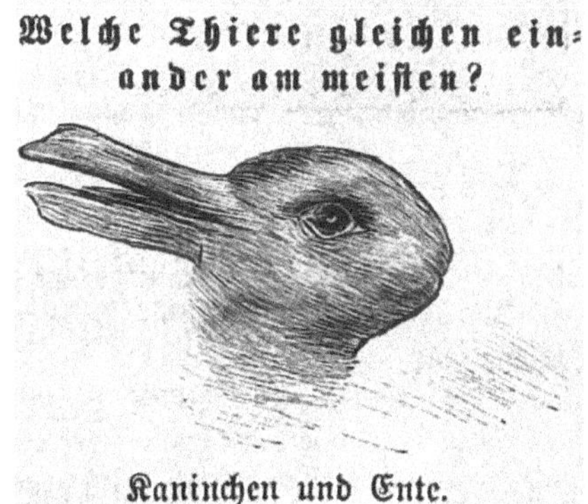

Fig. 1: Original Duckrabbit, engraving, artist unknown, in *Fliegende Blätter*, October 23, 1892, 17.

Debuting in the satiric magazine, the "duckrabbit" was embedded in a joke on likeness/similitude, framed by the trick question: "Which animals are most alike?" and the par-

---

1 "Welche Thiere gleichen einander am meisten?," *Fliegende Blätter*, October 23, 1892. Its first appearance in psychology came with Joseph Jastrow, "The Mind's Eye," *Popular Science Monthly* (January 1899). For a short overview of the different versions of the image and their uses see Peter Brugger, "One Hundred Years of an Ambiguous Figure: Happy Birthday, Duck/Rabbit!" *Perceptual & Motor Skills* 89, no. 3 (1999): 973–77, and I. C. McManus et al., "Science in the Making: Right Hand, Left Hand. II: The Duck–Rabbit Figure," *LATERALITY* 15 (2010): 166–68, doi:10.1080/13576500802564266.

adoxical answer, "Rabbit and duck."[2] Beyond being a mere punchline, the answer equally served as an aide in the discovery of both animals in the forms of the image. After all, context can heavily influence a first encounter with the notorious "duckrabbit," which may include semantic framing, additional visual cues, and even fuzzier information like viewer expectations. Unfamiliar beholders were clearly biased toward recognizing only a duck or some kind of bird in the month of October, while during easter, a rabbit was identified significantly more often.[3] If, as Ernst Gombrich suggested, one was to draw a duckpond around the image, or show test subjects slides of rabbits before projecting the ambiguous image, this would indubitably affect its spontaneous identification as well.[4] Attempting to account for all these different influences on human perception, neuropsychologist R. L. Gregory concluded, in radical terms, that therefore perceptions should be regarded as (often unconscious) *inferences*, based on miscellaneous signaled sensual data, both present and stored in memory.[5] "Perception is, in this sense, a kind of betting."[6]

Ludwig Wittgenstein, who in a similar vein recognized the inseparable entanglement of meaning with perception as well as that of sensuality with thinking, famously used a more simplified drawing of the *duckrabbit* scheme for his *Philosophical Investigations* (fig. 2).[7] It served him as his prime object for reflecting upon the paradoxical nature of 'seeing as' through aspect perception as opposed to common 'seeing'. To Wittgenstein, a 'change of aspect' brought about in the picture meant suddenly seeing a supposedly known quantity (i.e., the rabbit) *as* something else (i.e., the duck) in a flash:

> The change of aspect. 'But surely you would say that the picture is altogether different now!' But what is different? My impression? My point of view?—Can I say? I *describe* the alteration like a

---

2 See William J. T. Mitchell, *Picture Theory* (Chicago and London: The University of Chicago Press, 1995), 56. Therein he clarifies: "Certainly the rabbit and the duck don't 'resemble' each other: . . . they are 'nested' together—that is located, imagined or pictured *in the same gestalt*."

3 Peter Brugger and Susanne Brugger, "The Easter Bunny in October: Is it Disguised as a Duck?" *Perceptual and Motor Skills* 76 (1993), doi:10.2466/pms.1993.76.2.577. McManus et al. report that upon reviewing their article, which laid bare a significant age effect (children more often identifying the duck), Peter Brugger interestingly also suggested that pop-cultural phenomena, like the popularity of Donald Duck cartoons, might play a role (McManus et al., "Science in the Making," 185).

4 See Ernst H. Gombrich, *The Image and the Eye: Further Studies in the Psychology of Pictorial Representation* (Oxford: Phaidon, 1982), 36. A similar idea is already given by Ludwig Wittgenstein, *Philosophische Untersuchungen: Zweite Auflage / Philosophical Investigations: Second Edition*, trans. Gertrude E. M. Anscombe (Oxford: Blackwell Publishers, 1999 [1953]), 165/165e.

5 Richard L. Gregory, "The Confounded Eye," in *Illusion in Nature and Art*, ed. Richard L. Gregory and Ernst H. Gombrich (London: Duckworth, 1973), 54–55.

6 Gregory, "The Confounded Eye," 83. See Stewart E. Guthrie, *Faces in the Clouds: A New Theory of Religion* (New York and Oxford: Oxford University Press, 1993), 42, who erroneously ascribed this quote to Gombrich (in the same volume as Gregory).

7 "'Seeing as . . .' is not part of perception. And for that reason, it is like seeing and again not like." Wittgenstein, *Philosophische Untersuchungen*, 197e.; see also Sara Fortuna, *Wittgensteins Philosophie des Kippbilds: Aspektwechsel, Ethik, Sprache*, trans. Arnold A. Oberhammer (Vienna and Berlin: Verlag Turia + Kant, 2012), 45–46.

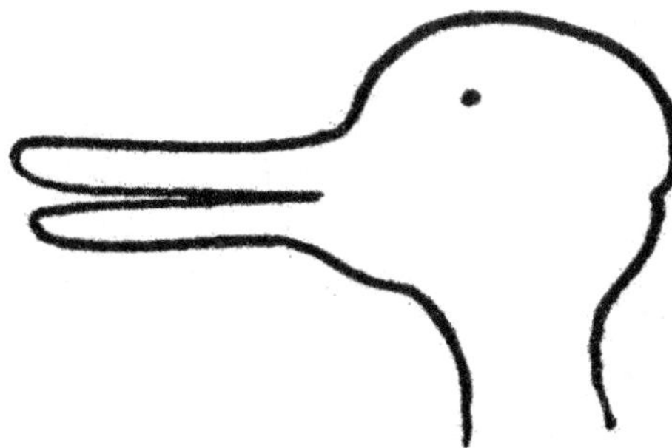

Fig. 2: Ludwig Wittgenstein, simplified Duckrabbit scheme, drawing, Ludwig Wittgenstein, *Philosophische Untersuchungen: Zweite Auflage / Philosophical Investigations: Second Edition*, trans. Gertrude E. M. Anscombe (Oxford: Blackwell Publishers, 1999 [1953]), 194.

perception; quite as if the object had altered before my eyes. . . . The expression of a change of aspect is the expression of a *new* perception and at the same time of the perception's being unchanged.[8]

The observation of a virtual transformation ("as if the object had altered before my eyes") that, according to Wittgenstein, happens instantaneously, in the way of the "dawning" or "flashing" of an aspect (*Aufleuchten*) points out a sudden, dynamic change from uniform stability via instability to bi-stability.[9] An entirely new property of the image emerges all while it formally and physically remains still and unchanged. Starting from this observation, I intend to advance from the often-discussed role of perception and cognition before the "duckrabbit" to their entanglement with the form and structural properties of the image. This attempt tries to reclaim the ambiguous *potential* of such images as participants within beholding, akin to what W. J. T. Mitchell called "a transaction between pictures and observers activated by the internal structural effects of multistability."[10]

Indeterminacy and Openness

When approaching ambiguous imagery, it must be borne in mind that they are first and foremost carriers of distinct pictorial properties as results of artistic practices, in order to be then susceptible to the active part of "the beholder's share."[11] This becomes clear upon revisiting the ample amount of different "duckrabbit" schemata that have circulated over the last hundred or so years (fig. 3). As Brugger and Brugger

---

8 Wittgenstein, *Philosophische Untersuchungen*, 195e–196e.

9 Wittgenstein, *Philosophische Untersuchungen*, 194e, 197e. See Ernst H. Gombrich, *Art and Illusion: A Study in the Psychology of Pictorial Representation* (London: Phaidon 1992), 4–5. The common German denomination *Kippbild* (literally: tipping image or tilting image) for a multistable/ambiguous image precisely accentuates this instability and the dynamic of change from one identification to the other in lieu of the supposed (be it multi- or bi-)stability of the different states of the image.

10 Mitchell, *Picture Theory*, 57.

11 See Gombrich, *Art and Illusion*, 154–203.

and McManus et al. have shown, a near equilibrium in the identifiability of both figures can hardly be achieved by most of them. We might go so far as to claim that a
truly "egalitarian" bi-stable "duckrabbit" is an entirely rare breed if not a theoretical
construct. The actual degrees of ambiguity that the different versions can elicit, are,
barring outer factors, contingent not only on their individual overall designs, but
will effectively hinge on formal minutiae.[12] The smallest differences in the contours,
the inclination of the ears/beak, or the positioning of the eye have the potential to
tip over the perception of the structural whole. After all, it is their composition that
sets formal constraints for the experience of which animal is preferred over the
other and that will decide over the degree of easiness of seeing each image respectively.[13]

At the same time, the twin-images are defined by a lack of comprehensiveness,
which introduces necessary degrees of abstraction and indeterminacy into the process
of their concretion. The "duckrabbit" is fundamentally rudimentary and necessitates
this built-in incompleteness. The condition of the possibility of multi-stability is its contingency on artistic vagueness and formal reduction, since neither figure can resemble
an actual rabbit nor a duck too closely. In this sense, producing ambiguity is as much a
process of cutting or filtering likeness as it is one of weaving it, both mentally and manually.

It is important to point out that, in traditional art history, Wolfgang Kemp's "aesthetic of reception" introduced a conceptual notion of indeterminacy and unfinished-
ness into the quality of all artworks *per se*.[14] Its "blanks" or joints would, in fact, constitute an "elementary matrix"[15] for the interaction with the artwork. Within art, this
potential openness becomes most glaringly apparent and is designed most effectively
within deliberately ambiguous imagery. Exceeding a mere status as "blanks," one
might say, that they lay out carefully devised "traps of the visible" that the observer
may fall into.[16]

---

12 Brugger and Brugger, "The Easter Bunny"; McManus et al., "Science in the Making."
13 McManus et al., "Science in the Making," 181–82, report among their findings that even Wittgenstein's figure is "very biased towards the duck."
14 For an English introduction, see Wolfgang Kemp, "The Work of Art and its Beholder: The Methodology of the Aesthetic of Reception," in *The Subjects of Art History: Historical Objects in Contemporary
Perspectives*, ed. Mark A. Cheetham (Cambridge: Cambridge University Press, 1998).
15 Kemp, "The Work of Art and its Beholder," 188.
16 See Michel Weemans, "L'Image Double: Piège et Révélateur du Visible," in *Voir Double: Pièges et Révélations du Visible*, ed. Michel Weemans, Dario Gamboni, and Jean-Hubert Martin (Paris: Éditions
Hazan, 2016).

TABLE 1

JASTROW'S (1899) DUCK/RABBIT FIGURE WITH ELEVEN PUBLISHED VARIANTS: DISPLAYED ARE 100 STUDENTS' RATINGS OF EASE OF SEEING A BIRD/RABBIT AND THE MEAN AMBIGUITY SCORE

Duck/Rabbit Variant[a,b] *		Reference	Rated Ease of Seeing a				Dominant Alternative Ambiguity Score[d]	
			Bird[c]		Rabbit[c,e]			
			M	SD	M	SD	M	SD
1	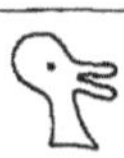	(7)	2.33	1.8	7.08	2.1	4.75	2.7
2	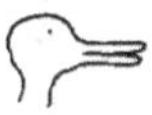	(20)	1.72	1.3	6.20	2.4	4.49	2.8
3	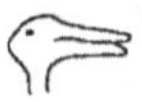	(4,5)	2.26	1.7	6.28	2.3	4.02	2.8
4	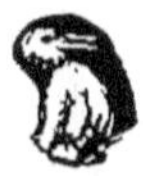	(21)	2.08	2.0	4.55	3.1	2.47	4.3
5	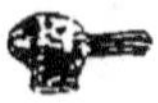	(15)	2.15	1.7	4.22	2.6	2.07	3.2
6	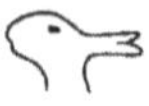	(8,9,10,19,22)	3.28	2.1	4.02	2.5	0.74	3.5
7	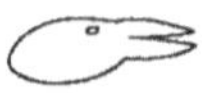	(3)	5.52	2.8	6.18	2.6	0.66	3.5
8	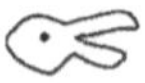	(18)	3.48	2.3	3.88	2.5	0.40	3.6
9	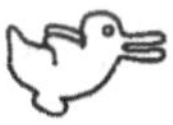	(25)	3.54	2.8	3.55	2.9	0.01	5.0
10	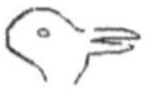	(6)	3.61	2.3	3.53	2.3	0.08	3.5
11	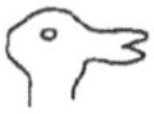	(12,23, cf. 16)	3.67	2.2	3.06	1.9	0.61	2.8
12	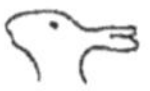	(1)	5.35	2.4	2.48	1.7	2.87	3.2

[a] Figure 5 is Jastrow's (1899) original drawing. [b] Figures 1, 2, 4, 5, 6 and 10 are mirror images of the drawings used in the original study. [c] For Figures 7 and 9 bird/rabbit ratings depended in part on beak/ear orientation (see text). [d] Absolute value of difference between bird and rabbit rating (a value of zero indicates perfect ambiguity). [e] Scale 1: easiest, 9: most difficult.
*In the order of most bird-dominant to most rabbit-dominant[c].

Fig. 3: Peter Brugger, overview of twelve Duckrabbit variants with ambiguity score, Peter Brugger, "One Hundred Years of an Ambiguous Figure: Happy Birthday, Duck/Rabbit!" *Perceptual & Motor Skills* 89, no. 3 (1999): 975, table 1.

Patterns and Potential

Using another inductive example, Gregory's tree (fig. 4), observers will predominantly see the abstracted outlines of a short trunk under a voluminous crown. Peripherally, three miniscule lines scribbled at the bottom right play their part in fleshing out the imagination as supposed blades of grass in the soil. However, with only a few more sketches, Gregory radically alters the probability of seeing a tree by adding a small attribute on the right side of the crown (fig. 5). The same forms now appear entirely transformed into the silhouette of neck, hair, and face of a cigarette smoker.[17] Arguably, there is some sort of "sleight of hand" involved between the two versions: upon revisiting fig. 4, we notice that Gregory only accentuated what he originally had shaped as vague anthropomorphic outlines that were there to be discovered and activated in the first place.

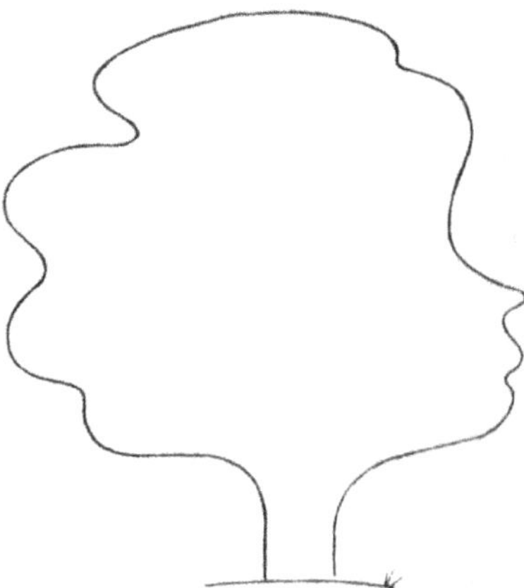

Fig. 4: R. L. Gregory, "Just a tree?," drawing, Richard L. Gregory, "The Confounded Eye," in *Illusion in Nature and Art*, ed. Richard L. Gregory and Ernst H. Gombrich (London: Duckworth, 1973), 81, fig. 29.

Gregory's experiment is reminiscent of an account by the early modern humanist and art theoretician Leon Battista Alberti, who, in his 1435 treatise *De Statua*, speculated about how early primitive sculptors would have been able to realize their first works of art through the discovery of anthropomorphic shapes in tree trunks or clods of earth: "Those [who were inclined to express and represent the bodies brought forth by nature] would at times observe in tree trunks, clumps of earth or other objects

---

17 Gregory, "The Confounded Eye," 81–82, figs. 29 and 30.

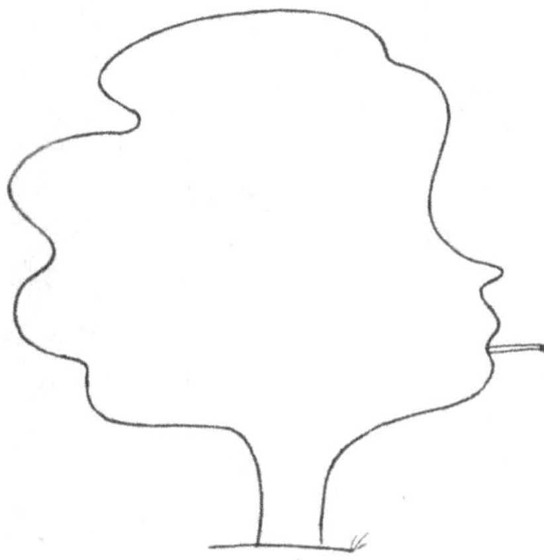

Fig. 5: R. L. Gregory, "The same tree with an addition," drawing, Gregory, *The Confounded Eye*, 82, fig. 30.

of this sort, certain outlines (*lineamenta*) which through some light changes could be made to resemble a natural shape."[18]

The supposed technique which Alberti had projected back onto human ancestry incorporates a cognitive bias into the stage of design, which has been much researched by recent neuropsychology as *pareidolia*.[19] It describes an inborn human tendency for the detection of significant structures in underdetermined visual stimuli that humans, as pattern seekers, are hardwired to look for.[20] This phenomenon can be counted as one of many interrelated human tendencies of anthropomorphism, which also include cognitive biases like "hyperactive agency detection."[21] We know that, while a conventional notion of agency is certainly untenable with regard to images, they are indubitably able to elicit some variants of a "living presence response" when beheld.[22]

---

18 Leon Battista Alberti, cited Horst W. Janson, "The Image Made by Chance in Renaissance Thought," in *De Artibus Opuscula, XL: Essays in Honour of Erwin Panofsky,* Vol. 1, ed. Millard Meiss (New York: New York University Press, 1961), 254.

19 The word was originally coined in 1866 by German Psychiatrist Karl Ludwig Kahlbaum in a paper on "delusions of the senses" (Karl L. Kahlbaum, "Die verschiedenen Formen der Sinnesdelirien," *Zeitschrift für Psychiatrie* 23 (1866): 56–78). It stems from the Greek words *para* (παρά: beside, alongside, instead) and *eidōlon* (εἴδωλον: image, form, shape).

20 The most common form, sometimes taken pars pro toto, is "face pareidolia." See, among many other studies, Orit Hershler et al., "The wide window of face detection," *Journal of Vision* 10 (2010), doi:10.1167/10.10.21. Gombrich, *Art and Illusion*, 87, put it bluntly: "Whenever anything *remotely* facelike enters our field of vision, we are alerted and respond."

21 For a comprehensive overview and an attempt in systematization, see Marco Antonio Correa Varella, "The Biology and Evolution of the Three Psychological Tendencies to Anthropomorphize Biology and Evolution," *Frontiers in Psychology* 9 (2018), https://doi.org/10.3389/fpsyg.2018.01839.

22 A pointed critique of notions of agency in art can be found in Matthew Rampley, "Agency, Affect and Intention in Art History: Some Observations," *Journal of Art Historiography* 24 (2021). On "living pres-

Not by accident, it appears to be a case of anthropomorphic pattern seeking when Wittgenstein describes, amid his analysis of the "duckrabbit" example, another (unknown) "puzzle picture" (*Vexierbild*). Therein, he reports, one would suddenly be able to discern, instead of branches, the "solution" of a human shape. Interestingly, the key to this change, which can be found beyond the recognition of color and shapes, lies in the revelation of what Wittgenstein calls "a quite particular 'organization'" that the image possesses.[23] This "quite particular organization" that transcends the mere form of the image cannot be taken literally as a physical property of the image, nor can it only be wholly dismissed as a subjective projection of *pareidolia*. Instead, it is clearly an intrinsic potential that the image has been imbued with and that had been waiting to be actuated by the beholder's cognition.

In this sense, one might point to what Dario Gamboni termed "potential images," which are "established—in the realm of the *virtual*—by the artist but dependent on the beholder for their realization."[24] This posits a quality of latency, waiting for its virtual manifestation, that lies neither in the form of the image nor in the beholder's cognition, but is activated in a hard-to-define virtual space during the process of collaboration between artist, work, and beholder.[25]

When faced with the problem of the pictorial dynamics at play in the apprehension of ambiguous imagery of various kind, one is, at some point, inevitably certain to hit these firm walls of a "black box." The change that occurs between different impressions of the very same form, such as Wittgenstein's "change of aspect," is as manifest virtually as it is undisplayable and physically untraceable within the forms of the image. To quote him again, once more:

---

ence response," see Alfred Gell, *Art and Agency: An Anthropological Theory* (Oxford: Oxford University Press, 1998) and Caroline van Eck, *Art, Agency and Living Presence*: From the Animated Object to the Excessive Object (Boston, Berlin, and Munich: De Gruyter, 2015), 45–66.

23 Wittgenstein, *Philosophische Untersuchungen*, 196e: "I suddenly see the solution of a puzzle picture. Before, there were branches there; now there is a human shape. My visual impression has changed and now I recognize that it has not only shape and color, but also a quite particular 'organization.'"

24 Dario Gamboni, *Potential Images: Ambiguity and Indeterminacy in Modern Art*, trans. Mark Treharne (London: Reaktion Books, 2002), 18 (emphasis the author's). See also Dario Gamboni and Richard Leydier (Interview), "Images Potentielles," *Artpress2* (2009) and Dario Gamboni, "Ambiguité Visuelle et Interprétation," in *Voir Double: Pièges et Révélations du Visible*, ed. Michel Weemans, Dario Gamboni, and Jean-Hubert Martin (Paris: Éditions Hazan, 2016), 42–45. Ambiguous pictorial phenomena have been approached by modern art history with a multitude of concepts, which each emphasize different qualities of the images in question. Horst Janson, for example, investigated them within a cultural tradition of "images made by chance" in nature (Janson, "The Image Made by Chance"; see also Giacomo Berra, "Immagini casuali, figure nascoste e natura antropomorfa nell'immaginario artistico rinascimentale," *Mitteilungen des Kunsthistorischen Instituts in Florenz* 43 (1999): 358–419. The notion of *hidden images* or *cryptomorphs* was most comprehensively and highly critically examined by James Elkins, *Why are Our Pictures Puzzles? On the Modern Origins of Pictorial Complexity* (London: Routledge, 1999). Felix Thürlemann's representational concept of "double mimesis" is closest to the more traditional idea of bi-stability of two representations (Felix Thürlemann, *Dürers Doppelter Blick* [Konstanz: UVK, 2010]).

25 Gamboni, *Potential Images*, 19.

> My visual impression has changed;—what was it like before and what is it like now?—If I represent it by means of an exact copy—and isn't that a good representation of it?—no change is shown. And above all do *not* say "After all, my visual impression isn't the drawing; it is this––which I can't show to anyone."––Of course, it is not the drawing, but neither is it anything of the same category, which I carry within myself.[26]

No marks were left, no molecule budged. Yet, at the same time, a dynamic encounter occurred between an observer and an artifact, which is transgressing its own form in an image act.[27]

---

26 Wittgenstein, *Philosophische Untersuchungen*, 196e.
27 See comprehensively: Horst Bredekamp, *Image Acts: A Systematic Approach to Visual Agency*, trans. Elizabeth Clegg (Boston, Berlin, and Munich: De Gruyter, 2017), 209–64.

Léa Perraudin

Surface Tension: Venice and the Undercurrents of Mediation

How to move through a city that comes from the water as privileged frame of reference, that thinks every aspect of itself through the waterfront as access point, that has agreed on liquid grounds of supply and disposal, of spatial scarcity and temporal submergence. A city that has challenged its inhabitants and visitors to move and to sense otherwise while undeniably and irreversibly turning itself into everything that comes and goes with being one of the top destinations of globalized tourism. Who lives here, who walks here, who dares to swim, who needs to be carried?

Luce Irigaray famously challenged Martin Heidegger's notion of dwelling as critically reiterated toward aerial matter.[1] Echoing Astrida Neimanis and their queering of Irigaray's elemental conversations, Venice urges me to ask what is at stake when dwelling on and with the liquid element.[2] The city facilitates water in its omnipresence as an infrastructural preset and as a resource. But not only in terms of it being the vital factor for carbon-based life, but rather of it being the solvent for an imaginary. That is the resource of water in contemporary Venice. A figuration of a picture opportunity and a brief but riskless adventure in getting lost in the lagoon maze. The cruise ships have been banned—so, fingers crossed, that at least Piazza San Marco is flooded while we are here.

This is how my field notes from October 2021 during the Anthropocene Campus in Venice themed "Water Politics in the Age of the Anthropocene" began.[3] I would like to take

---

1 See Luce Irigaray, *The Forgetting of Air in Martin Heidegger* (Austin: University of Texas Press, 2001); See Martin Heidegger, "Building Dwelling Thinking," in *Poetry, Language, Thought*, ed. and trans. Albert Hofsdater (New York: Harper & Row, 1971), 145–61.

2 See Astrida Neimanis, *Bodies of Water: Posthuman Feminist Phenomenology* (London: Bloomsbury Academic, 2017).

3 The Anthropocene Campus Venice was organized by the Center for the Humanities and Social Change, Ca' Foscari University of Venice and the Max Planck Partner Group "The Water City" within the framework of the Anthropocene Curriculum by Haus der Kulturen der Welt (Berlin) and Max Planck Institute for the History of Science (Berlin), supported by the Federal Foreign Office of Germany. I would like to thank the organizers Cristina Baldacci, Shaul Bassi, and Pietro Daniel Omodeo for hosting a thought-provoking program in Venice and for gathering a vast variety of researchers, artists and practitioners. I remain utterly grateful for Heather Contant, L. Sasha Gora, and Ifor Duncan, the instructors of the seminar "Venice is Leaking: Interventions in the Lagoon-City Continuum" for providing a truly unique, engaging, and critical space of embodied research and exchange and for the generosity to make us participants part of their Lagoon ecosystem for a week that exceeded in many ways what one could wish for in any academic context. I would also like to thank the Cluster of Excellence "Matters of Activity" for the support and funding of my research stay in Venice by the German Research Foundation. After eighteen months of remote work and online conferencing due to the ongoing global COVID-19 pandemic, I followed the invitation to an in-person attendance in Venice that came with both

them as a starting point to contextualize some broader questions when it comes to tipping points—a notion to indicate a critical and oftentimes irreversible change in a process, system, or network (choose your preferred relational vernacular) that brings about an unprecedented loss of stability. The following is an attempt to rephrase such tipping points through the frictions they engender and how we might conceive of the waters of Venice as a particular mediator of said frictions. As I will show, they allow me to tackle some aspects of the aquatic specificity of the Venetian lagoon as a matter of surface tensions and in terms of a logic I call compromised abundance.

Compromised Abundance

As epitome of the heteronormative romantic getaway, Venice provides a vast imaginary of what lovers can do on or by the water in order to make their trip worthwhile. The city has in fact capitalized on the liquid element by turning it into an asset that allows for yet another photo opportunity couples will take in gondolas, on bridges, during sunsets, and at outdoor dinners.[4]

While in Venice, I was in search of another coupling—the coupling of processes that bring about a complicated and troubling surplus, which I would like to introduce as "compromised abundance." Being one of the most visited tourist destinations in the world, Venice engenders such abundance on conflicting terms. Yet, it is responsible for a particular, commodified experience of the lagoon environment. What allows one to float? In the case of water, the properties that play out at the interface of two phases— liquid and gaseous—cause the water to minimize its surface area and prevent light objects with a higher density than water from submerging. This molecular phenomenon is called surface tension. While minimizing something (the surface), something else that allows for the stability and integrity of this surface (the tension) is increased. In tandem with the macroscopic principle of buoyancy that indicates that the upward force on a (completely or partially) submerged body is equal to the weight of the liquid displaced, floating is made possible.[5]

---

the convenience and environmental impact of taking a flight from Berlin to Venice, renting an Airbnb and using vaporetti, the motorized water vehicles for getting around in the city as well as other touristic commodities. I decided to come to Venice because of the number of site-specific activities on, through, and with the water planned by the seminar instructors and co-conducted by Venetian residents and seminar participants.

4 In the following, I will focus on particular experiences of contemporary Venice that are engendered by touristic imaginaries and spatial circumstances that are site specific to the current anthropocenic lagoon environment. Therefore, I will not further investigate, e.g., the historical dimensions of power, trade, and the challenges of building a city on water nor elaborate on the multispecies communities of the lagoon. Nevertheless, they are as well matters of surface tension.

5 "Surface tension," *Encyclopedia Britannica*, https://www.britannica.com/science/surface-tension; "Archimedes' principle," *Encyclopedia Britannica*, https://www.britannica.com/science/Archimedes-principle.

This brief digression into fluid dynamics is far from being a mere metaphor for the waters of Venice. It allows me to trace how floatability is granted through displacement and how it prioritizes the neatness of a surface and while doing so, causes tensions to maximize and unfold as compromised abundance. In attuning to the material multitudes that make up Venice, this exercise of tracing floatability spells out a practice of care that intends to perceive of care as something beyond maintenance in that it lets go of the idea of material integrity.[6] In that sense, compromised abundance is both the diagnosis of a harmful dynamic bearing witness in the Anthropocene as well as a motive to unlearn predominant paradigms of materiality. Caring as brought forth by María Puig de la Bellacasa is a non-normative form of material engagement.[7] It entails acknowledging our complicity in the status quo but also the commitment to embrace unruliness, vulnerabilities, and impurities in an attempt at aspirational solidarity as Alexis Shotwell insists.[8]

At this point, a closer look into another imaginary of contemporary Venice is indicated—an imaginary that feeds from climate crisis and the realization of planetary urgency: tropes of globalized tourism, extinction, and ecocide, and consequently the attempt to defeat the elements through geo-engineering. Some of them can be subsumed under the denominator "toxic sublime," a term proposed by Jennifer Peeples in that they illustrate the contaminated yet visually conflicted sites of the Anthropocene: polluted waters, damaged landscapes, scenes of ecological disasters.[9] What should be added here and what also seems crucial in terms of toxicity is that these tropes are brought forth by a highly invasive and harmful—in that sense toxic—understanding of occupying and defining spaces through reinforcing the privileged modes and motives of the passivated materials of modernity. They themselves urge us to critically address our own research practices and their institutionalized presets as questions of extraction.

Critical theory work has come to terms with these definitory practices of Western universalism and how they attribute activity and passivity to matter, to bodies, and environments and the way they prioritize ways of worldmaking by operations of ignoring, displacing, and erasing other practices and knowledges. Kathryn Yusoff insists that "extractable matter must be both passive (awaiting extraction and possessing of prop-

---

6 Together with Gina Caison, I have argued elsewhere for such a practice of care as related to the waters of Venice through a site-specific walkshop intervention. See Gina Caison, Léa Perraudin, "To Carry Water: An Invitation to Move and Sense Otherwise," in *Venice and the Anthropocene: An Ecocritical Guide* ed. Lucio De Capitani et al. (Venice: Wetlands Books, 2022).

7 See María Puig de la Bellacasa, *Matters of Care in Technoscience: Speculative Ethics in More Than Human Worlds* (Minneapolis: University of Minnesota Press, 2017).

8 See Alexis Shotwell, "Complexity and Complicity: An Introduction to Constitutive Impurity," in *Against Purity: Living Ethically in Compromised Times* (Minneapolis: University of Minnesota Press, 2016), 1–20.

9 See Jennifer Peeples, "Toxic Sublime: Imaging Contaminated Landscapes," *Environmental Communication* 5, no. 4 (2011): 373–92.

erties) and able to be activated through the mastery of white men."[10] Emphasizing the obscuring operations of such acts of passivation, Macarena Gómez-Barris exposes the extractive view as facilitator of the "reorganization of territories, populations, and plant and animal life into extractible data and natural resources for material and immaterial accumulation."[11]

The following three vignettes on the aquatic specificity of the lagoon city are characterized by surface tensions that allow me to further operationalize what I mean by compromised abundance: they capitalize on an experience of flow and floating I am inclined to remain suspicious of.

Surface Tension: Displacement

As the world kept pouring visitors into Venice by docking cruise ships in the city center over the last decades, the conflicting scales that have been reinforcing the vulnerabilities of the water-land continuum became increasingly apparent. In pre-pandemic times, the city witnessed over twenty million visitors annually. On busy days the 50,000 permanent residents share the historical city center with up to 120,000 tourists who are roaming the squares, narrow streets and canals.[12] Through their presence, their gaze and their action they largely define how the urban space is used and navigated, creating a spatial matrix that intersects pollution and preservation, (material) exhaustion and prosperity. Consequently, the cruise ship and its wider infrastructural complex is a matter of surface tension. With the ultimate ban of larger cruise ships entering the city area as of August 2021 as well as the pandemic lockdowns that made global tourism come to a halt, the waters of Venice cleared up and temporarily looked different. Clear water in this case is not necessarily a sign of better water quality, it is rather due to the lack of water traffic that would usually cause the sediments to whirl up. Movement causes turbidity—and stagnation makes things surface in a particular way: The deserted streets, abandoned hotels, Airbnbs, and Biennale venues during lockdown unmistakably revealed that the city, by turning itself into a place that accommodates twice as many tourists than residents, had ultimately encouraged gentrification and privileged the use of space toward touristic needs while tolerating displacement of various other bodies.[13] The lack of touristic influx during the COVID-19 pandemic

---

10 Kathryn Yusoff, *A Billion Black Anthropocenes or None* (Minneapolis: University of Minnesota Press, 2019), 14.

11 Macarena Gómez-Barris, *The Extractive Zone: Social Ecologies and Decolonial Perspectives* (Durham, NC: Duke University Press, 2017), 5.

12 In comparison to a population of approximately 175,000 residents in 1951 and approximately 93,000 in 1981.

13 This kind of displacement equally applies to the transformation of local business spaces into temporal venues for the Venice Biennale and collateral events. Recently, the Austrian Pavilion of the 18th International Venice Architecture Biennale 2023 has critically commented on these processes of

due to global travel restrictions didn't cause much ease for the living situation of residents. As Venice's tourism dependent economic infrastructure stagnated, the strategically overlooked labor at the heart of making the lagoon city a consumable experience —cleaning, maintenance, and other services for tourist sites, transport, restaurants, and accommodation—revealed the presets of its invisibilization. COVID-19 reinforced precarious conditions of dwelling and working in Venice, with residents losing jobs and facing the threat of becoming unhoused while touristic accommodations remained abandoned during lockdown.[14] Due to the particular circumstances of the strict lockdowns that required residents to shelter in place and consequently caused a steady use of the domestic water piping system while the piping in restaurants and touristic accommodations remained mostly unused, a significant disturbance in the infrastructure of supply and disposal appeared.[15]

Another cause of displacement obviously is the liquid element itself that puts crucial challenges at the organization of residential and public spaces through increasing land scarcity and effects of climate gentrification. At the same time, the water carries with it the conflicting temporalities of these vulnerabilities. "Time" and "tide" have a shared linguistic history, the German *Gezeiten* (tide) is a remnant to this etymological proximity. An electronic display in the front window of Farmacia Morelli in Campo San Bartolomeo keeps track of the total amount of permanent residents in the historical city center. The constantly decreasing number is eerily connected to the slow creep of sea level rise and the steady touristic influx that caters to the logic of short-term stays, day trippers, Biennale visitors, historic architecture appreciators, and weekend honeymooners.[16] At the end of May 2022, leaflets with nothing but the number 49,999 printed on them appeared in highly frequented areas of Venice. They served as an anonymously invoked warning sign for the decline in number of permanent residents

---

urban displacement and exclusion instigated by touristic monoculture and the spatial demands of Biennale infrastructures. See AKT and Hermann Czech, *Partezipazione* (Vienna: Luftschacht, 2023).

14 Public housing is a contested subject in the residential politics of Venice. The project OCIO monitors the city's housing situation and its development since the mid 1990s. Currently, approximately two thousand public housing units are vacant, as ATER (the local government agency for residential building) reports. Most of them are deliberately left unoccupied or will be made available for tourist accommodations eventually. The complicated urban logistics of the lagoon, accompanied with high cost in maintenance and modernization in the historical city center as well as the status of Venice as UNESCO world heritage site pose a challenge at policy making. See "L'Osservatorio CIvicO sulla casa e la residenza – Venezia," OCIO, https://ocio-venezia.it/; See "Azienda Territoriale per l'Edilizia Residenziale," ATER, https://www.atervenezia.it/.

15 Many thanks to Rosella Alba for pointing that out during a conversation at the Anthropocene Campus Venice.

16 From April 1, 2024, day-trippers will have to pay a fee between 3 € and 10 € to enter the historic city center, depending on the season and crowdedness of the city. However, the introduction of this fee has been postponed several times.

that was estimated to happen during July 2022.[17] The admonitory 49,999 can be read as another tipping point for the city, that reduces the likelihood and possibility for an actual lived urban experience beyond touristic tropes.

In addition to these socio-political and spatial consequences that occur when things come to a standstill and cause displacement, surface tension in the aquatic specificity of Venice is closely tied to the phantasm of flow and frictionless transmission.

Surface Tension: Floating on Friction

Much has been said about the elemental forces of water. Hans Blumenberg's understanding of the open sea as sphere of the unreckonable and lawless, Gaston Bachelard's elemental musings on fluids as hormones of imagination, Michel Serres's philosophy of vortex and passages. They all stem from the idea of an unfathomable, yet universal aqueous space "out there" that—although being sensed and channeled—must fail to be entirely mapped and measured.[18] I share with them the sentiment that even when stagnant, water does not cease to be in transformation. However, it is neither my intention to reiterate the elemental universality of water, nor attempt to get hold of what water *is* but to seek for the frictions being carried out in the Venetian lagoon.

Most water bodies have been entangled in ongoing efforts of domestication or are at least significantly affected by human activity. These impacts include the accumulation of plastic waste in oceans, the increase of anthropogenic pollutants in oceans, the weaponization and privatization of water resources, as well as the reorganization and destruction of entire ecosystems through practices like damming, river engineering, and overfishing. One could easily make an argument about the multiple twisted naturalizations at work here.[19] It is striking, however, that it also seems to come naturally to make the connection between the ubiquity and existential dimension of water as compared to the ubiquity and existential dimension of data flow in technological infrastructure, of their supply and disposal. Scholarship has made sense of the apparent metaphorical and material connection between water and contemporary technological culture. John Durham Peters traces the cultural histories of flow and

---

17 See Campaign for a Living Venice, "49,999 residents: The blitz sounding the alarm about Venice," https://campaignforlivingvenice.org/2022/05/27/49999-residents-the-blitz-sounding-the-alarm-about-venice/.

18 Gaston Bachelard, *Water and Dreams: An Essay on the Imagination of Matter,* trans. Edith R. Farrell (Dallas: Dallas Institute Publications, 1983); Michel Serres, *Hermès V, Le passage du Nord-ouest* (Paris: Éditions de Minuit, 1980); Hans Blumenberg, *Schiffbruch mit Zuschauer: Paradigma einer Daseinsmetapher* (Frankfurt: Suhrkamp, 1979).

19 I have argued elsewhere for the conflicting temporal taxonomies of plastics as related to water, waste, and futurity, see Léa Perraudin, "Tales from the Great Pacific Garbage Patch: Speculative Encounters with Plastic" in *Müll: Interdisziplinäre Perspektiven auf das Übrig-Gebliebene* ed. Christiane Lewe, Tim Othold, and Nicolas Oxen (Bielefeld: transcript, 2016), 143–70, https://doi.org/10.1515/9783839433270-007.

connectivity in his philosophy of elemental media; Thomas Sutherland points out the ontologies of flux central to accelerated digital capitalism.[20] The liquid element proves to spread out evenly, self-consistent, transparent. Its materiality embodies the phantasm of absolute transmission. But things are becoming turbid as Venice itself reveals the messy backsides of the flowlike imperative of connectivity. Surface tension in tandem with buoyancy, as introduced earlier, allows for the pleasant feeling of floating. As I will show, it also constitutes a particular understanding of friction as commodity in the Venetian lagoon.

In her seminal book on global connection, Anna Tsing sheds light on what gives grip to worldly encounter: friction. She presents frictions as the overlooked and oftentimes contingent but constitutive factors of "interconnection across difference" in globalized capitalism.[21]

> Commodities seem so familiar that we imagine them ready made for us throughout every stage of production and distribution, as they pass from hand to hand until they arrive at the consumer. Yet the closer we look at the commodity chain, the more every step – even transportation – can be seen as an arena of cultural production. Global capitalism is made in the friction in these chains as divergent cultural economies are linked, often awkwardly. Yet the commodity must emerge as if untouched by this friction.[22]

Consequently, frictions must be considered integral for any production of commodities, yet this fact remains strategically disguised. So, how do frictions and commodities meet in Venice?

Most visitors begin their journey through Venice with the visceral realization that the vast majority of transport and travel is guided by waterways; while the only feasible means of land transportation is by foot—there are no motorized vehicles, bicycles, or other means of land transportation available in the historic city center. Bridges, stairs and narrow cobblestone alleys guide the path of trolley-equipped tourists to their accommodation. Some of them are organized along the waterfront, but oftentimes they are rather intersecting with the maze of waterways. Consequently, Venice greets its visitors with the bliss of a riskless adventure, a noticeable but harmless exhaustion that differs from the comfort of other well-organized city trips and renders friction itself into a commodity. This friction is characterized by a temporary deviation of the flow that water is intrinsically linked to. It serves as the opportunity to engage in an experience of an ambiguous spectacle all over the city.

---

20 John Durham Peters, *The Marvelous Clouds: Toward a Philosophy of Elemental Media* (Chicago: University of Chicago Press, 2015); Thomas Sutherland, "Liquid Networks and the Metaphysics of Flux: Ontologies of Flow in an Age of Speed and Mobility," *Theory, Culture & Society* 30, no. 5 (2013): 3–23, http://doi.org/10.1177/0263276412469670.
21 Anna Lowenhaupt Tsing, *Friction: An Ethnography of Global Connection* (Princeton, NJ: Princeton University Press, 2004), 4.
22 Ibid.

Flooded plazas and alleys are a common occurrence in Venice. Moreover, they are a welcomed iteration of consumable friction for many visitors to document an "authentic" experience of Venice. Walking through the city in body contact (or at least in uncommon proximity) with what makes this city unique—its waters—grants a photo opportunity not to be missed. Being considered as adventurous activities reframes these events of climate emergency as sideshow of disaster tourism. Colorful disposable rubber boots that are sold where the *acqua alta*[23] occurs and that serve as a quirky accessory further emphasize the transactional logic of the frictions (figs. 1, 2). The gadgets used to document such experiences—predominantly smartphones—themselves are subject to the unattainable imperative of connectivity attributed to both water and digital culture.

Street signs are rare in Venice. Additionally, the narrow, mazelike alleys increase the chance of getting lost; GPS works imprecisely and leads visitors who rely on gadget-based mobile navigation oftentimes to dead-ends or uncrossable waterfronts. Consequently, going astray has become part of the "authentic" touristic experience of the city—a favorable detour, another spectacle that feeds from friction. Images, goods, bodies of flesh and of water are migrating through the city. As of July 2022, the hashtag #Venice lists 16.3 million posts on Instagram. The energy consumption and data traffic for navigating and documenting the spectacle of the lagoon on the go is directly linked to the question of what is surfacing in this city. With water being both the asset for the romantic vista cliché as well as a commodity snapshot of friction, other attunements remain overlooked.

Surface Tension: Filtering the Tide

When it comes to accounting for the highly invasive anthropogenic impacts on the planet, technological fixes are oftentimes presented as a convenient and seemingly self-evident solution. However, in many cases practices of geo-engineering follow a biased logic of scalability. As Anna Tsing argues, scalability presents the promise "to expand without distorting the framework" while at the same time "by its design, covers up and attempts to block the transformative diversity of social relations."[24]

Venice resides on instable grounds in subsidence. Due to natural subsiding of the area and anthropogenic groundwater overexploitation in the twentieth century, paired with the continuous rise of sea level, over the last century Venice has lost approximately twenty-five centimeters in elevation.[25]

---

23 *Acqua alta* refers to the occurrence of unusually high tides resulting in flooding, particularly in Venice and other coastal regions along the Adriatic Sea.
24 Anna Lowenhaupt Tsing, "On Nonscalability: The Living World Is Not Amenable to Precision-Nested Scales", *Common Knowledge* 18, no. 3 (2012): 505–24, 523, https://doi.org/10.1215/0961754X-1630424.
25 Luigi Tosi, Pietro Teatini, and Tazio Strozzi, "Natural versus anthropogenic subsidence of Venice," *Scientific Reports* 3, 2710 (2013): 1–9, 1, https://doi.org/10.1038/srep02710.

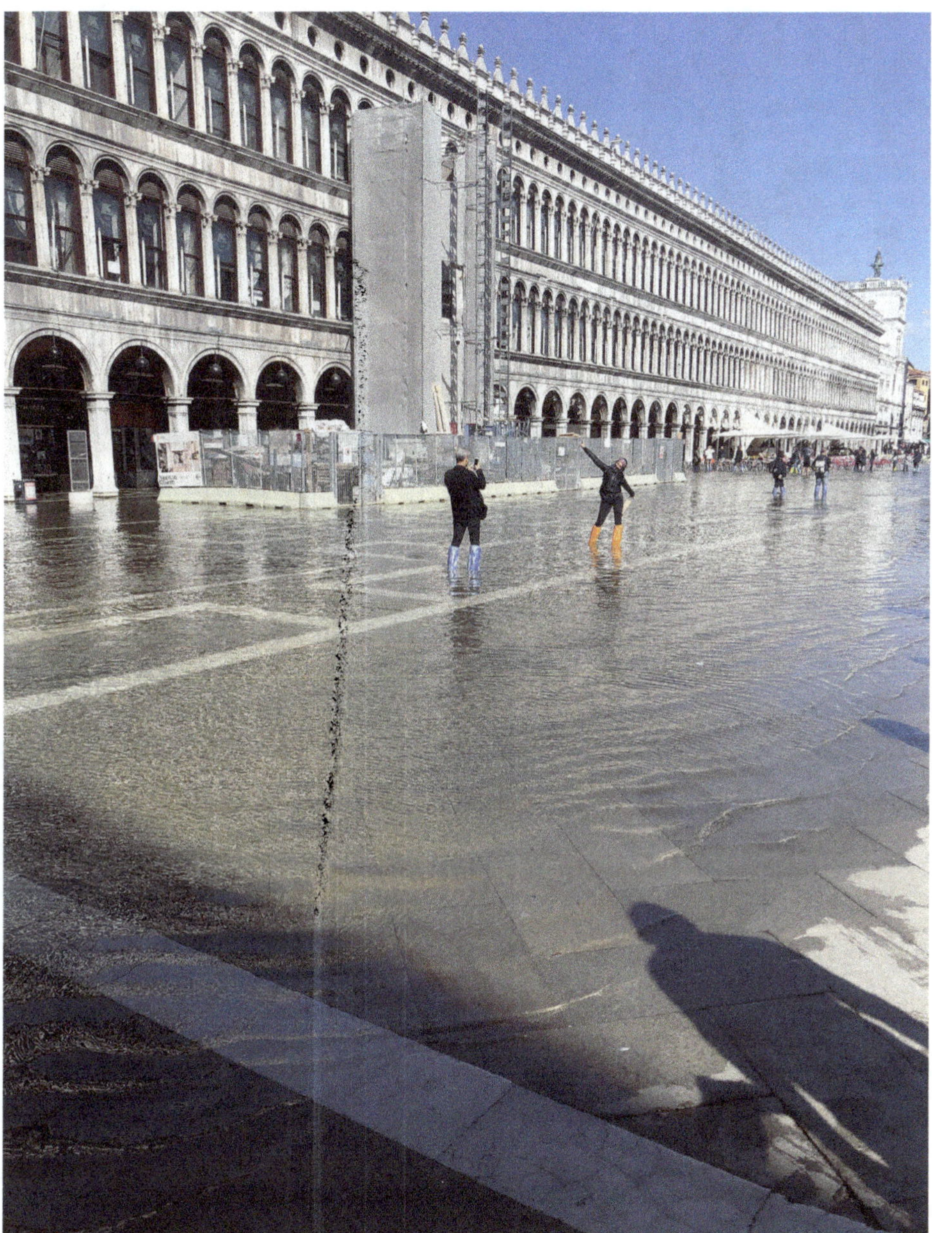

Fig. 1: Posing in disposable rubber boots at Piazza San Marco.

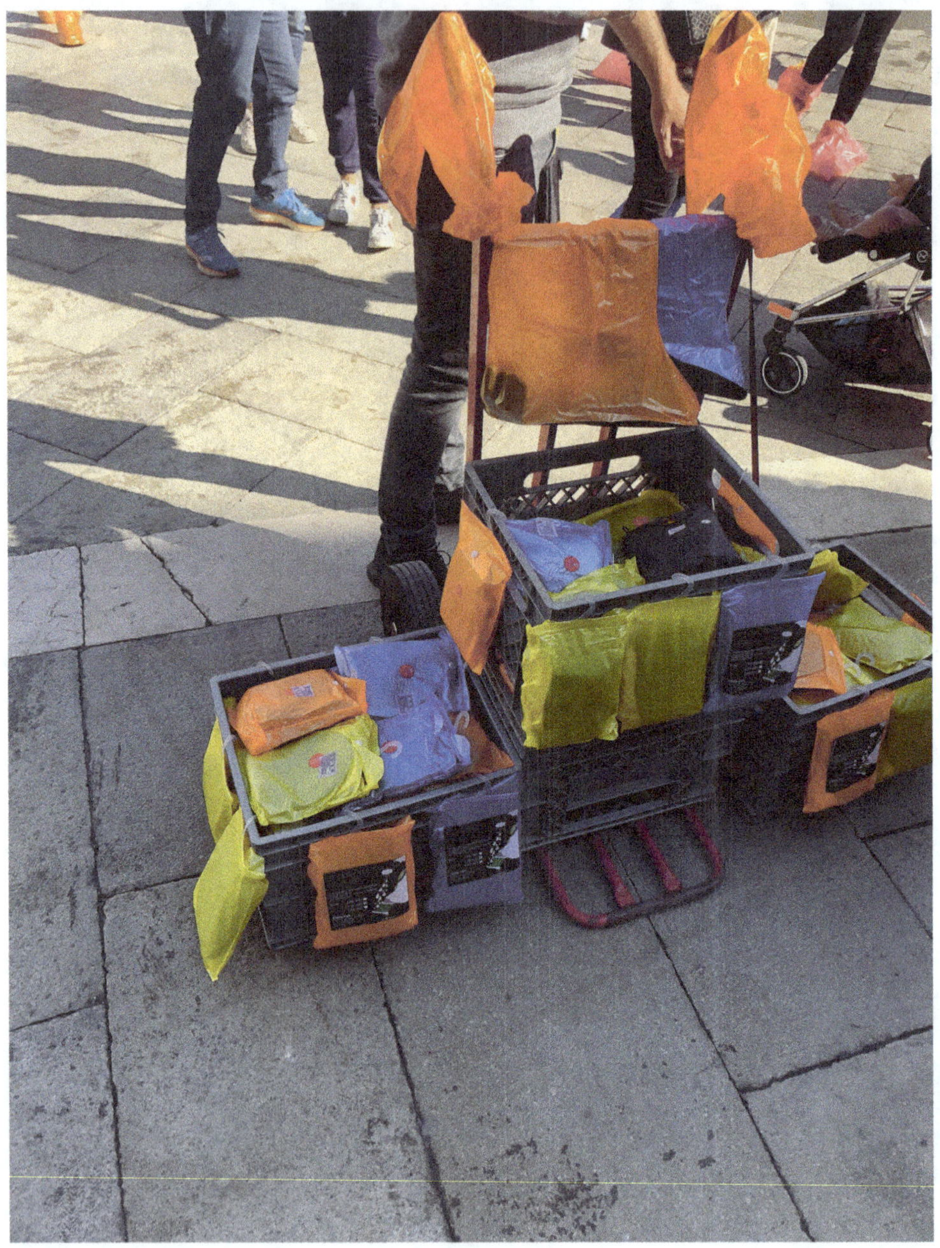

Fig. 2: Disposable rubber boots, ten euros per pair, Piazza San Marco.

Various conceptual models in geo-engineering over the last two decades attempt to make use of the aquifers below the city, in order to "inflate" the porous sediments with up to 150 billion liters of sea water and consequently to elevate the city.[26] Other concepts include liquified carbon dioxide as the substance to be injected that would additionally create a massive carbon sink below the city to store emissions.[27] Plans like these appeared over the years but submerged again. Currently, Venice is grappling with questions of uplifting mostly by creating ad-hoc micro elevations to manage pedestrian traffic, cat walks that are set up when the *acqua alta* hits.

Venice's prime example of a long-term project in local geo-engineering is MOSE (Modulo Sperimentale Elettromeccanico), an integrated system of rows of mobile gates that are installed at three inlets of the Venetian lagoon. The system is intended to protect the city temporarily from the high tides of the Adriatic Sea. The project has been in development for almost thirty years and is considered a questionable infrastructural and ecological endeavor with conflicting bureaucratic and economic interests. Various test runs have provided poor results in reliably preventing flooding, yet after the devastating effects of the *acqua alta* and the Sirocco storm surge in late 2019 the city worked toward having the system fully implemented by the end of 2021.

MOSE allows to filter water—albeit not primarily of its components but rather a filtering of the tide itself, its distribution and possibility condition. Therefore, MOSE is a filter along the lines of chaos and order, risk and convenience. This consequently results in filtering the lagoon environment and modifying it in terms of its biome as well as the distribution and use of land. This in itself has serious implications for how the city and its surrounding area of islands and salt marshes will ultimately be challenged by the crucial question of what is above and what is below water level. In my reading, MOSE imposes a hylomorphic understanding of matter while at the same time disregarding the fact of being exposed to the elemental force in its capacity to take and change form. With it being in the making for three decades, MOSE appears to be a sluggish symbol of the ecological crisis that needs to be accounted for. It presents practices of separation as enablers of change and understands them through the imposed action of the hand, as the story of Moses itself is framed in Exodus:

"Then Moses stretched out his hand over the sea, and all that night the Lord drove the sea back with a strong east wind and turned it into dry land. The waters were divided, and the Israelites went through the sea on dry ground, with a wall of water on their right and on their left."[28] What does such an understanding of the imposed action of the hand imply?

---

26 Giuseppe Gambolati and Pietro Teatini, "Anthropogenic Uplift of Venice by Using Seawater," *Venice Shall Rise Again*, ed. Giuseppe Gambolati and Pietro Teatini (Amsterdam: Elsevier, 2014), 57–77, https://doi.org/10.1016/B978-0-12-420144-6.00005–4.

27 See Bernhard Schrefler and Bonacina, Cesare, "Possible CO2 Injection in Aquifers below Venice," *Revue européenne de génie civil* 9 (2005): 809–16, https://doi.org/10.1080/17747120.2005.9692785.

28 Exodus 14: 21–22.

Hands In and Against the Anthropocene

Throughout the three vignettes I have traced some aspects of the aquatic specificity of the Venetian lagoon as matters of surface tension. All three of them also present a particular understanding of conceptual and actual hands responsible for creating, maintaining, or resisting such surface tensions. The invisible labor that lies at the core of the infrastructural complex of globalized tourism is carried out by actual hands that clean, serve, care, repair, distribute, and transport all across the city. While taking snapshots and performing acts of navigating and documenting their surroundings through technological devices, visitors rely mostly on their actual hands to perform the task, engaging with conceptual hands in an exercise of deixis. Whereas geo-engineering and attempts at defying the elements heavily rely on the conceptual hand as a means of problem solving and mastery.

Manipulation, maintenance, handling—these terms are etymologically derived from the Latin, French, or English word for the upper extremity. They are intrinsically linked to a particular framing of technological evolution of the human species through tool use and the skilled deployment of the hand as brought forth by Ernst Kapp in 1877 in his early accounts of a philosophy of technology.[29] As paleoanthropologist André Leroi-Gourhan pointed out, technologies themselves can be understood as mechanisms of hominization. They occur through the liberation of the hand from being merely a means of transportation into one of grasping and manipulation and the liberation of the mouth from a means of grasping and manipulation into one of expression. By Leroi-Gourhan's account, in tandem they give rise to stabilized practices of shaping, handling, translating, and categorizing the surroundings and bring forth increasingly complex systems and conventions of language, knowledge, and archival.[30]

If we think of technology-environment-human relations in this particular way, we come about multiple practices of sensing, making, and dwelling in order to make a hostile place intelligible, stabilized, profitable—livable in the broadest sense. Livable for whom though? Critical research on the Anthropocene has tackled this conflicted story of proto-promethean control and mastery attached to a universalist understanding of humankind paired with an evolutionary and therefore oftentimes teleological understanding of technology.[31] It is deeply inscribed in the colonial-imperialist complex to apply a vocabulary of terrestrial-infrastructural progress that engendered the technological advancements of modernity. This vocabulary is then utilized in order to dis-

---

29 See Ernst Kapp, *Grundlinien einer Philosophie der Technik: Zur Entstehungsgeschichte der Cultur aus neuen Gesichtspunkten* (Braunschweig: G. Westermann, 1877).

30 See André Leroi-Gourhan, *Gesture and Speech*, trans. Anna Bostock Berger (Cambridge, MA: MIT Press, 1993).

31 See Jason W. Moore, *Anthropocene or Capitalocene? Nature, History and the Crisis of Capitalism* (Oakland: PM Press, 2016); Joni Adamson, "We Have Never Been Anthropos: From Environmental Justice to Cosmopolitics," in *Environmental Humanities: Voices from the Anthropocene*, ed. Serpil Oppermann and Serenella Iovino (New York: Rowman & Littlefield, 2016), 155–73.

guise the highly invasive and violent appropriation and exploitation of territories, bodies, and materials. The ongoing neo-colonial technocapitalist logic of extractivism reinforces these asymmetries, while *anthropos* appears as the well-trained planetary engineer whose success is granted by a vast tableau of system stabilizers at the cost of other bodies (and their labor, oftentimes conducted through hands).[32] But what about stability after all? There is in fact leaking all over the place. In that sense, many accounts of these histories of technology can be reframed as a historiography of cosmetics, in which a particular narrative of human mastery is accentuated. But the shortcomings of this kind of manipulation usually remain disguised, superficially concealed in yet another achievement of either skillful engineering or technological epiphany and therefore remain insensitive to other kinds of openings.

Liquid Handshake

Stacy Alaimo vigorously claims: "The Anthropocene is no time to set things straight."[33] What she calls "the impermeability of the western human subject"[34] directly feeds into the narrative of alleged detached mastery and its respective practices of stabilization. Being exposed and exposing oneself through other mediums and multitudes consequently entails a queering of these presupposed fixed and stable set-ups toward acknowledging environmentally embodied vulnerabilities in action. Think of leaking not as a threat but as occurring, as watery digits that might be reached out to (fig. 3). So, what does it take to engage in that kind of liquid handshake? That is, to think of the relation of surrounding and surrounded in a novel way. The troubling surplus in and along the Venetian waters that I have introduced as "compromised abundance" equally operates as a means of Anthropocene diagnostics as well as a motive to challenge the paradigm of passivated materiality. That also entails a perspective on care that emphasizes caring as, through, and for exposure and wetness. Obviously, this takes more than just showing open palms and taking action by applying quick solutions that reproduce the very same mode of operating that has caused the surface tensions in the first place. Obviously, this takes more than assuming the universal validity of other practices and departure points as they themselves are by definition activities in precarity and don't reside on stable grounds.

What could a hand do instead? Let's start with the willingness and commitment to carry. Together with Gina Caison, Maud Canisius, Camilla Bertolini, and Ifor Duncan, I

---

32 See Anna Tsing, "Earth Stalked by Man," *The Cambridge Journal of Anthropology* 34, no. 1 (2016): 2–16.; Donna J. Haraway, "Tentacular Thinking: Anthropocene, Capitalocene, Chthulucene," *e-flux Journal* #75 (2016): 1–17, http://www.eflux.com/journal/75/67125/tentacular-thinking-anthropocenecapitalocene-chthulucene/.
33 Stacy Alaimo, *Exposed: Environmental Politics and Pleasures in Posthuman Times* (Minneapolis: University of Minnesota Press, 2016), 1.
34 Ibid., 5.

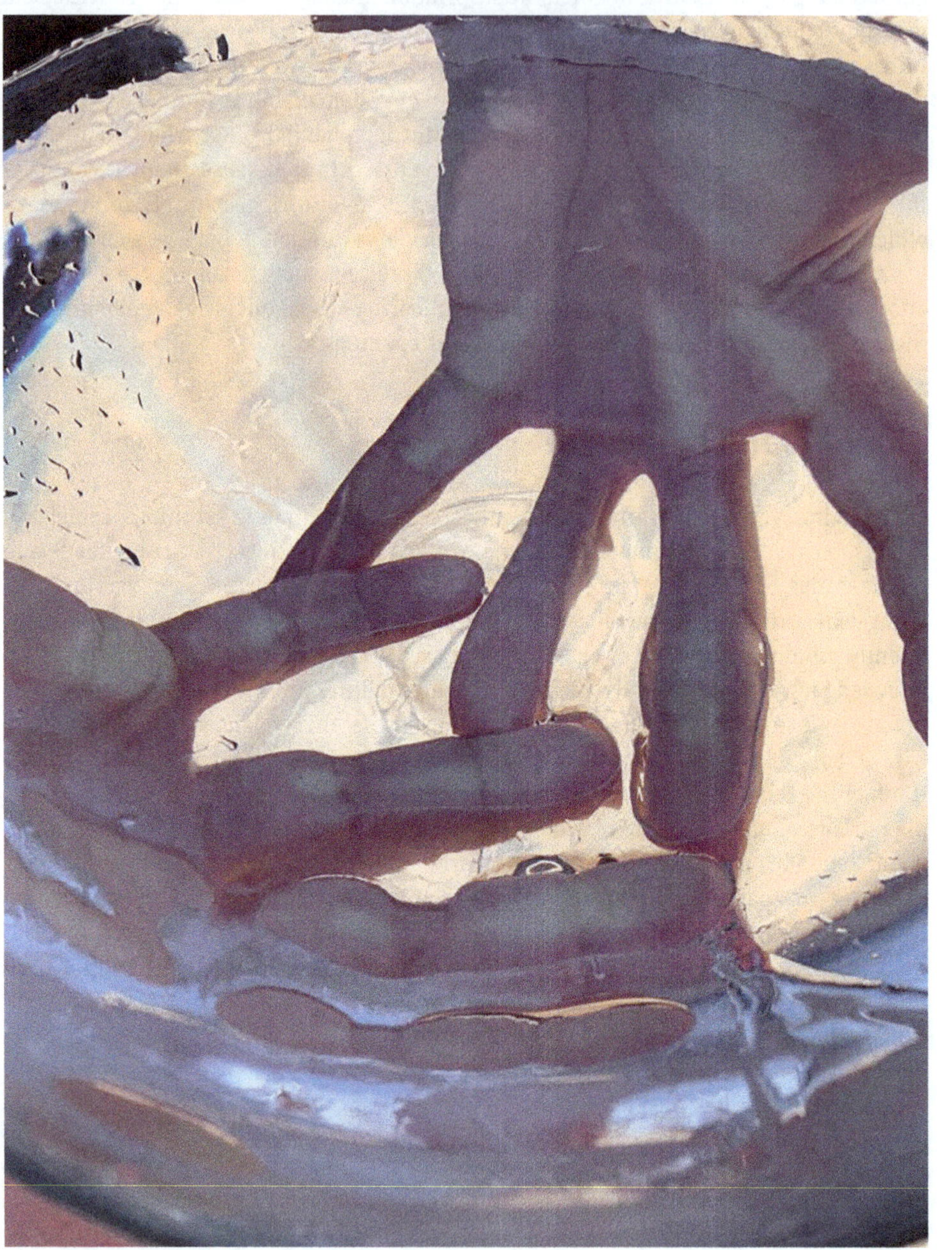

Fig. 3: Watery digits at the Venetian *sestiere* of Dorsoduro.

developed a walkshop exercise on carrying open water through Venice to engage in the activity of *walking as research*—a collaborative, material, and political practice.[35] As transcorporeal encounter, the mundane activity of carrying allowed us to tune into the effects of such a simple gesture on our movement, our senses, and the experience of our surroundings and vice versa—in short: to explore the way we carry ourselves while carrying water being carried by the waters of Venice. The communal act of carrying an open vessel filled to the brim with Venetian water through the city would soon lead to spillage as the vessel was passed between hands: liquid handshakes among the group, residents of Venice, and fellow nonhumans. This exercise in hydro-feminist kinship revealed that practices of displacement and filtering stick to the multiple bodies (whether made of flesh, compromised water, or creeping climate) in Venice. They manifested infrastructurally and transcorporeally as we decided to dissolve the tropes of blissful spectacle of this city (whether romantic or catastrophic) and simultaneously ushered a nuanced experience of friction in the lagoon environment.[36]

Carrying is caring. If care is thought of in this way, we also come to acknowledge via Neimanis that we ourselves are leaking bodies of water.[37] In that sense, any caring activity has to be conceived as something beyond maintenance, but an articulation, a way of thinking as Puig de la Bellacasa puts it, a joint exposure that makes a plea for spillage and wetness.[38]

I brought three bottles back to Berlin that I filled with different bodies of water available in Venice: lagoon, tap, store bought. Just from looking, one is unable to tell the difference between them and assess what their contents might cause when ingested. Two of them are considered drinkable, the other one is a polluted mix of salt water and fresh water. I assume that few would be willing to take their chances in picking the safe one (fig. 4).

---

35 This collaborative walkshop drew inspiration from a conversation with Astrida Neimanis during the Anthropocene Campus Venice and their foundational work on walkshopping together with environmental artist Perdita Phillips. See Astrida Neimanis and Perdita Phillips, "Postcards from the Underground," *Journal of Public Pedagogies*, no. 4 (November 2016). https://doi.org/10.15209/jpp.1181. For the broader methodological framework of walking as research and its material, performative and political implications as a queer, more-than-representational, and radical practice, see Stephanie Springgay and Sarah E. Truman, *Walking Methodologies in a More-than-Human World: WalkingLab* (New York: Routledge 2017).
36 For a comprehensive account of the walkshop activities in terms of spillage, exposure, and the futurity of post-Anthropocene thinking, see Caison and Perraudin, *To Carry Water.*
37 See Neimanis, *Bodies of Water,* 29.
38 See Puig de la Bellacasa, *Matters of Care,* 72. Together with Clemens Winkler, I have discussed elsewhere the uptake of care as a practice, motive, and commitment in humanities and design research discourse in terms of their shared *and* conflicting ways of worldmaking, institutional and material legacies, and argued for a nuanced reading of care as related to questions of li(v)ability. See Léa Perraudin and Clemens Winkler, "Designing with Care? A Pending Question," in *Material Trajectories: Designing with Care?* ed. Léa Perraudin, Clemens Winkler, Claudia Mareis, and Matthias Held (Lüneburg: Meson, 2023), 15–29.

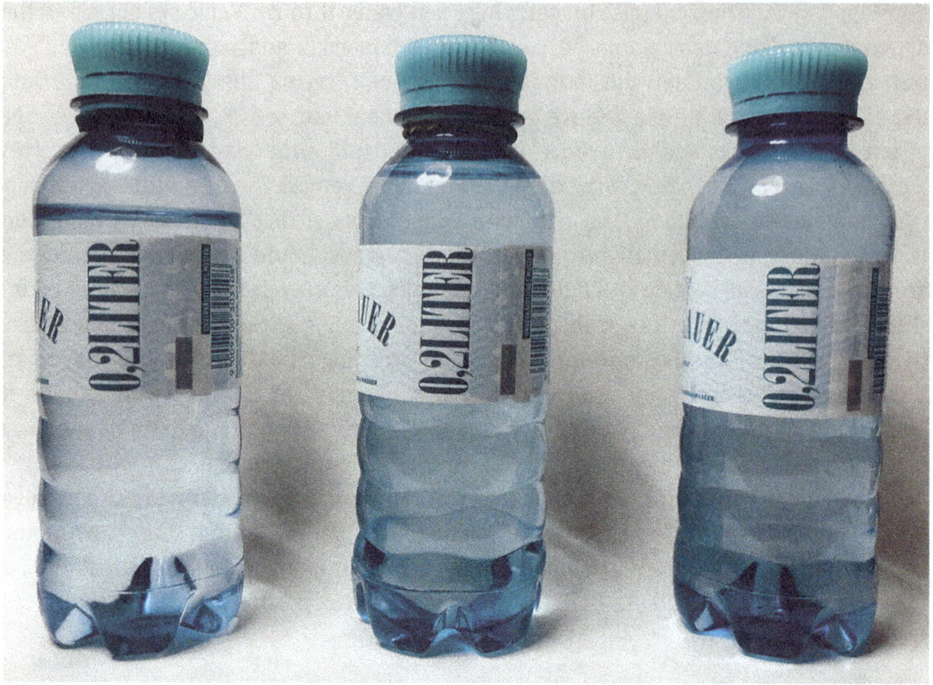

Fig. 4: Three bottles: lagoon, tap, store bought

The question of what it actually means to be refreshed and replenished by a body of water, entails asking about availability, access, safety—immediate questions of privilege that in turn apply to myself, as I am residing in institutionalized spaces of academia, above water level, being minorly affected by pollutants in my surroundings. I am involved in another kind of surface tension during a research stay that is funded by the German Research Foundation while engaging in academic discussions and workshop activities conducted by Venetians and long-term residents in Venice. My time spent in the lagoon city is characterized by the attempt to move and sense otherwise.[39] *Yet, how deeply might one get submersed in a limited time; what is the measure of said depth and what is the medium of submergence? How do the waters of Venice mediate temporal frictions as they play out at the fringes of academic knowledge, material poetics and geopolitic realities? After rowing in the lagoon and swimming at Isola di Sant'Erasmo, after eating Salicornia and after taking a sip of compromised water, after writing about pollution, displacement, subsidence, care, and embodied practices of research, I still know little about the lived experience of Venice and speak from a position of convenience and choice of what I address and what I strategically or involuntarily overlook during a temporal stay. Through the well-planed framework of the program, I was granted access to many experiences that tourists and residents might never engage in. My first time in*

---

39 See Caison and Perraudin, *To Carry Water.*

Venice has been shaped by those experiences and simultaneously has left me mostly insensitive to this point of what makes up the aquatic specificity of the Venetian lagoon in terms of its history of wealth and power, and how water was rendered a mediator of practices of separation, violence, and extraction as they significantly contributed to what I encountered and need to be addressed as surface tensions themselves.

I would like to thank Gina Caison for the collaborative work on Carrying Water that served as a point of departure for the broader framework of this paper. My gratitude also is expressed to Yoonha Kim for our valuable conversations on filtering, frictions and fieldnotes and her careful editorial advice as well as Ise Schubert for her insightful comments and expertise on the Venetian imaginaries in action.

Material Memories

Claudia Mareis

Rubber Frictions: Material Legacies beyond the 1851 Great Exhibition

Introduction

The notion of "active matter" has gained momentum in a variety of disciplines, ranging from materials science to cultural studies, philosophy, architecture, and design.[1] On the one hand, new digital fabrication technologies and the development of "smart" materials have increased interest in materiality and blurred the boundaries between the biological and the built environment, between analogue and digital structures and production techniques.[2] On the other hand, in relation to these technoscientific developments, neo-materialist movements, especially feminist and post-humanist theories, provided important impulses for a novel understanding and critical investigation of matter.[3] The understanding of "active materials"[4] and "vibrant matter,"[5] as proposed by new materialist scholars, has broadened the focus "to the non-human or more-than-human, and the biological and ecological dimensions of life matters" and has challenged us "to consider how non-human bodies or matters might contribute to their own actualization."[6] Ultimately, questions of material epistemologies and material onto-politics have gained further importance through the far-reaching post- and decolonial critique of the hegemonic Western knowledge system and its attempts of (un)making worlds within and through colonialist-modernist power systems.[7] This in-

---

1 I thank Michaela Büsse, Sria Chatterjee, and the reviewer of this text, whose valuable comments helped improve my argumentation.

2 On these developments see, e.g., Paola Antonelli, ed., *The Neri Oxman Material Ecology Catalogue* (New York: The Museum of Modern Art, 2020); Seetal Solanki, *Why Materials Matter: Responsible Design for a Better World* (Munich: Prestel, 2019); Skylar Tibbits, ed., *Active Matter* (Cambridge, MA: MIT Press, 2017); Rivka Oxman and Robert Oxman, *The New Structuralism: Design, Engineering and Architectural Technologies* (Hoboken, NJ: Wiley, 2010).

3 For an overview on new materialist writings and positions see, e.g., Iris van der Tuin and Rick Dolphijn, ed., *New Materialism: Interviews & Cartographies* (Ann Arbor: Open Humanities Press, 2012); Diana Coole and Samantha Frost, "Introducing the New Materialisms," in *New Naterialisms: Ontology, Agency, and Politics*, Diana Coole and Samantha Frost, ed. (Durham, NC: Duke University Press, 2010), 1–46.

4 Peter Fratzl, Michael Friedman, Karin Krauthausen, and Wolfgang Schäffner, ed., *Active Materials* (Berlin: De Gruyter, 2021).

5 Jane Bennett, *Vibrant Matter: A Political Ecology of Things* (Durham, NC: Duke University Press, 2010).

6 Astrida Neimanis, "Material Feminisms," in *Posthuman Glossary*, Rosi Braidotti and Maria Hlavajova, ed. (London: Bloomsbury, 2018), 243.

7 Walter D. Mignolo, *The Darker Side of Western Modernity: Global Futures, Decolonial Options* (Durham, NC: Duke University Press, 2011); Aníbal Quijano, "Coloniality and Modernity/Rationality," *Cultural Studies* 21, 2–3 (2007): 168–78.

cludes the specific ways in which matter—both human and more-than-human bodies —has been treated as "cheap thing" and "raw material."[8]

In this article, I will approach material histories and legacies in a way that understands so-called "raw materials" not just as "naturally" given, passive entities or as starting points for industrial production. Rather, I will consider this complex as "active matter," that is, as frictional socio-material processes of becoming.[9] By revisiting the historical event of the Great Exhibition of 1851 in London and contemplating the vulcanized rubber exhibits on display there, I will discuss and problematize "the conceptual and material apparatuses that *make* matter raw," as historian Adelene Buckland put it.[10] These apparatuses link the vulcanization of rubber, a chemical intervention at the molecular level, not only to industrial commodification and capitalist free trade in the mid-nineteenth century, but even more so to colonial power, forced labor, and the "culture of terror"[11]of the Amazonian rubber boom around 1900. In what follows, I will argue that vulcanized rubber, more than simply being a technically enhanced "raw material," must be seen as a site of conflicted material onto-politics, of exploitation, and oppression, where living bodies are to be transformed into passive industrial commodities. Drawing on my background in cultural history and design, I will pay particular attention to the ways in which the raw material of rubber has been discursively framed in discussions of style and taste to attempt moral comparison and cultural othering.[12] I am less interested, however, in what "new" knowledge or industrial "inventions" were generated by Western rubber technicians or designers at the time than in what *other* Indigenous knowledges were expropriated and marginalized, and how the rubber industry's violent past lives on both in intergenerational trauma and transformative futures.[13]

---

8 Patel Raj and Jason W. Moore, A History of the World in Seven Cheap Things: A Guide to Capitalism Nature and the Future of the Planet (London: Verso, 2018).

9 Claudia Mareis, Helen Pinto, Amanda Winberg, and Emile De Visscher, "Rubber Violence," in *Material Legacies. DESIGN LAB #13*, Michaela Büsse, Amanda Winberg, and Helen Pinto, ed. (Berlin: Kunstgewerbemuseum Berlin, 2023), 6–7.

10 Adelene Buckland, "Introduction to Volume 1: Raw Materials," in *Victorian Material Culture. Volume 1: Raw Materials*, ed. Adelene Buckland (London: Routledge, 2023): 18–34, 22–23 (original emphasis).

11 Michael Taussig, "Culture of Terror – Space of Death: Roger Casement's Putumayo Report and the Explanation of Torture," *Comparative Studies in Society and History* 26, no. 3 (1984): 495.

12 Although I examine a somewhat different constellation in my work, I would still like to refer to art historian Debora L. Silverman's important work on Belgian Art Nouveau as "imperial modernism," created with raw materials from the Congo and inspired by the local culture there. See Debora L. Silverman, "Art Nouveau, Art of Darkness: African Lineages of Belgian Modernism," *West 86th: A Journal of Decorative Arts, Design History, and Material Culture*, Part I: 18, no. 2 (2011), 139–81; Part II: 19, no. 2 (2012): 175–95; Part III: 20, no. 1 (2013), 3–61.

13 As a white, Central European scholar, I am aware of my situated perspective and privileged position when it comes to referring to non-Western, Indigenous knowledge in an academic context. Although the reference to this knowledge in what follows is made in order to disrupt the heroic narratives of progress in European and North American industrial history, it still runs the risk of being reappropriated and

Though historical in nature, my text is guided by some conceptual reflections by ethnographer Anna Tsing. In her ethnographic study on the multiple uses of the Indonesian rainforest, she alludes to the conflicted history of rubber by building her argument around the dynamics of what she calls "friction." As she elaborates: "Consider rubber. Coerced out of indigenous Americans, rubber was stolen and planted around the world by peasants and plantations, mimicked and displaced by chemists and fashioned with or without unions into tires and, eventually, marketed for the latest craze in sports utility vehicles. ... Industrial rubber is made possible by the savagery of European conquest, the competitive passions of colonial botany, the resistance strategies of peasants, the confusion of war and technoscience, the struggle over industrial goals and hierarchies, and much more that would not be evident from a teleology of industrial progress. It is these vicissitudes that I am calling friction."[14]

In essence, Tsing claims that the concept of friction "refuses the lie that global power operates as a well-oiled machine," it rather indicates "where the rubber meets the road."[15] That is to say, friction is the moment when *matter*, both physical and social, collides in a seemingly universal system of both resistance and commodification and leaves its traces behind. In this sense, the introduction of vulcanized rubber at the Great Exhibition of 1851 and its promotion as essential raw material for Western industrialization and mass production is less characterized by the linear logic of techno-scientific progress, or the "smooth operation of global power,"[16] but by the causes and effects of friction. Engaging friction as a tool to navigate material histories and legacies might not only lead to a different way of looking at the sticky matter of rubber, but might also help to unsettle a *transformative* potential that is clinging to the very idea of raw material as incommensurable "active matter."[17]

---

objectified within the "modernist/colonialist" knowledge-power system that produced these narratives in the first place.

14 Anna Lowenhaupt Tsing, *Friction: An Ethnography of Global Connection* (Princeton: Princeton University Press, 2004), 6.

15 Ibid.

16 Ibid.

17 I owe this thought to Adelene Buckland's great introduction on raw materials in Victorian material culture, which inspired my writing a lot. In this regard I would also like to thank Kaja Ninnis for pointing me to this text. See Buckland, "Introduction to Volume 1: Raw Materials."

Revisiting the Great Exhibition and Its Classification System

The Great Exhibition of the Works of Industry of All Nations of 1851 is considered such a thoroughly researched event that new insights are hardly to be expected.[18] However, nineteenth-century historians Louise Purbrick and Adelene Buckland have recently called for the reexamination of this (and similar) international trade fairs from the perspective of material culture, especially with regard to so-called raw materials. According to Purbrick, "reflecting on raw materials offers an alternative to many of the existing histories of [the] Great Exhibition of 1851" since it allows for "a writing of a world history that re-evaluates the significance of the substances of which the earth itself is comprised and the consequences of their movement across its surface."[19] In doing so, she argues, one might be able to expand and disrupt the imperial, industrial, and capitalist narratives installed by and around this exhibition.[20]

Buckland, on the other hand, deconstructs the category of the "raw material" as such in her highly illuminating text on Victorian material culture. Borrowing from new materialist scholar Karen Barad,[21] she argues that "the raw material is a particularly explicit example of the 'material-discursive' construction of all matter," which "can only *exist as a raw material* through the 'intra-actions' of imperial and industrial networks of power."[22] She thus problematizes the fact that the category of the raw material is usually taken for granted (even in new materialist theories), instead of being revealed as a discursive construct, or more precisely, as the colonial and capitalist "fantasy" of turning living bodies into passive matter. "[The] fantasy of the raw material", she argues, "is a fantasy of the earth and of almost all living things as at least potentially devoid of agency, purpose or narrative of their own, but on the cusp of a narrative transformation into capital – a transformation always to be enacted from without."[23] Before I elaborate further on this point with regard to the raw material of rubber and its frictional material legacies, I would like to briefly introduce the histor-

---

18 On this point, see Alexander C. T. Geppert, "Welttheater: Die Geschichte des europäischen Ausstellungswesens im 19. und 20. Jahrhundert: Ein Forschungsbericht," *Neue Politische Literatur*, 1 (2002): 10–61, 40.

19 Louise Purbrick, "The Raw Materials of World History: Re-visiting the Great Exhibition's Objects," *World History Connected* 13, no. 1 (2016), https://worldhistoryconnected.press.uillinois.edu/13.3/forum_01_purbrick.html.

20 A recent example of such a re-examination of Victorian material culture along raw materials such as coal is the following article: Colin Fanning, "'The Indispensable Agent': Coal and Its Displacements in Victorian Britain," *Journal of Design History* 34 no. 3 (2020): 212–226.

21 See Karen Barad, Meeting the Universe Half Way: Quantum Physics and the Entanglement of Matter and Meaning (Durham, NC: Duke University Press, 2007).

22 Buckland, "Introduction to Volume 1: Raw materials," 22.

23 Ibid.

ical site of the Great Exhibition by focusing on its symbolic and material configuration.[24]

Organized by the Royal Society for the Encouragement of Arts, Manufactures and Commerce under the patronage of Prince Albert, The Great Exhibition of the Works of Industry of All Nations was the first world trade fair of its kind:[25] a massive effort to showcase the achievements of the industrialized nations of the time while ostensibly promoting free trade, knowledge and cultural exchange, as well as "the unity of mankind."[26] Over six million people came to witness the spectacular event, many of them using the newly built railway system for the first time in their lives.[27] The exhibition, which lasted from May to October 1851, took place in the Crystal Palace: a monumental modular building almost entirely made of cast iron and glass sheets, designed by garden architect Joseph Paxton.[28] Being the largest greenhouse of the time,[29] the Crystal Palace quickly became a symbol for the triumph of modern industry and engineering as well as for Victorian culture (and its decay) (figs. 1, 2).[30]

Although the exhibition was propagated under the auspices of international pacifism and free trade it was also shaped, according to historian Jeffrey Auerbach, by "strident nationalism and even racism,"[31] "fierce nationalistic competition and xenophobia in international affairs; the persistence of aristocratic power; the increasing division of British society into classes; the beginnings of British economic decline; and, the entrenchment of racial attitudes that would characterize the expanding British em-

---

24 Overviews on the Great Exhibition include Louise Purbrick, *The Great Exhibition of 1851: New Interdisciplinary Essays* (Manchester: Manchester University Press, 2001); Jeffrey Auerbach, *The Great Exhibition: A Nation on Display* (New Haven: Yale University Press, 1999); John E. Findling and Kimberly D. Pelle, *Historical Dictionary of World's Fairs and Expositions 1851–1988* (Westport: Greenwood Press, 1990). Paul Greenhalgh, *Ephemeral Vistas. The "Expositions Universelles," Great Exhibitions and World's Fairs, 1851–1939* (Manchester: Manchester University Press 1988); Utz Haltern, *Die Londoner Weltausstellung von 1851: Ein Beitrag zur Geschichte der bürgerlich-industriellen Gesellschaft im 19. Jahrhundert* (Münster: Aschendorff, 1971).

25 For the origins and prehistory of the World's Fair, see Auerbach, *The Great Exhibition*, 9–31; Greenhalgh, *Ephemeral Vistas*, 3–26.

26 *Great Exhibition of the Works of Industry of all Nations, 1851: Official Descriptive and Illustrated Catalogue* Vol. 1 (London: Spicer Brothers, 1851), 3, https://www.e-rara.ch/zut/content/structure/6726808.

27 Hermione Hobhouse, "3. The Legacy of the Great Exhibition," *RSA Journal* 143, 5459 (1995): 49.

28 For more details on the Crystal Palace, see Folke T. Kihlstedt, "The Crystal Palace," *Scientific American*, 251, 4 (1984): 132–43; Dustin Valen, "On the Horticultural Origins of Victorian Glasshouse Culture," *Journal of the Society of Architectural Historians*, 75, 4 (2016): 403–23.

29 "It covered 772,824 square feet (about 19 acres) in plan. It was 1,848 feet long by 408 feet wide and had an addition on the north side measuring 936 by 48 feet. Its longitudinal central aisle, the 'main avenue,' was 72 by 66 feet high, and its vaulted transept was 72 by 108 feet high." See: Kihlstedt, "The Crystal Palace," 133.

30 On the changing reception and memory culture of the Great Exhibition and the Crystal Palace, see Jeffrey Auerbach, "The Great Exhibition and Historical Memory," *Journal of Victorian Culture* 6, 1 (2001): 89–112.

31 Ibid., 99.

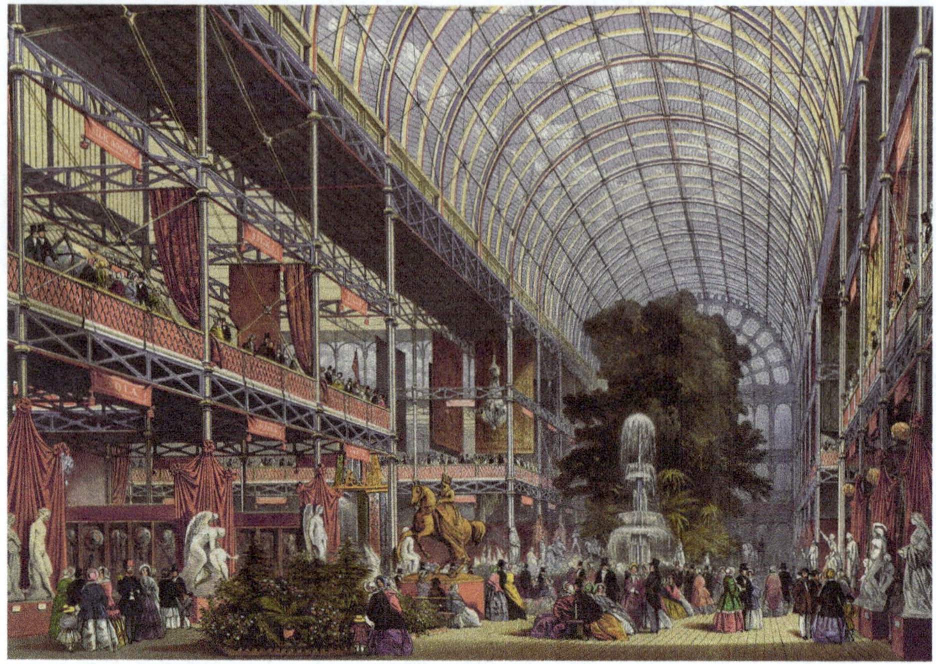

Fig. 1: Inside the Great Exhibition of 1851 in London. *Dickinsons' Comprehensive Pictures of the Great Exhibition of 1851.* Dickinson Brothers, Her Majesty's Publishers, 1852. https://doi.org/10.5479/sil.495268. 39088008102741.

pire."[32] Moreover, with the strong presence of British colonies and dependencies, the Great Exhibition has been characterized as an "imperial display,"[33] a battleground for cultural othering and moral comparison. In particular, the large Indian court highlighted the importance of the colonies to British success in trade and industry (fig. 3).[34] The Great Exhibition presented itself as a dynamic, yet conflicted site, to quote cultural historian Catherine Hall, where "identities were . . . constructed in a process of mutual constitution" and "the making of self" was realized "through the making and marking off of others."[35]

At the center of these cultural negotiations were the material exhibits—how they were produced, selected, classified, displayed, and evaluated. But more importantly—

---

32 Ibid., 107–108.

33 See: Greenhalgh, *Ephemeral Vistas*, 52–81.

34 On the Indian representation at the Great Exhibition, see Carol A. Breckenridge, "The Aesthetics and Politics of Colonial Collecting: India at World Fairs," *Comparative Studies in Society and History* 31, 2 (1989): 195–216; Saloni Mathur, *India by Design: Colonial History and Cultural Display* (Berkeley: University of California Press, 2007), 14–17.

35 Catherine Hall, "Culture and Identity in Imperial Britain," in *The British Empire: Themes and Perspectives*, ed. Sarah E. Stockwell (Oxford: Blackwell Publishing, 2008), 203.

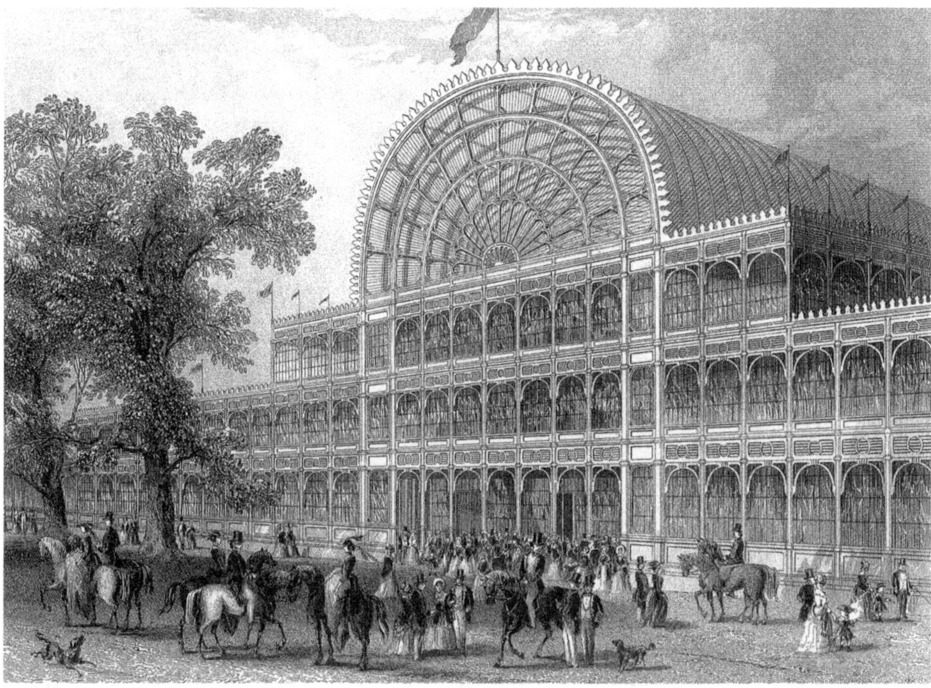

Fig. 2: Exterior of the north transept of the Crystal Palace, London, built for the Great Exhibition, 1851, artist unknown.

what was not shown, suppressed, marginalized, or hidden from the glittering halls of the Crystal Palace. In bringing together almost 14,000 exhibitors and more than 100,000 exhibits from commerce, industry, arts, and science,[36] the Great Exhibition introduced a new perceptual regime for the eye and the senses.[37] It created a "phantasmagoria of capitalist culture,"[38] to use Walter Benjamin's words. While linked to an educational idea,[39] a new form of public sphere emerged, in which the gesture of exhibiting was intended to prepare the visitors for their role as consumers in the industrial modernity that was to come. In terms of design history, the Great Exhibition had a lasting impact on how de-

---

36 Auerbach, The Great Exhibition of 1851, 91.

37 On this point, see Monika Wagner, "Die Erste Londoner Weltausstellung als Wahrnehmungsproblem, " in *Ferrum: Nachrichten aus der Eisenbibliothek 66* (Schaffhausen: Stiftung der Georg Fischer AG, 1994), 38. See also the basic text on this subject: Tony Bennett, "The Exhibitionary Complex", *New Formations*, 4 (1988): 73–102.

38 Walter Benjamin, "Paris, the Capital of the Nineteenth Century," in *Walter Benjamin: The Writer of Modern Life. Essays on Charles Baudelaire*, ed. Michael W. Jennings (Cambridge, MA: The Belknap Press of Harvard University Press, 2006), 37.

39 On public accessibility of institutions such libraries, museums, and world fairs in the nineteenth century, see Jürgen Osterhammel, *The Transformation of the World: A Global History of the Nineteenth Century* (Princeton: Princeton University Press, 2014), 7–15.

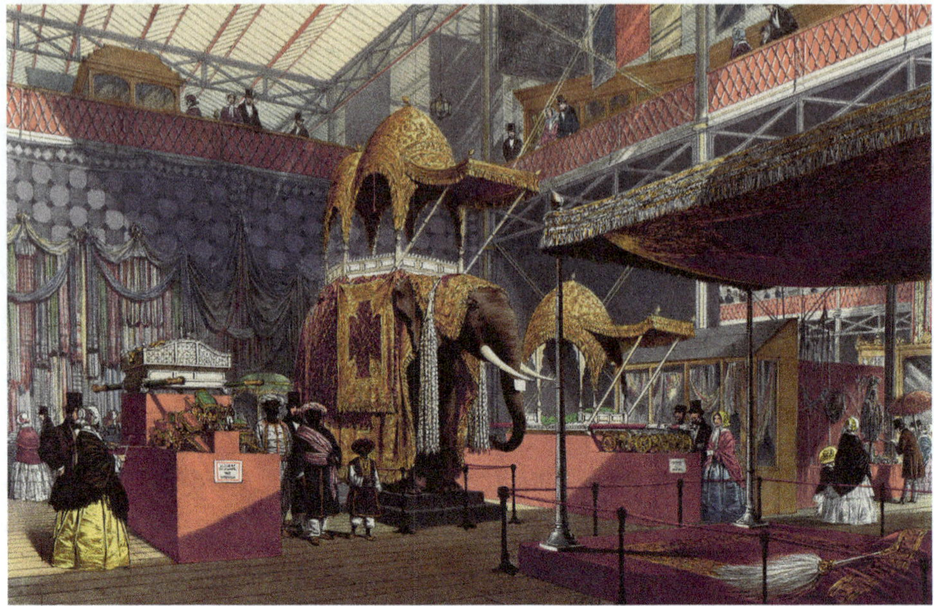

Fig. 3: Exhibition gallery representing India. *Dickinsons' Comprehensive Pictures of the Great Exhibition of 1851.* Dickinson Brothers, Her Majesty's Publishers, 1852.

sign was conceived as an aesthetic practice and commercial activity. It introduced, for the first time, "a broader lay audience to the industrial arts of the globe" and made design a site for "enchantment, edification, and entertainment," design historian Lara Kriegel says.[40] Being called "a lesson in taste" by contemporary observers,[41] the event sparked controversial discussions on the aesthetic and material quality of the manufactured products on display and helped set the standards of modern Western design education.[42]

The exhibits included industrial and agricultural machinery (such as steam engines, hydraulic presses, looms), technical instruments and pharmaceutical procedures, weapons, sculptures, domestic appliances, furniture, textiles, furs, stuffed animals, plant seeds, food, and many more things. Also shown were specimens of so-called raw materials, such as rubber pieces, coal lumps, cotton bales, or limestones, which

---

40 Lara Kriegel, *Grand Designs: Labor, Empire, and the Museum in Victorian Culture* (Durham, NC: Duke University Press, 2007), 87–88.

41 Ralph Nicholson Wornum, "The Exhibition as a Lesson in Taste," in *The Art Journal Illustrated Catalogue: The Industry of All Nations 1851* (London: George Virtue, 1851), I–XXII.

42 One direct legacy of the Great Exhibition was the founding of the South Kensington Museum in 1852, which later became the Victoria and Albert Museum, one of the world's leading museum of crafts and design today. It also set the standards for modern Western design education in Great Britain and beyond.

were considered the starting point for various industrial products and processes (fig. 4).[43] While this mixture of exhibits might appear random at first glance, the organizers, together with scientists, had worked out a more or less coherent classification system for a large part of the objects on display (fig. 5).[44] It consisted of thirty classes of objects, allocated to four main sections, that were Raw Materials, Machinery, Manufactures, and Fine Arts. It applied to the exhibits from Great Britain and its colonies and dependencies at the time, which together formed by far the largest part of the exhibition.[45] All the other national courts, however, did *not* follow this system (apparently for reasons of feasibility and logistics), but were individually organized and set up by local committees according to their own aesthetic and pragmatic criteria. Thus, although the classification system did not affect the exhibition as a whole, due to British nationalist and imperialist dominance of the event it played an important role in defining hegemonic material culture in terms of its industrial-commercial usefulness, cultural value, and moral superiority. In doing so, it "went far beyond the older classifications of natural history to unify nature, culture, and industry in one grand system," as historian Jürgen Osterhammel states.[46] It is therefore worth taking a closer look at its basic assumptions and modes of operation.

First of all, the classification system used for Great Britain and its colonies depicted the logic of the ideal manufacturing process: beginning from the raw materials of the earth to their mechanical treatment by machines, the skilled manufacturing process and the artistic refinement.[47] It thus symbolized the material and moral transformation of crude nature into refined cultural objects and industrial commodities. According to Purbrick, raw materials, such as metals, minerals, vegetable and animal substances, belonged "to the most lowly category" within the classification system, since "they begin, and remain behind the event of industrialization."[48] They transcend the logic of industrialization in a peculiar way, given their different geological or biological ontologies and temporalities, and had to be made fit through numerous epistemic, technical, and logistical maneuvers. However, as Purbrick states, the act of creating such a classification system placed its authors in an omniscient position for identifying and mastering the various stages from "the earth's simple substances transformed into sophisticated objects."[49]

The classification system of Great Britain and its colonies also reflected the great enthusiasm for scientific taxonomy typical of the Victorian age. It was a clear reminiscent of earlier universalist projects of the eighteenth century, such as the taxonomy

---

43 See Auerbach, The Great Exhibition of 1851, 92.

44 Ibid.

45 *Great Exhibition of the Works of Industry of all Nations, 1851: Official Descriptive and Illustrated Catalogue* Vol. 1 (London: Spicer Brothers, 1851), 89–106, https://www.e-rara.ch/zut/content/structure/6726808.

46 Osterhammel, The Transformation of the World, 14–15.

47 See Auerbach, The Great Exhibition of 1851, 92.

48 Purbrick, "The Raw Materials of World History", unpaginated.

49 Ibid.

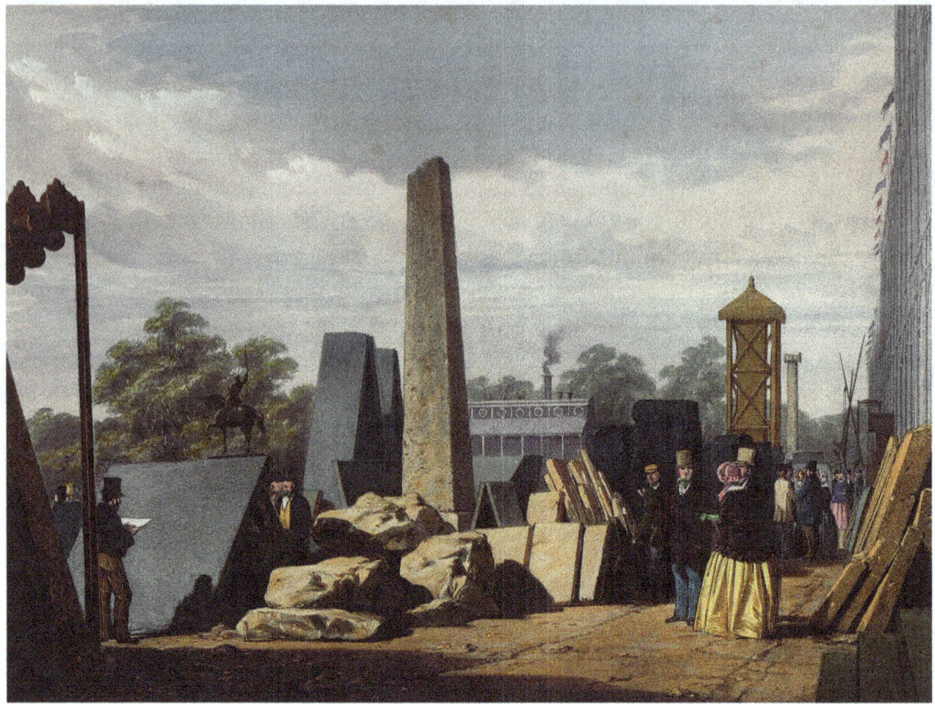

Fig. 4: The exterior of the Great Exhibition, Hyde Park, London. *Dickinsons' Comprehensive Pictures of the Great Exhibition of 1851.* Dickinson Brothers, Her Majesty's Publishers, 1852.

Swedish botanist and zoologist Carl Linnaeus, accused today for laying the grounds of scientific racism,[50] or the *Encyclopédie ou Dictionnaire raisonné des sciences, des arts et des métiers* by French philosophers Denis Diderot and Jean d'Alembert. With the main difference, however, that the classification system of a large part of the Great Exhibition served above all commercial interests, as Auerbach argues. To him, it was "a testament to the power and status of commerce" and showed "that everything in the world could be organized along commercial lines."[51] The colonies in particular served a variety of commercial interests: as source of cheap 'raw material' and human labor, as inspiration for new styles and aesthetic expressions, and as market for products manufactured in the Northern Hemisphere. Paul Greenhalgh summarizes the strong imperial-commercial nexus as following: "Like everything else at the *Great Exhibition*, Empire was a commodity, a thing more important than but not dissimilar to shawls, ironwork, flax, or indeed, sculpture."[52]

---

50 On this point, see Mary Louise Pratt, *Imperial Eyes: Travel Writing and Transculturation* (London: Routledge, 2003), 24–37.

51 Auerbach, The Great Exhibition of 1851, 94.

52 Greenhalgh, *Ephemeral Vistas*, 54.

CLASSIFICATION OF SUBJECTS IN THE THIRTY CLASSES INTO WHICH
THE EXHIBITION IS DIVIDED.

RAW MATERIALS.

CLASS.

I. Mining, Quarrying, Metallurgical Operations, and Mineral Products.
II. Chemical and Pharmaceutical Processes and Products generally.
III. Substances used for Food.
IV. Vegetable and Animal Substances, chiefly used in Manufactures, as Implements, or for Ornament.

MACHINERY.

V. Machines for direct use, including Carriages and Railway and Naval Mechanism.
VI. Manufacturing Machines and Tools.
VII. Civil Engineering, Architectural, and Building Contrivances.
VIII. Naval Architecture and Military Engineering; Ordnance, Armour, and Accoutrements.
IX. Agricultural and Horticultural Machines and Implements.
X. Philosophical Instruments and Processes depending upon their use; Musical, Horological, and Surgical Instruments.

MANUFACTURES.

XI. Cotton.
XII. Woollen and Worsted.
XIII. Silk and Velvet.
XIV. Manufactures from Flax and Hemp.
XV. Mixed Fabrics, including Shawls, but exclusive of Worsted Goods (Class XII.).
XVI. Leather, including Saddlery and Harness, Skins, Fur, Feathers, and Hair.
XVII. Paper and Stationery, Printing and Bookbinding.
XVIII. Woven, Spun, Felted, and laid Fabrics, when shown as specimens of Printing or Dyeing.
XIX. Tapestry, including Carpets and Floor-cloths, Lace and Embroidery, Fancy and Industrial Works.
XX. Articles of Clothing for immediate personal or domestic use.
XXI. Cutlery and Edge Tools.
XXII. Iron and General Hardware.
XXIII. Working in precious Metals, and in their imitation, Jewellery, and all articles of Virtu and Luxury, not included in all other Classes.
XXIV. Glass.
XXV. Ceramic Manufactures, China, Porcelain, Earthenware, &c.
XXVI. Decoration Furniture and Upholstery, including Paper-hangings, Papier Maché, and Japanned Goods.
XXVII. Manufactures in Mineral Substances, used for building or decoration, as in Marble, Slate, Porphyries, Cements, Artificial Stones, &c.
XXVIII. Manufactures from Animal and Vegetable Substances, not being Woven or Felted, or included in other Sections.
XXIX. Miscellaneous Manufactures and Small Wares.

FINE ARTS.

XXX. Sculpture, Models, and Plastic Art.

Fig. 5: Classification of subjects in the thirty classes into which the exhibition was divided. *Great Exhibition of the Works of Industry of all Nations, 1851: Official Descriptive and Illustrated Catalogue, Vol. 1* (London: Spicer Brothers, 1851), 89.

Perhaps more than anything, the classification system introduced to present the material culture of Great Britain and its colonies served as an epistemological apparatus for demonstrating Western superiority and effecting colonial power. It not only played a key role in the attempt to organize, systematize, compare, and evaluate the various national exhibits in terms of their industrial usefulness and advancement, but also in measuring and making (supposed) cultural progress evident on a global scale. After all, a cultural evolutionist view prevailed at the Great Exhibition, according to which the various cultures of the world were still at different stages of civilization—with the industrialized countries, in particular, the British Empire at the forefront.[53] This view, which also applied to the category of the raw materials, was linked to a moral mandate, as Buckland explains: "If Britain was the most successful empire in the world, this was precisely because . . . it had learned to unleash the raw materials of the earth on a scale that had never yet been witnessed and to the supposed benefit of all mankind. In this sense, the raw material as concept materialised an imaginative fusion of both industry and empire, grafting them into a shared narrative of the transformation of the earth to its highest potential."[54]

In the new industrial and capitalist world order, which the Great Exhibition both manifested and enacted, the classification of material culture was shaped through a complex interplay of commercial interests and moral comparison, cultural evaluation, and colonial power. According to anthropologist Arturo Escobar, this was made clear by the exhibition design: "As visitors made their way through the glass cathedral, it became clear to them that not all peoples in the world had achieved the same level of 'development,' for there was no way the arts from 'the stationary East' nor the handicrafts from 'the aborigines' could ever match the 'progress' of the West."[55] In that sense, to him, the Great Exhibition must be seen, not just as an epistemological apparatus, but as a machine "for effecting . . . *coloniality*."[56]

Rubber as Projection Site for Cultural Othering

Universal epistemologies, such as the Great Exhibition's classification system, are powerful tools in rendering invisible other modes of knowing and being that have no place within them. In this sense, the exhibition itself not only symbolized the triumph of Western industry and technology, but contributed with its quasi-scientific setup to the exclusion of "other, non-scientific forms of knowledges and . . . the subaltern social groups whose social practices were informed by such knowledges," as Boaventura de Sousa Santos, João Arriscado Nunes, and Maria Paula Meneses remind us.[57] To elaborate on this point, it is worth considering *rubber* as a particularly exclusive raw material and a site of exploitation and oppression. Although it attracted the attention of many visitors, retailers, and the council's jury members at the Great Exhibition, the

---

53 On this point see, e.g., Paul Young, "Mission Impossible: Globalization and the Great Exhibition," in *Britain, the Empire, and the World at the Great Exhibition of 1851*, Jeffrey A. Auerbach and Peter H. Hoffenberg, ed. (Hampshire/Burlington: Ashgate, 2008), 3–25.
54 Buckland, "Introduction to Volume 1: Raw materials," 19.
55 Arturo Escobar, *Designs for the Pluriverse: Radical Interdependence, Autonomy, and the Making of Worlds* (Durham, NC: Duke University Press, 2018), 31.
56 Ibid. (original emphasis).
57 Santos, Boaventura de Sousa, João Arriscado Nunes, and Maria Paula Meneses, "Opening Up the Canon of Knowledge and Recognition of Difference," in *Another Knowledge Is Possible: Beyond Northern Epistemologies*, ed. Boaventura de Sousa Santos (New York: Verso, 2008 [2007]), xviii.

way this material was staged and received represented only a very limited part of the "pluriversal" material knowledge and ontologies that came with it.[58]

Rubber seemed to be everywhere at the Great Exhibition.[59] The Crystal Palace's exhibition halls were literally flooded by products made out of rubber and ebonite, a hardened form of rubber that served as substitute for the more expensive ebony and ivory.[60] At the time, natural rubber, also known as *caoutchouc* or *India rubber*, was not yet cultivated in plantations in Southeast Asia or Africa, as this was the case at the end of the nineteenth century, but was still harvested from wild rubber trees in the Amazon basin. It was gained from the milky sap of the Pará rubber tree, the *Hevea brasiliensis*, native to the Amazon rain forest by cutting the trunk of the rubber tree and collecting the escaping sticky fluid, called latex, in a vessel. Rubber then is basically a coagulated form of the latex, formed into lumps on site and transported to Europe and North America for further processing and production purposes (fig. 6).

According to the *Official Descriptive and Illustrated Catalogue* these exhibits included, for instance, water taps made of rubber, various wheels and tires, buffers for railway carriages, water-pillows for hot and cold water, waterproof coats and mackintoshes or rubber boats and pontoons (fig. 7). It was "such a general collection of rubber manufactures as the world had never before seen," British rubber manufacturer Thomas Hancock attested.[61] In addition, visitors could also see objects made of gutta percha, a material quite similar to rubber, yet endemic in Malaya.[62] What made rubber so attractive for manufactures, designers, and retailers at the time were its material properties, in particular its elasticity and impermeability to water. Also, rubber proved to be a poor conductor of energy and heat, which is why it was used at the time for the in-

---

58 On the topic of pluriversality, see Blaser, Mario and Marisol de la Cadena, "Pluriverse: Proposals for a World of Many Worlds," in *A World of Many Worlds*, Marisol de la Cadena and Mario Blaser, ed. (Durham, NC: Duke University Press, 2018), 6. See also: Arturo Escobar, *Designs for the Pluriverse: Radical Interdependence, Autonomy, and the Making of Worlds* (Durham, NC: Duke University Press, 2018); Arturo Escobar, *Pluriversal Politics: The Real and the Possible* (Durham, NC: Duke University Press, 2020). On the topic of pluriversal politics see also Arturo Escobar, Designs for the Pluriverse: Radical Interdependence, Autonomy, and the Making of Worlds (Durham, NC: Duke University Press, 2018); Arturo Escobar, Pluriversal Politics: The Real and the Possible (Durham, NC: Duke University Press, 2020).
59 John Tully, *Devil's Milk: A Social History of Rubber* (New York: Monthly Review Press, 2011), 43.
60 It should be noted, however, that the terms "India rubber" or "caoutchouc" were not explicitly mentioned in the classification system under the section Raw Materials, only "Gum Elastics and Gutta Percha" were listed there. See *Great Exhibition of the Works of Industry of all Nations, 1851: Official Descriptive and Illustrated Catalogue*, Vol. 1 (London: Spicer Brothers, 1851), 91, https://www.e-rara.ch/zut/content/structure/6726808.
61 Quoted in John Loadman, *Tears of the Tree. The Story of Rubber: A Modern Marvel* (Oxford: Oxford University Press, 2005), 70.
62 See John Tully, "A Victorian Ecological Disaster: Imperialism, the Telegraph, and Gutta-Percha," *Journal of World History* 20, 4 (2009): 559–79.

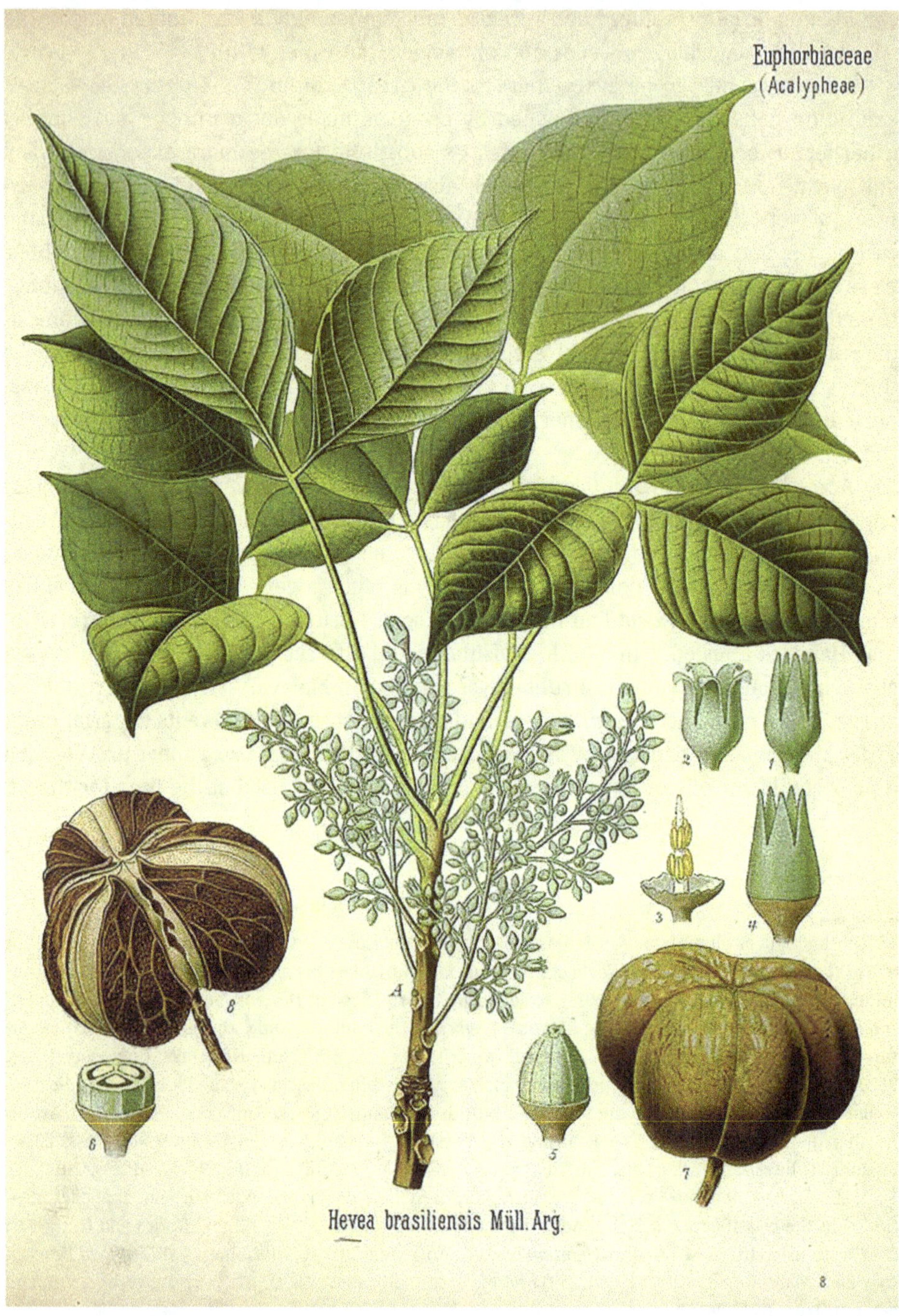

Fig. 6: Hevea Brasiliensis, color lithography after a botanical illustration of Hermann Adolph Koehler medicinal plants, 1887. Published by Gustav Pabst, Koehler, Germany.

sulation of submarine telegraph cables.[63] Last but not least, it seemed to be almost infinitely shapeable and applicable with regard to potential designs and commercial uses.

From an arts and crafts perspective, many Western visitors unfamiliar with rubber wrongly perceived it as a "raw material" without history and tradition. Since rubber came from the so-called New World, to them it was not charged with any known historic Western style or established arts or crafts tradition, but offered allegedly "universal" design possibilities instead. Rubber thus became a surface for ambiguous cultural and moral projections and imaginings. Against the background of industrialization and the associated social and ecological problems in the Northern Hemisphere, discourses around rubber on the one hand reflected a longing for an intact pre-industrial world and a healthy nature. On the other hand, these discourses also created and perpetuated problematic stereotypes about the supposedly pristine, uncultivated state of the rainforest and the Indigenous people's lack of culture and history. Construed as the antithesis of the "civilized," "modern" Western world, rubber became the material embodiment of the "foreign," "wild," "primitive," and "unknown."

Especially for art critics visiting the Great Exhibition, rubber was a material that fascinated and irritated them at the same time. Although rubber proved extremely valuable to Western industry and commerce, and was valued for its many uses, it was not considered a sublime or superior material, like marble, for example, but rather as crude and inferior, amorphous and compliant. Famous German architect Gottfried Semper, who reported on the Great Exhibition from his London exile, called rubber the "monkey among useful materials," and criticized the "peculiar submissiveness" with which this elastic material gives itself away for all purposes, especially for imitations.[64] Semper was probably referring to the expression "the art as the monkey of nature," an old metaphorical figure from the early modern period, which symbolizes the mimetic function of poetry.[65]

---

63 Rolf E. Hummel, *Understanding Materials Science: History, Properties, Applications* (New York: Springer 2004 [1998]), 271.

64 "Ein wichtiger Naturstoff hat erst in neuester Zeit auf dem ganzen weiten Gebiete der Industrie eine Art von Umwälzung hervorgebracht, und zwar vermöge seiner merkwürdigen Gefügigkeit, mit welcher er sich zu allen Zwecken hergibt und leiht. Ich meine das Gummi elasticum oder den Kautschuk, wie er auf Indisch benannt wird, dessen stilistisches Gebiet das weiteste ist, was gedacht werden kann, da seine fast unbegrenzte Wirkungssphäre die Imitation ist." Gottfried Semper, "Die textile Kunst für sich betrachtet und in Beziehung zur Baukunst," *Der Stil in den technischen und tektonischen Künsten oder praktische Ästhetik: Ein Handbuch für Techniker, Künstler und Kunstfreunde*, vol. 1 (Frankfurt: Verlag für Kunst und Wissenschaft, 1860), 112, https://digi.ub.uni-heidelberg.de/diglit/semper1860/0157. The quotes in the text are by the author.

65 See: Roland Borgards, "Affenpoesie: Literatur und Wissen in der Frühen Neuzeit," in *Animali: Tiere und Fabelwesen von der Antike bis zur Neuzeit*, Luca Tori and Aline Steinbrecher, ed. (Milan: Skira, 2012), 261–62. See also Horst Bredekamp, "Der Affe der Natur: Zur Utopie des Spiels," in *Spielwissen und Wissensspiele: Wissenschaft und Game-Branche im Dialog über die Kulturtechnik des Spiels*, Thomas Lilge and Christian Stein, ed. (Berlin: transcript Verlag, 2018), 24.

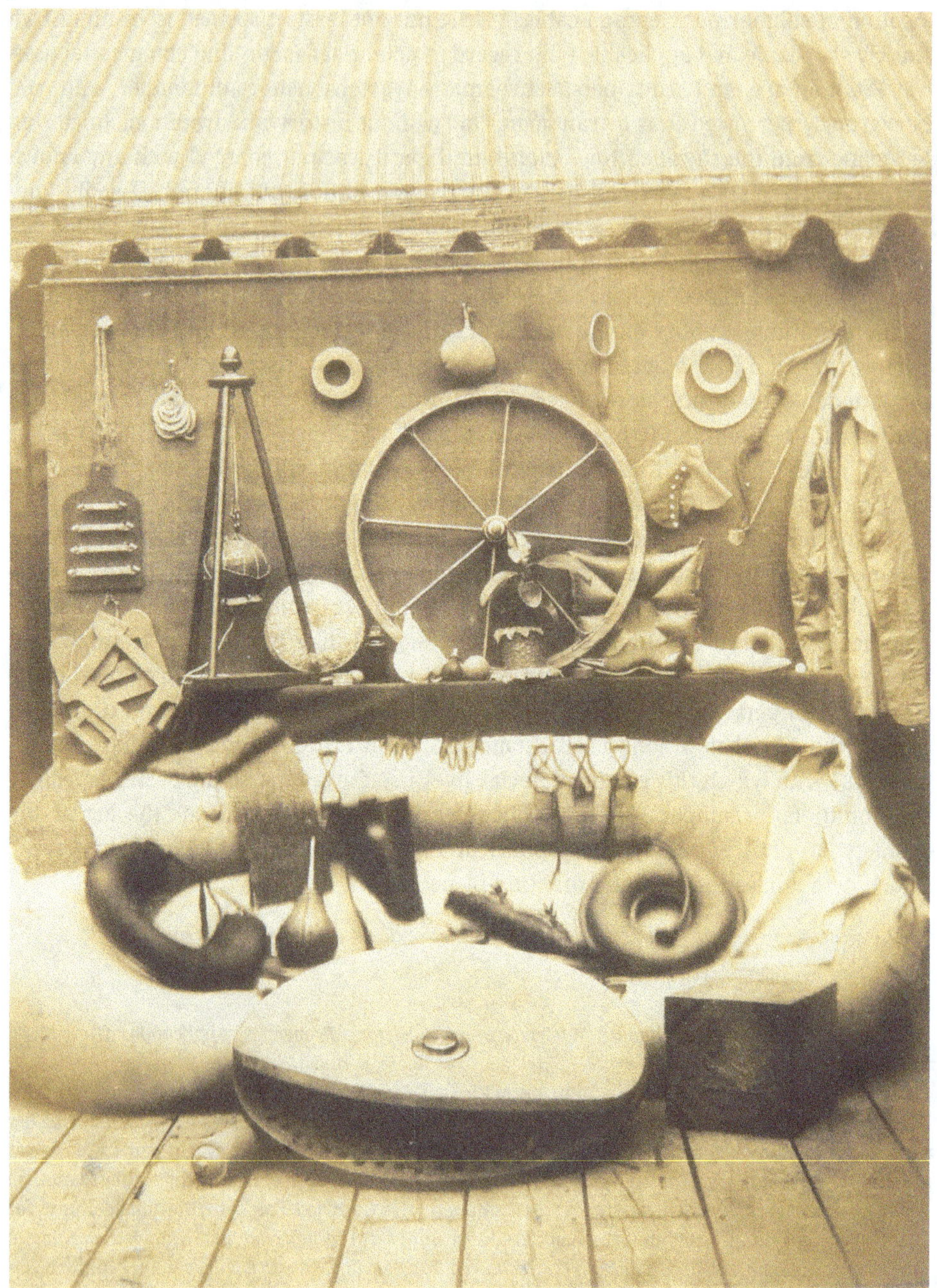

Fig. 7: Articles in Indian rubber by Mackintosh, The Great Exhibition, 1851, salted paper photoprint by Claude-Marie Ferrier.

However, from a cultural history perspective, the image of the monkey—understood as an uncanny borderline creature—also functions as "a strategic element of a differentialist anthropology," as a way to maintain the divine boundary between humans and animals.[66] In this sense, Semper's statement about rubber being the "monkey among useful materials," along with its exoticizing undertone, could well be read as the negotiation of anthropological boundaries and cultural othering, where the monkey as the uncanny "wild other" ensures one's own "civilized humanity."

After having visited the Great Exhibition, Semper was especially concerned about the trend to imitate precious materials, such as ebony and ivory, with cheaper surrogate materials such as rubber: "Rubber and gutta-percha are vulcanized and utilized in a thousand imitations of wood, metal, and stone carvings, exceeding by far the natural limitations of the material they purport to represent," he wrote.[67] Rubber did not come off well in this reading either, as it did not stand for high quality and originality, but rather for inferior imitation and bad taste. However, discussions about the proper artistic handling of a particular material were not limited to rubber but reflected a general interest in the subject of materiality and material culture at the time. A key question in the material debates of the mid-nineteenth century was whether mankind in the age of industrialization—like other important historical epochs before—was able to develop an own epochal style, a distinctive aesthetic language of form and use of material.

Based on aesthetic theories that saw in every material an inherent "gestalt," a predestined form, many artists, architects, and critics searched for an artistic and industrial style that would do "justice" to the laws of the material. Belgian architect and designer Henry van de Velde, for instance, argued that things were beautiful only if they "were made according to logic, according to reason, according to the principles of the rational being of things and according to the exact, necessary and natural laws of the material used for this purpose."[68] Also the German discussion around the term *Materialgerechtigkeit*, which can be translated as "truth to material," arose from the criti-

---

66 "Der Mensch setzt den Affen als Grenze, mittels derer er sich von den Tieren unterscheidet. Der Affe als Grenzwesen ist demnach ein strategisches Element einer differentialistischen Anthropologie, in der sich der Mensch gegenüber den Tieren seiner selbst versichert." Borgards, *Affenpoesie*, 261–62. The quote in the text is translated by the author.

67 Gottfried Semper, "Science, Industry, and Art," in *The Four Elements of Architecture and Other Writings*, trans. Harry F. Mallgrave and Wolfgang Herrmann (Cambridge, UK: Cambridge University Press, 2010), 134.

68 "Ich . . . bemerke, dass es Menschen in unserer Zeit gibt, die Schönes geschaffen haben, einzig deshalb schön, weil die Dinge nach der Logik, nach der Vernunft, nach den Prinzipien des vernünftigen Seins der Dinge und nach den genauen, notwendigen und natürlichen Gesetzen des dazu verwandten Materials hergestellt wurden." Henry van de Velde, "Kunstgewerbliche Laienpredigten (1902)," in *Materialästhetik: Quellentexte zu Kunst, Design und Architektur,* Dietmar Rübel, Monika Wagner, and Vera Wolff, ed. (Berlin: Reimer, 2005), 169. The quote in the text is translated by the author.

cism of the lack of an "appropriate" style for the industrial age and the search for more honest and rational design criteria.[69]

Against the historical backdrop of industrialization, nationalism, and imperialism, material became a "dazzling battle term," according to art historians Dietmar Rübel, Monika Wagner, and Vera Wolff: "In the name of material, both artisanal working methods and more 'honest' machine work or purist ornamentlessness could be fought for."[70] These debates were often linked to a normative pedagogical impetus, in that questions of material were simultaneously used to negotiate questions of national identity, moral superiority, and cultural progress. Especially non-Western raw materials, such as rubber, had a strong symbolic dimension in those material debates. Similar to the way the concept of *Terra nullius*, the supposedly "unowned" and "lawless" land, was used to justify colonization,[71] also the raw materials deported from the colonies were often described this way: as having no considerable history, no indigenous ownership, no proper cultural or aesthetic characteristics, and no material qualities. Rubber was treated as if it were *Materia nullius*, if you will.[72] Stripped off its Indigenous histories, knowledges, and territories, of its cultural and aesthetic attributions, and even of its material properties it seemed to be just a lawless, passive matter waiting to be "discovered" and made morally and commercially "productive" through Western ingenuity.

From Vulcanization to a "Culture of Terror"

The commercial use of rubber in the Northern Hemisphere and its activation as essential raw material for the era of industrialization was only made possible, to recall Adelene Buckland, through "conceptual and material apparatuses that *make* matter raw."[73] This process, however, was not at all a linear process, or a "smooth operation of global power," as Anna Tsing argues, but emerged from frictions caused by slavery, resistance,

---

69 On this point, see Monika Wagner, "'Materialgerechtigkeit': Debatten um Werkstoffe in der Architektur des 19. und frühen 20. Jahrhunderts," in *Historische Architekturoberflächen: Kalk – Putz – Farbe*, ed. Jürgen Pursche (Munich: ICOMOS – Hefte des Deutschen Nationalkomitees 39, 2003), 135–38; Nadine Rottau, *Materialgerechtigkeit: Ästhetik im 19. Jahrhundert*, Forschungsberichte Kunst + Technik, Vol. 2 (Aachen: Shaker, 2012).

70 Dietmar Rübel, Monika Wagner, and Vera Wolff, "Vorwort," in *Materialästhetik: Quellentexte zu Kunst, Design und Architektur*, Dietmar Rübel, Monika Wagner, and Vera Wolff, ed. (Berlin: Reimer, 2005), 10. The quote in the text is translated by the author.

71 See: Stuart Banner "Why Terra Nullius? Anthropology and Property Law in Early Australia," in *Law and History Review* 23, 1 (2005): 95–131.

72 In contrast to the Aristotelian concept of *Materia prima*, the unformed matter, which still contains all the possibilities of becoming form, I would like to use this term *Materia nullis* to emphasize colonialized, racialized modes of material becoming.

73 Buckland, "Introduction to Volume 1: Raw materials," 22–23 (original emphasis).

assertive power, and interpretive hierarchies.[74] Proof of all this is the technology of rubber vulcanization, which is the subject of the following section. In the Crystal Palace, among all rubber manufacturers one stood out in particular. It was inventor Charles Goodyear, who contributed to the exhibition court of the United States of America with a small array of award-winning rubber and ebony products, consisting of black boats and pontoons.[75] They were set up on the middle of an iron bridge constructed by engineer Nathaniel Rider,[76] draped in the shape of a black mountain, and adorned by a red flag (figs. 8, 9). What the visitors were supposed to see, with a little imagination, was a volcano made out of rubber goods with a little flame on top. More important than the rubber goods themselves, however, was a specific chemical treatment of the material, called "vulcanization," which Goodyear presented on the Great Exhibition's world stage.

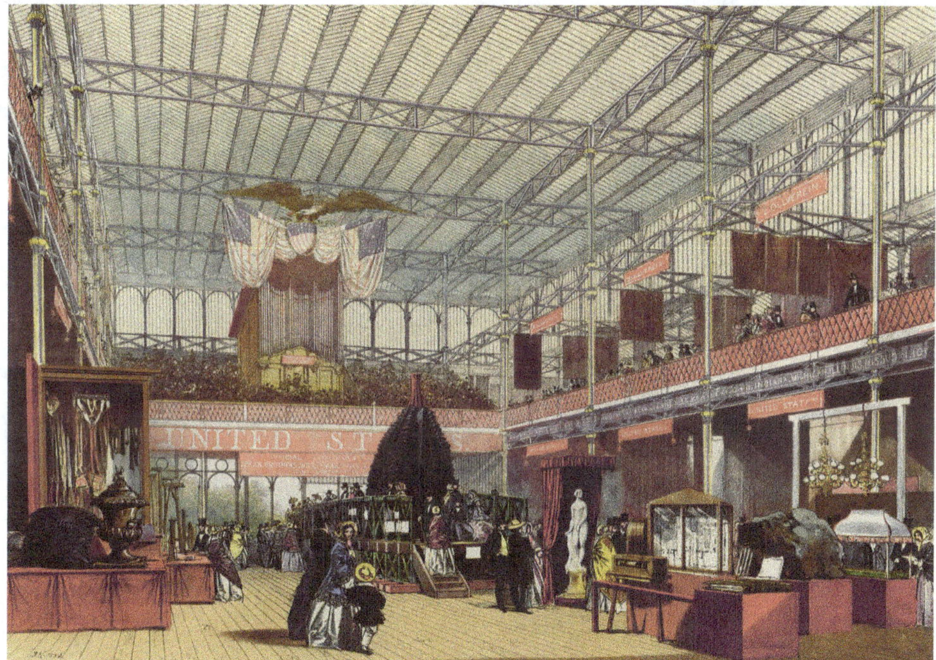

Fig. 8: Exhibition gallery representing America. *Dickinsons' Comprehensive Pictures of the Great Exhibition of 1851.* Dickinson Brothers, Her Majesty's Publishers, 1852.

Vulcanization technology, named after the Roman god of fire, proved to be a veritable game changer in the industrial use of rubber. Before the "invention" of vulcanization

74 Tsing, *Friction*, 6.
75 Marcus Cunliffe, "America at the Great Exhibition of 1851," *American Quarterly* 3, no. 2 (1951): 122.
76 Victor C. Darnell, "The Pioneering Iron Trusses of Nathaniel Rider," *Construction History* 7 (1991): 69–81.

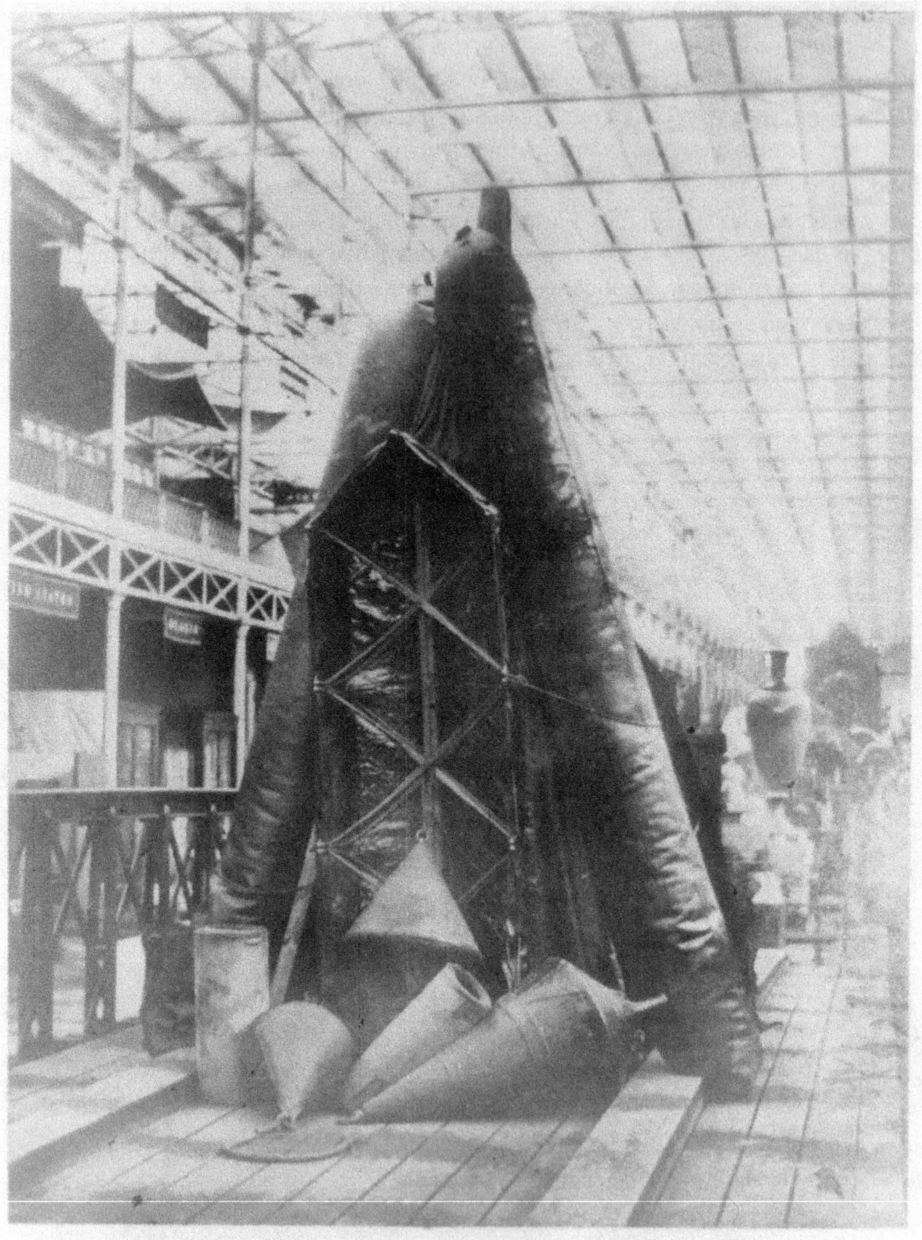

Fig. 9: Charles Goodyear exhibit of vulcanized India rubber products (boat, pontoons, and buoys), 1851, salted paper photoprint by Claude-Marie Ferrier or Hugh Owen.

most attempts to process rubber for larger industrial and commercial purposes in Europe and North America from the mid-eighteenth century onward failed. Although there had been a few rubber applications in the early nineteenth century, such as erasers, elastic bands, mackintoshes or rubber boots, the commercial production and use of this foreign material remained complicated.[77] Despite the fact that latex had been successfully used for thousands of years by the Indigenous peoples in Mesoamerica and South America to make boats and shoes waterproof.[78]

What proved to be challenging for European and North American manufacturers, however, was the inherent activity of this material, or more precisely, their unfamiliarity in handling this material. Due to its active thermoplastic properties, rubber becomes soft and sticky in hot temperatures and brittle in cold. Goods manufactured of untreated rubber thus tended to change their consistency in rather unwanted and unpleasant ways: they became sticky or leaking, started to stink, or even rotted in warehouses. After a series of previous experiments by Friedrich Lüdersdorff in Germany and Nathaniel Hayward in the United States, in 1839 Goodyear eventually found a way to successfully treat rubber with sulfur and heat to eliminate unwanted material properties.[79]

Vulcanization technology was the key to control material activity and transform rubber into a stable, manageable commodity essential to Western industrialization and mass production (fig. 10). In the process of vulcanization, the polyisoprenes of the rubber crosslink with the help of sulfur molecules. Compared to the original material, after vulcanization rubber has permanent elastic properties, this means, it will return to its original position after mechanical stress (fig. 11). In addition, it has a higher tensile strength, can be stretched more easily and is more resistant to weathering and ageing. What vulcanization does to rubber, is in short, that it "increases elasticity while it decreases plasticity."[80] Another name for vulcanization, that is still in use in material science and engineering today, is to "cure" or "curing." To me, this term seems to refer to the idea that the material's inherent natural activity is seen as an "illness" or a "defect" that needs to be "cured" by modern man.

In combination with the masticator, a machine invented by Thomas Hancock in 1821 for shredding rubber waste, and other rubber-related technologies, the technology of vulcanization helped to launch the rubber industry on a large scale in Europe and

---

77 See Tully, *Devil's Milk*, 36–38.

78 Ibid., 29–33; Harp, *A World History of Rubber*, 12–13, Furthermore, see Dorothy Hosler, Sandra L. Burkett, and Michael J. Tarkanian, "Prehistoric Polymers: Rubber Processing in Ancient Mesoamerica," *Science* 284, 5422 (1999): 1988–91; Joan C. Long, "The History of Rubber – A Survey of Sources about the History of Rubber," *Rubber Chemistry and Technology* 74, no. 3 (2001): 493–508.

79 On the genealogy of vulcanization see Tully, *Devil's Milk*, 40; Loadman, *Tears of Trees*, 232–33. On the material science of vulcanization see A. Y. Coran, "Vulcanization," in *The Science and Technology of Rubber*, fourth edition, James E. Mark, Burak Erman, and Mike Roland ed. (Amsterdam: Elsevier, 2013), 337–81.

80 A. Y. Coran, "Vulcanization," 338.

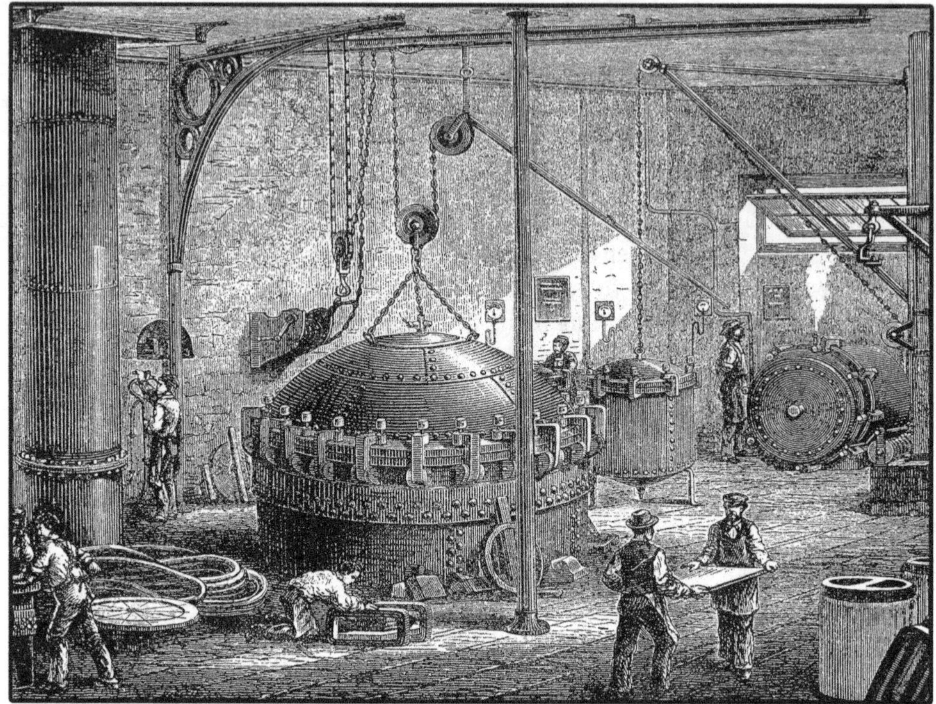

Fig. 10: Historical French illustration from the nineteenth century of a rubber vulcanization machine to convert rubber into durable material through a chemical process.

North America and made rubber the backbone of industrialized Western societies. "Up to this point, the latex had generally been smoke-cured into objects, such as boots or balls, on site in South America, then shipped to Europe or North America," historian of rubber Stephen L. Harp explains.[81] "Now, however, solid balls of smoked rubber could be broken down, easily dissolved, and made into objects in the Northern Hemisphere. Like cotton manufacture, which the British empire largely removed from India over time and installed in Britain, the base of rubber manufacture would similarly move from South America to Europe and the United States."[82]

Over time, the rubber industry radically transformed the modern Western world of consumption, communication, and transportation from submarine cables to pneumatic tires and rubber condoms. Also in the Amazon basin, home to the Pará rubber tree, rubber industry changed everything it came in touch with: it shaped entire habitats and ecologies, caused the rise and decay of extensive networks of modern infrastructure, including splendid trade cities such as Manaus or Belém. It brought fabulous wealth to those who ruled the rubber market and unspeakable suffering to those who

81 Harp, A World History of Rubber, 13.
82 Ibid.

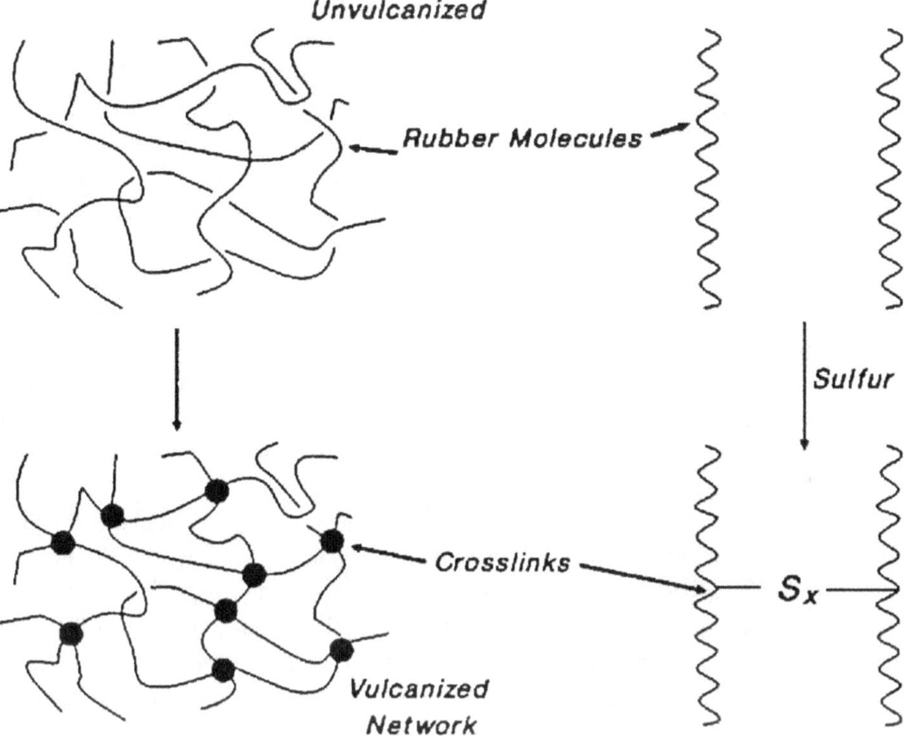

Fig. 11: Molecular network formation through vulcanization. A. Y. Coran, "Vulcanization," in *The Science and Technology of Rubber*, ed. James E. Mark, Burak Erman and C. Michael Roland (Amsterdam et al.: Elsevier, 2013), 337–381, 338.

did the dirty work of manually tapping latex in the forests or processing rubber in factories.[83] The emergent automobile industry, in particular, generated such an enormous demand for natural rubber, that it could only be met by forced labor and exploitative cultivation. In the period from 1860 to 1910, the insatiable greed for rubber in Europe and North America led to the infamous speculative "rubber boom,"[84] with the Amazon rain forest being "the primary source of the finest grades of crude rubber in the world."[85] The interest in Amazonian rubber, however, ended by World War I when Bra-

---

83 Tully, *Devil's Milk*, 23.

84 Barbara Weinstein, *The Amazon Rubber Boom, 1850–1920* (Stanford: Stanford University Press, 1983). See also Tully, *Devil's Milk*, 65–76.

85 Bradford Barham and Oliver Coomes. "Wild rubber: Industrial Organisation and the Microeconomics of Extraction During the Amazon Rubber Boom (1860–1920)," *Journal of Latin American Studies* 26, no. 1 (1994): 40.

zil lost its rubber monopoly to the British colonies in Southeast Asia due to illegally deported rubber seeds, among other reasons.[86]

The "great transformation"[87] brought about by the rubber industry would not have been possible without an extremely violent regime of slavery and forced labor, dispossession, torture and killing.[88] Within the colonial-imperial system of forced labor and slavery, the flesh and the bodies of Black and Indigenous people in particular were treated "as the raw material for infinite forms of use."[89] (This is not to say, however, that in the midst of violence there were not also incommensurable acts of Black and Indigenous resistance as well as spaces of evasion.) Anthropologist Michael Taussig has described the brutal regime of the Amazon rubber industry as "culture of terror. It aimed not only at exploiting the labor of enslaved indigenous and Black rubber tappers but was fueled by a mutual mimesis of cultural othering and fear. Taussig speaks in this context of a "colonial mirror which reflects back onto the colonists the barbarity of their own social relations, but as imputed to the savage or evil figures they wish to colonize"[90] (fig. 12).

Under the brutal regime of *Peruvian Amazon Rubber Company* alone in the Putumayo area, a conflicted site between now Peru and Colombia, estimated 30,000 Indigenous people, mostly Boras and Witotos, died in the late nineteenth and early twentieth century of disease and abuse.[91] Colombian ethnohistorian Juan Alvaro Echeverri recalls how rubber affected the Putumayo as follows: "The total Indian population was reduced to perhaps less than a tenth between 1900 and 1930, and the surviving ones were forcefully resettled on the Putumayo River and further south. A few managed to escape north or to hide in the forest. Their social, political and ceremonial organization was severely shattered, and their territory was depopulated, as the forest regrew in what had been a densely populated region."[92] Also in the Congo Free State, ruled by Belgium's King Leopold II, estimated ten million people died due to rubber violence between 1885 and 1908.[93]

---

86 See: Tully, *Devil's Milk*, 72.

87 Ibid., 41–44.

88 On the interplay of race, migration, and labor within the rubber industry, see Harp, *A World History of Rubber*, 10–39.

89 On this point, see Tiffany Lethabo King, "The Labor of (re)reading Plantation Landscapes Fungible(ly)," *Antipode*, 48, no. 4 (2016): 1022–39, 1028. I owe this reference to Adelene Buckland's introduction to "raw materials" in *Victorian Material Culture*.

90 Taussig, "Culture of Terror," 495.

91 Luisa Abad González, *Etnocidio y resistencia en la Amazonía peruana* (Cuenca: Ediciones de la Universidad de Castilla-La Mancha, 2003), 177 On the rubber related genocides in the Putumayo area as well as in the Congo Free State see, e.g., Tully, *Devil's Milk*, 77–121.

92 Juan Alvaro Echeverri, "The Putumayo Indians and the Rubber Boom," *Irish Journal of Anthropology* 14, no. 2 (2011): 13.

93 An important, albeit controversial, account of the Congo atrocities is Adam Hochschild, *King Leopold's Ghost: A Story of Greed, Terror, and Heroism in Colonial Africa* (Boston: Houghton Mifflin, 1999). A more recent critical discussion of this book is Sarah De Mul, "The Holocaust as a Paradigm for the Congo Atrocities: Adam Hochschild's 'King Leopold's Ghost'," *Criticism* 53, no. 4 (2011): 587–606.

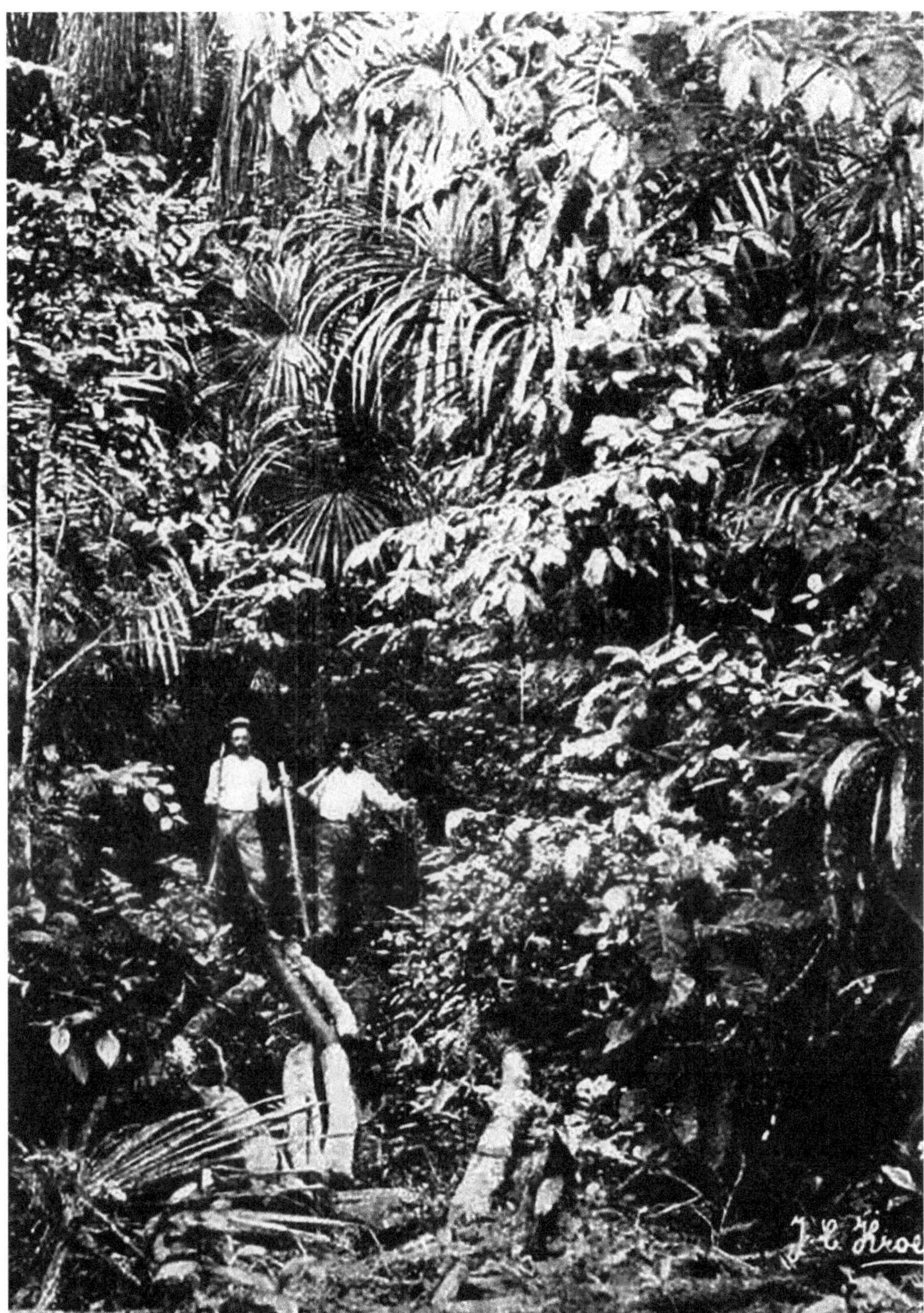

Fig. 12: Historical photograph from the Peruvian Amazon, in Walter Ernest Hardenburg, *The Putumayo: The Devil's Paradise. Travels in the Peruvian Amazon Region and an Account of the Atrocities Committed upon the Indians Therein* (London: Fischer Unwin, 1913), 74.

The transformation of natural rubber into a useful raw material for Western industries must be seen as the result of a long process of effecting colonial power and the displacement of Indigenous knowledge, starting already in the late fifteenth and sixteenth centuries. "Columbus and other Europeans 'discovered' rubber, much as they 'discovered' so much of the flora of the Western Hemisphere and sub-Saharan Africa," Stephen Harps notes, rather "indigenous people introduced it to them."[94] Indigenous rubber knowledge always encompassed many more aspects and different types of knowledge than just its mere utilization. Based on a holistic and spiritual understanding of the living, animated nature,[95] it included precise botanical expertise, biogeological knowledge of the location of latex-bearing trees in the jungle, and a vast technical knowledge of handling latex and producing high-quality latex products.[96] Many of the commercial goods later produced by the European rubber industry, such as raincoats, rubber boots, and rubber toys were "invented" first by Indigenous people in Mesoamerica and South America and only later adopted by Europeans and North Americans.[97] Until the mid-nineteenth century, the Amazon region hosted a sophisticated local rubber industry that in many ways surpassed its European counterpart. At the time of the Great Exhibition, hundreds of thousands of rubber boots were still manufactured in the Pará region and exported to Europe and North America, as they were superior in quality to the products manufactured there.[98]

Long before the invention of vulcanization by Goodyear, there also existed an Indigenous, that is, biological method of curing rubber, comparable in its antibacterial and stabilizing effects.[99] This method consisted of smoking dry latex over a smoldering fire made of tropical branches and nuts: "Smoked indigenous rubber products were both highly elastic and durable. They did not have the 'diseases' that the rubber products made in Europe and North America had," chemist and historian Jens Soentgen says.[100] In fact, many of the material problems that the rubber industry in the Global North had struggled with, such as the stickiness of rubber, resulted from a lack of knowledge in handling this delicate organic material.[101] Moreover, the long shipping to Europe and North America meant that rubber had to be conserved in a way that made further processing much more difficult, compared to the direct handling of fresh latex onsite. In other words, vulcanization was not so much a "cure" for a defective "raw material," but an answer to a problem created by industrialization itself. Seen in this light, vulcanization technology, much-praised at the Great Exhibition of 1851 in London, was less of "a

---

94 Harp, A World History of Rubber, 11.

95 Tully, *Devil's Milk*, 32.

96 Jens Soentgen, "Die Bedeutung Indigenen Wissens für die Geschichte des Kautschuks," *Technikgeschichte* 80, no. 4 (2013): 301.

97 Ibid., 306, 321.

98 Ibid., 313.

99 Ibid., 311–19.

100 Ibid., 315. The quote in the text is translated by the author.

101 Ibid., 319.

grand exemplification of the present state of human industry, and of the efforts of mind."[102] Instead, it was simply a site-specific solution to a problem that had already been solved elsewhere before, with disastrous consequences however.

On the Baskets of Darkness and Light

The case of vulcanization clearly shows that Western rubber technology needed both the appropriation and degradation of non-Western, Indigenous knowledge to stabilize its supposed superior state of cultural progress and civilization. In this regard, Santos, Nunes, and Meneses remind us, that the "production of the West as hegemonic knowledge required the creation of an Other," which was "constituted as an intrinsically disqualified being . . . available for use and appropriation."[103] With the introduction of vulcanization technology in Europe and North America and the industrial rubber boom, Indigenous rubber knowledge and technologies seemed to fade into the background, despite it having been crucial for its rise in the beginning. But this conception is misleading. While the "High-Tech Plundering" and destruction of the Amazon rainforest has continued on a grand scale to this day,[104] Indigenous rubber knowledge has been preserved on site and further developed in the direction of alternative, cooperative and sustainable models of production.[105] The initiative Couro Vegetal da Amazônia, for instance, "brought together more than 200 local and indigenous families in three forest communities, providing training in an innovative processing method to produce sheets of vulcanized rubber."[106] Worth mentioning is also the work of Brazilian feminist environmentalist Bia Saldanha, who engaged over the last decades with local producers in the Chico Mendes reserve in the production of ethically and sustainably sourced rubber. Her work resulted, most famously, in a collaboration with the French company Veja for an eco-friendly vegan sneaker.[107]

---

102 Robert Hunt, "Introduction," in *Hunt's Hand-book to the Official Catalogues: An Explanatory Guide to the Natural Productions and Manufactures of the Great Exhibition of the Industry of all Nations, 1851* (Cambridge, UK: Cambridge University Press 2011 [1851]), vi.

103 Santos, Nunes, and Meneses, "Opening Up the Canon of Knowledge and Recognition of Difference," xxxv.

104 See, e.g., Laymert Garcia dos Santos, "High-Tech Plundering, Biodiversity, and Cultural Erosion: The Case of Brazil," in *Another Production Is Possible: Beyond the Capitalist Canon*, ed. Boaventura de Sousa Santos (London: Verso, 2007), 151–81.

105 For a more comprehensive approach on this topic, see Boaventura de Sousa Santos, ed., *Another Production Is Possible: Beyond the Capitalist Canon* (London: Verso, 2007).

106 "Couro Vegetal da Amazônia," Equator Initiative, https://www.equatorinitiative.org/2017/05/28/couro-vegetal-da-amazonia/.

107 Anna Schnuck, "Der Regenwald ist kurz davor seine Regenerationsfähigkeit zu verlieren," VIERTEL \ VOR (August 23, 2019), http://viertel-vor.com/2019/08/23/regenwald-brand-bia-saldanha-rubber-veja-klima/.

However, in the prevailing Western narratives of rubber, the people who crafted and cared about this "proto-biocapital" in Mesoamerica and South America in the first place,[108] long before the Europeans even knew about it, are still reduced to either primitive savages," "submissive rubber tapers," or "victims" of the rubber industry. But that is just another devaluation of Indigenous knowledge and memory. "In all those tales," Juan Alvaro Echeverri argues, Indigenous people "are not actual subjects but objects of compassion, fear or observation," they "do not have voice but are the objects of disputes among Whites."[109] In contrast, Echeverri attempts to capture the traumatic memory of the genocide perpetrated in the Putumayo region during the rubber boom through the voices of the surviving Indigenous Muinane people and their struggle to heal from the violent past.[110]

To the them, their and their ancestor's territory "is not just a tract of land that can be mapped or legally titled," Echeverri says, but it's a source of memory that "had remained amputated" ever since the rubber industry with its slaughters and massacres hit.[111] From the perspective of Muinane's older and younger generations, Echeverri reflects on how the rubber violence of the past had condensed into a transgenerational trauma, where memory and healing collide. In remembering the rubber violence, the older generation distinct between a "Basket of Darkness," full of locked away painful memories, and a "'Basket of Life," "where the seeds of the future are placed, looking forward to the growing of new generations and leaving behind the dangerous memories of violence and sorcery of the past."[112] However, with the locking away of trauma comes shame and silence, too. The difficulty of not even knowing how to think or talk about the traumas of the past is complicated by the fact that the Indigenous culture is still confronted with the enduring power of the colonial apparatus. The younger generation, Echeverri says, "functions like a mirror – a reflective space that allows them to face the past in an indirect way"; however "[this] reflective space is configured paradoxically by purely foreign devices: writing, schooling, use of the Spanish language, state recognition, and so forth."[113] So, it's a vicious colonial cycle from which there seems to be no escape—and yet it's the new generation that needs to be trusted

---

108 This expression comes from Quito-based Ecuadorian artist Adrián Balseca. "White Wake: On the Trail of 'Rubber Fever' in the South American Amazon," Collecíon Cisneros, https://coleccioncisneros. org/editorial/featured/white-wake-trail-rubber-fever-south-american-amazon.

109 Echeverri, "The Putumayo Indians and the Rubber Boom," 13. See also Juan Alvaro Echeverri, "To Heal or to Remember: Indian Memory of the Rubber Boom and Roger Casement's 'Basket of Life'," *ABEI Journal* 12 (2010): 49–64.

110 The legal restitution of territory in the 1980s by the Colombian government to the indigenous groups of the region, including the Witoto, Bora, Muinane, Miraña, Ocaina, Nonuya, and Andoque Indians was a moment of confrontation, since it became necessary "to revisit the territory and to face its memories." See: Echeverri, "The Putumayo Indians and the Rubber Boom," 13–14.

111 Ibid., 14.

112 Ibid., 13.

113 Ibid., 14.

with hope and healing: "[It] is to them we owe true truth and true justice. They are our actual true mirrors of memory."[114]

Sticky Rubber Matters

The Great Exhibition of 1851 in London, which was the starting point of this text, aimed to capture the industrial knowledge of its time in a supposedly universal gesture. And yet, much of the diversity and dignity of human ingenuity has remained untold and unseen. A closer look at the exhibits in the glittering halls of Crystal Palace, especially at the "raw materials" sections, reveals that things were much gloomier than promised at first sight. The very constellation, in which vulcanized rubber was staged at the exhibition, makes apparent the friction: To get to the Goodyear's rubber volcano in the US sections, visitors had to pass two sculptures. These were Hiram Power's well received sculpture the *Greek Slave* and Peter Stephenson's *Dying Indian*.[115] Although these two sculptures were intended at the time as symbols of humanity and freedom, in retrospect they appear as cynical comments on the cruel reign of terror of the rubber industry as well as the violent system of oppression that settler colonialism introduced to the Americas.

Obviously, there was no straight path, no "smooth operation of global power,"[116] leading from the milky sap of the Amazon Pará tree to the manufactured rubber pieces on display at the Crystal Palace in London. Rather, there were fierce collisions and contested battlefields fought in resistance by Indigenous and Black bodies, from which rubber emerged as a passivated and naturalized raw material for Western industries and mass production. However, it is important to understand that the very concept of 'raw material was never limited to the exploitation of vegetable or mineral materials, such as rubber or coal, but always depended on the degradation and exploitation of human bodies. Being a "raw material," as Adelene Buckland put it, was less about being "in a state of nature," but rather about politicizing or racializing "a hitherto meaningless plant or rock or animal or person," imagining them "as actualised by the mighty intellectual, physical and spiritual powers of industry and empire."[117]

The distinction between *active* and *passive* matter that I want to bring into play here has been constitutive of the entangled systems of colonialism, extractivism, slavery, and their racist legitimation. Racialization, according to inhuman geographer Kathryn Yusoff, "belongs to a material categorization of the division of matter (corporeal and mineralogical) into active and inert. Extractable matter must be both passive (awaiting extraction and possessing of properties) and able to be activated through

---

114 Ibid., 18.
115 Cunliffe, "America at the Great Exhibition of 1851," 119.
116 Tsing, *Friction*, 6.
117 Buckland, "Introduction to Volume 1: Raw Materials," 20.

the mastery of white men."[118] "Historically, both slaves and gold have to be material and epistemically made through the recognition and extraction of their inhuman properties," she concludes.[119] The distinction of *active* and *passive* matter thus lies at the very heart of a racialized "historical regime of material power."[120]

However, engaging friction as a tool to navigate material histories and legacies not just reveals exploitation, oppression, and trauma, but might also help to unsettle a *transformative* potential that is always already clinging to the very idea of "raw material" as incommensurable "active matter," as Adelene Buckland suggests.[121] Building on Tiffany Lethabo King's thoughts on "Black fungibility,"[122] that is, the speculative reimagination of "Black flesh and Black spaces … as counter-sites where new possibilities for (more than) humanness and freedom are articulated,"[123] Buckland urges for reframing "the raw material as itself a sign of 'fungibility'. . . 'that cannot be contained' by the violently intertwined histories of empire, industrialisation and labour."[124] Read against the grain, the very logics of commodification and extraction may thus open up other-than-exploitative ways of thinking through the violent repression of colonial and imperial raw material politics and the narratives and futurities that emerge from them. "What other histories and stories have been 'cut' from their materials the moment they are conceptualised as 'raw'?" she then asks.[125] This is precisely what we should be asking about the frictional material legacies of rubber and its inhumane appropriation as naturalized "raw material" for Western industrialization and mass production: What other incommensurable, resistant histories and stories of bodies, people, animals, plants, landscapes, habitats, and ecologies are attached to the sticky matter of rubber instead or beyond? What other futures might (or might not) be kneaded out of its formless mass yet?

---

118 Kathryn Yusoff, *A Billion Black Anthropocenes or None* (Minneapolis: University of Minnesota Press, 2018), 2–3.

119 Ibid., 3.

120 Ibid., 4.

121 Buckland, "Introduction to Volume 1: Raw Materials," 28.

122 Building on the foundational thoughts of Hortense Spillers and Saidiya Hartman, Tiffany Lethabo King frames "Black fungibility" as "the treatment of the Black enslaved body as an open sign that can be arranged and rearranged for infinite kinds of use". However, in her/their speculative reading, "Black fungibility . . . is conceptualized as the capacity of Blackness for unfettered exchangeability and transformation within and beyond the form of the commodity, thereby making fungibility an open-ended analytic accounting for both Black abjection and Black pursuits of life in the midst of subjection." See: King, "The Labor of (re)reading Plantation Landscapes Fungible(ly)," 1023–1025. See also Hortense Spillers, "Mama's baby, papa's maybe: An American grammar book," *Diacritics* 17, no. 2 (1987): 65–81; Saidiya Hartman, *Scenes of Subjection: Terror, Slavery, and Self-making in 19th Century America* (New York: Oxford University Press, 1997).

123 King, "The Labor of (re)reading Plantation Landscapes Fungible(ly)," 1036.

124 Buckland, "Introduction to Volume 1: Raw Materials," 29.

125 Ibid., 28.

Emile De Visscher

Crafting Time: Speculative Archaeology and Inorganic Becoming

Introduction

The practice of design is among the disciplines generating the most dramatic consequences for our environment. The will to build stable, perennial objects crossing time, combined with a strategy for constant renewal, programmed obsolescence, and trend generation is at the heart of the waste problem as much as the legacy of designers' actions. How to question this fate and develop alternative strategies? If a circular economy and bio design is the current answer, different approaches could be explored. The ecological crisis is essentially a temporal issue. It forces us to analyze the consequences of our actions over thousands of years. At the same time, however, it is a great acceleration, where every year counts, and the choices for drastic change have become stressfully urgent. The project presented in this article tries to take a different stance toward our legacy to future generations with a desire to act in the present. Entitled "petrification," this research has given rise to more than seven years of material and conceptual explorations. Started as a very pragmatic method to generate a new local craft, it then led to question the relations between organic and inorganic matter, the symbolic power of petrification as a material and cultural transformation, and the relations between technicity and culture through an interpretation of Gilbert Simondon's theory of technology. This research project has been essentially an investigation, full of discoveries, twists and turns. In the following pages, I will give an account of this situated investigation and its multiple facets.

Weaving, Folding, Cutting—Without Translations

The project begins in 2015 in the context of the maker movement. The idea of bringing production to as many people as possible, of designing objects or repairing them in local workshops, of a *democratic making*, was then developing in communities of designers, engineers, hackers, and companies. I was involved in several collective projects of this type—and would be considered a "maker." I was often invited to carry out innovation projects over short periods of time, also called hackathons, in schools, companies, or for public institutions. In this context, it was essential to find materials and tools that would allow everyone, whatever their degree of knowledge or sensitivity to DIY, to express their ideas, to shape them, and to refine them. In these hackathon contexts, we would bring 3D printers, laser cutters, or other digital fabrication tools. Of course, these machines required skills, knowledge, or mediation that were often inappropriate. I then realized the power and democratic aspect of paper and cardboard. Much more than modeling clay or drawing, these cellulosic materials were known to everyone. They implied tools and techniques that every actor could easily understand, and at the same time allowed to generate forms surprisingly sensitive and complex. Over and over, paper and cardboard were becoming the central material and came to be, in my experience, the most shared, universal, and democratic tool to materialize ideas. Cutting, making patterns, assembling, folding, creating structural lines, tensioning—all these operations are experienced from an early age, and remain known to all levels of society (fig. 1).

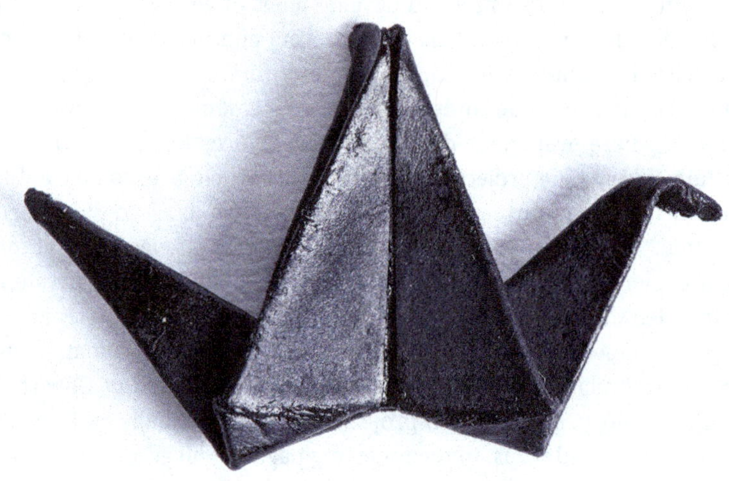

Fig. 1: Emile de Visscher, petrified paper origami, 2017. Experiments conducted with students from the PIG workshop at Chimie ParisTech, Paris.

However, the paper or cardboard prototype is, most of the time, only a temporary initial form. In a design process, paper has a long tradition.[1] This first handmade shape then requires a series of translations: it is measured, modeled on a computer, translated into technical parts, reproduced by a series of various processes that, after manufacturing, can finally be assembled and become functional objects. In all these translation processes, the tools of modeling, manufacturing, and their constraints intervene, and modify the initial idea. They may lose the immediacy and the power

---

1 Konstantin Grcic, "I Love to Hear Someone Cutting Cardboard," *The Art Newspaper*, January 1, 2010, https://www.theartnewspaper.com/2010/12/01/interview-with-designer-konstantin-grcic-i-love-to-hear-someone-cutting-cardboard.

of the first forms and the immediate intelligence of hand manipulations. How to skip these translation processes? Is it possible to directly transform paper and cardboard experiments into functional objects without all these biases? I realized that I had to find a way to transform the organic into the inorganic, that is, to find a way to petrify these materials. This process exists in nature. It can be found in the deserts of Arizona, where whole trees, preserved from decomposition, are progressively infiltrated by minerals that crystalize and transform organic matter into stone. This geological manifestation became the center of my research to understand its mechanisms (fig. 2).

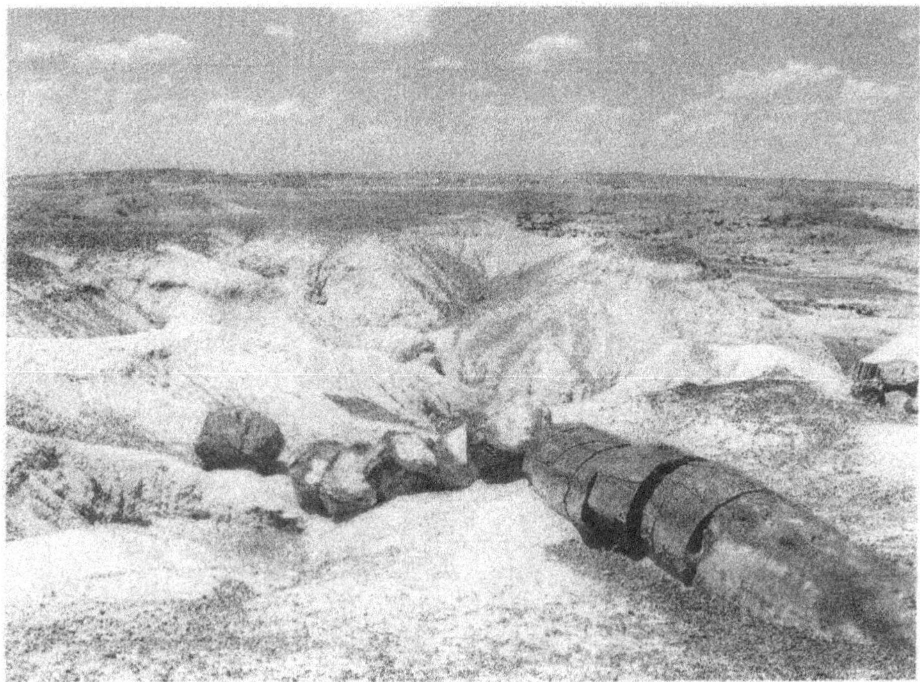

Fig. 2: Stefan Pauli, petrified tree in Petrified Forest National Park, United States, 2001.

Technophany: Myths, Labs, and Video Games

During an initial literature overview, I stumbled upon a popular science news reporting on a Californian laboratory that had succeeded in *petrifying* cellulose: "Achieving what would take millions of years in only a few days, scientists have drastically sped up the process of petrifying wood."[2] Of course, the scientific article of the chemis-

---

2 Brandon Miller, "Presto! Instant Petrified Wood Created in Lab," *Life Science*, January 27, 2005, www.livescience.com/110-presto-instant-petrified-wood-created-lab.html.

try lab itself did not talk about petrification, but rather about *bio-mineralization templating*. The method used involved a soaking in acid to remove the lignin of the wood cubes, followed by an infusion of silica to finally cook the carbon/lignin composite in an atmosphere furnace at 1,400 °C (fig. 3). During this firing, the carbon and silica could be transformed into silicon carbide (SiC), whose mechanical properties are close to diamond. The aim of the process was to produce silicon carbide filters, the filtration quality of which would depend on the porosity of the chosen wood. Several aspects were interesting here: the use of the term "petrification," the history of silicon carbide, or the transformation of biological structures for technical applications.

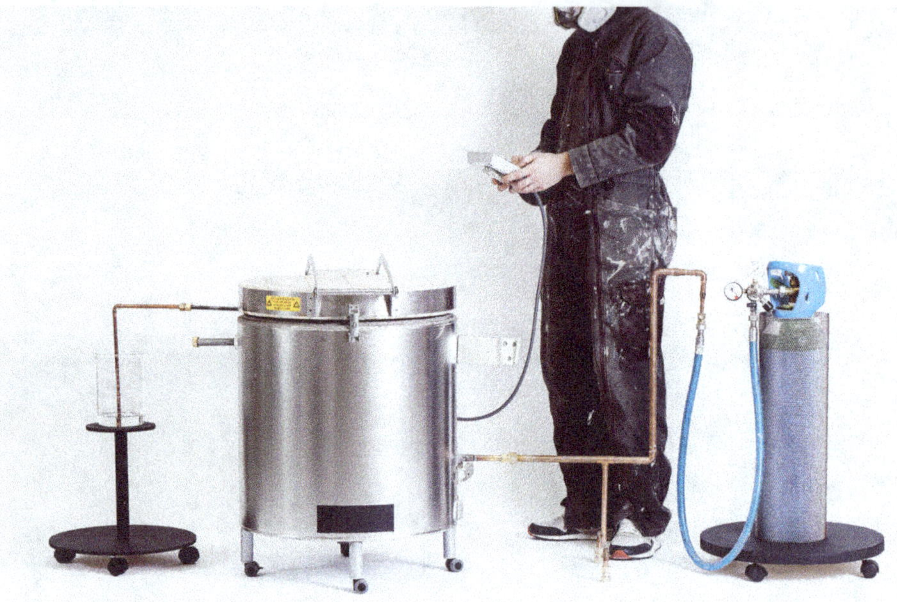

Fig. 3: Emile de Visscher, atmosphere furnace used for the petrification process, 2016

While materials science is usually looking for performance and control of its properties, the use of biological materials in engineering, mechanics, and design remains, in my experience, a problem. Too composite, too anisotropic, too tied to the ways in which it was grown, or the context of its production, biological matter is considered irregular, impossible to predict, and unscalable. In my engineering studies, they were not studied—we were limited to metals, plastics, and industrial ceramics. All of them were "solid," isotropic, scalable, stable, resistant to their direct environment, and thus allowed a prediction of the behaviors, calculation of their properties, and design with CAD software. But in this article, the strategy was different. Instead of using fully controlled and artificial materials, and then try to construct a complex foam system to generate filters out of them, these scientists use the structural intelligence of a natural material to obtain complex shapes, and transform the material into a stable and rigid structure.[3] Bio-templating, as a form of making, is in this sense closer to ancestral craft traditions than to modern engineering, as it relies on the intelligence, properties, and singularities of organic matter rather than reconstructing complex structures from passive and amorphous materials. It is, in a way, a matter of "making with,"[4] of telling other geo-stories than the *tabula rasa* and hylomorphic ones (fig. 4).

The crystal formed by this process, silicon carbide, is not without interest and history. A ceramic form that exists in nature, moissanite, was named after the French chemist Henri Moissan, who first discovered it in the Diablo Canyon in Arizona. It is present in very small quantities on our planet's crust and is always the result of an external contribution by meteors (fig. 5). Natural silicon carbide is older than our solar system and is considered a "pristine interstellar material."[5] It has extensively been studied to understand the nuclear and chemical processes occurring in the formation of stars. However, the synthesis of such a rock precedes its discovery in craters. Suggested by Franz Jakob Berzelius in 1824, the first produced SiC material was eventually achieved by Eugene Acheson in 1892, which he called "Carborandum" and started to produce industrially. Its mechanical properties are impressive, and its heat resistance and lightness make it useful for many applications, in aerospace, heating systems, or grinding tools. As such, silicon carbide could be considered the cursed brother of diamonds, quite similar in many applications, but never considered as valuable, precious, or metaphorically charged in society and common applications.

But to me, the most interesting discovery was the use of the term "petrification" in Miller's article. Why use this term? What does it refer to? The technical process, beyond the mechanical qualities, was calling for a deeply rooted collective metaphorical reference. As I dug into the literature, I realized that petrification was present everywhere in society. It can be found in video games, science fiction movies, comic books: stories about the power to transform living beings into stone, and about stones that come to life. But in a much deeper sense, it comes from myths all over the world. Whether in Japanese, Australian, Indian, South American, or Central African mythology, and even in the Bible, there are stories of humans turning into stone.[6] Celtic and Icelandic mythology, for example, associates rock formations with trolls petrified by the sun. In the European context, Medusa is without doubt the figure that petrification is constantly referring to. Medusa is one of the three Gorgons, those who gaze into her eyes turn into stone, her hair is made of snakes, and she is defeated

---

3 To learn more about these strategies: Paris Oskar, Ingo Burgert, and Peter Fratzl, "Biomimetics and Biotemplating of Natural Materials," *MRS Bulletin* 35, no. 3 (2010): 219–25.
4 "Making with" can be read through the lens of "Sym-poiesis," a concept developed by Donna Haraway. See Donna Haraway, "It Matters What Stories Tell Stories; It Matters Whose Stories Tell Stories," *a/b: Auto/Biography Studies* 34, no. 3 (2019): 565–75.
5 To know more about the exact origins of the "cosmic chemical memory" of silicon carbide: Jim Kelly, "The Astrophysical Nature of Silicon Carbide," *UCL online*, img.chem.ucl.ac.uk/www/kelly/history.htm.
6 "Petrifaction in Mythology and Fiction," Wikipedia, last modified July 16, 2023, https://en.wikipedia.org/wiki/Petrifaction_in_mythology_and_fiction.

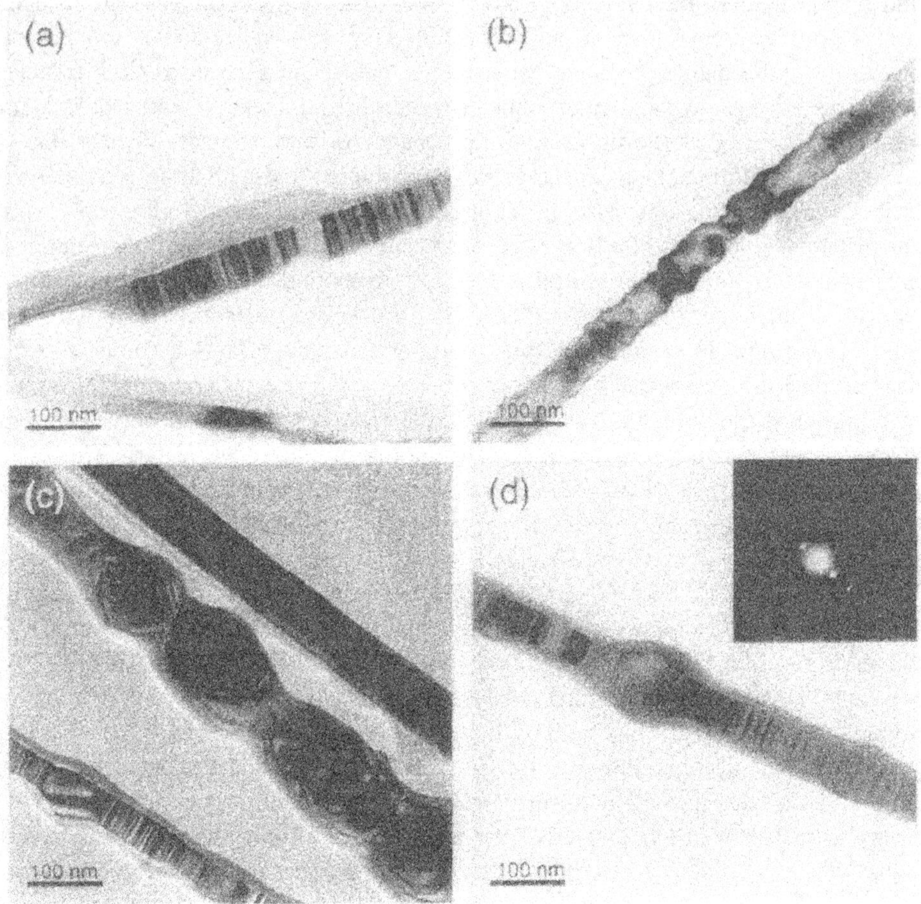

Fig. 4: Scanning electron microscopic images of SiC materials. In Yongsoon Shin, Chongming Wang, Gregory J. Exarhos, "Synthesis of SiC Ceramics by the Carbothermal reduction of Mineralized Wood with Silica," *Advanced Materials* 17, no. 1 (January 2005): 75.

by Perseus who uses a shield to protect himself and cut off her head, from which will come Chrysaor and Pegasus (fig. 6). Perseus will then take Medusa's head with him to defeat his enemies and eventually offer it to Athena, who will place it on her shield and thus obtain the power of petrification as a weapon to protect the Athenians. Symbol of a violent, uncontrollable, and dangerous femininity, at the antipodes of the strong and protective Athena, Medusa gives an account of an ambiguous position, powerful but enticing, seductive but dangerous, uncontrollable but potentially useful.

The petrification of wood we are talking about here is not only a technical process opening new manufacturing possibilities. It is also a cultural process bringing into play collective knowledge, dreams, and fundamental fears. The philosopher Gilbert Simondon had a term to qualify this quality: "technophany." In a series of conferences given

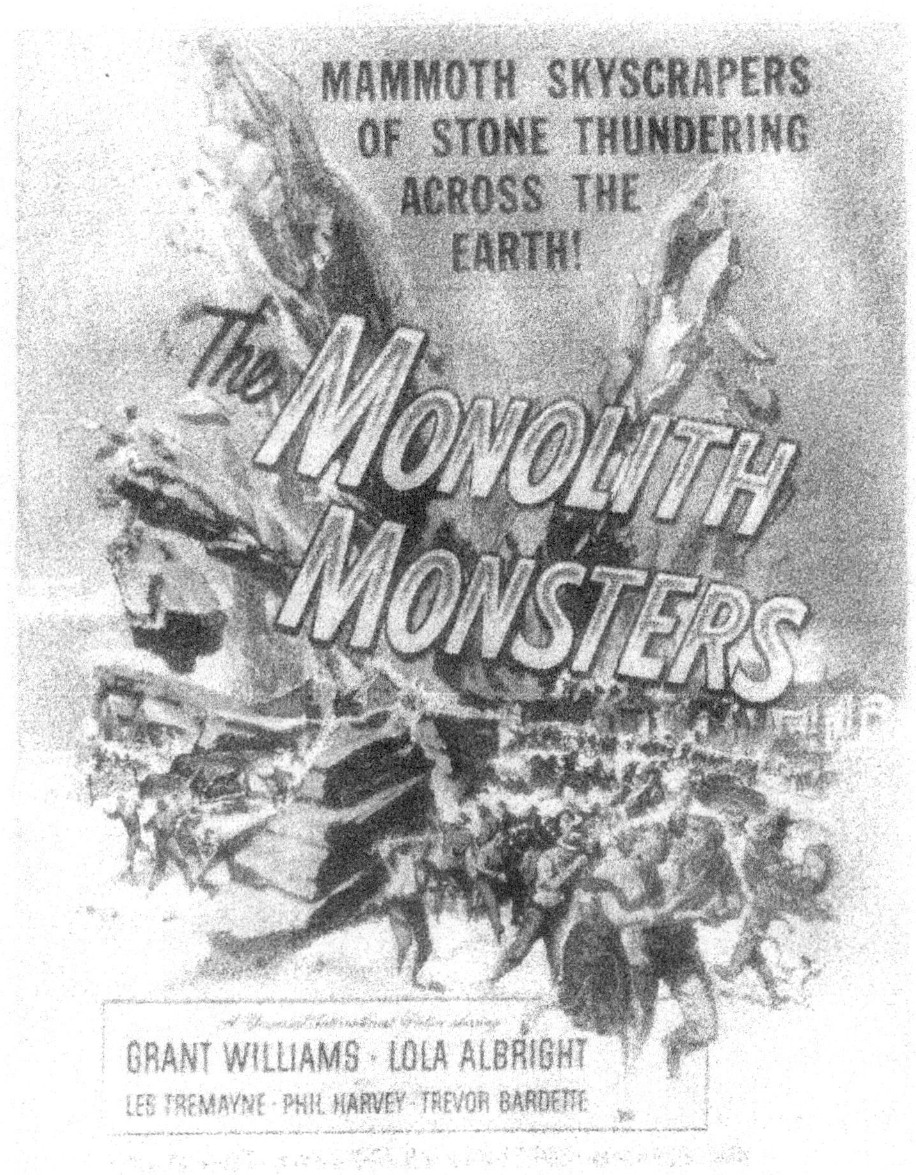

Fig. 5: William Reynold Brown, *The Monolith Monsters*, 1957, poster art.

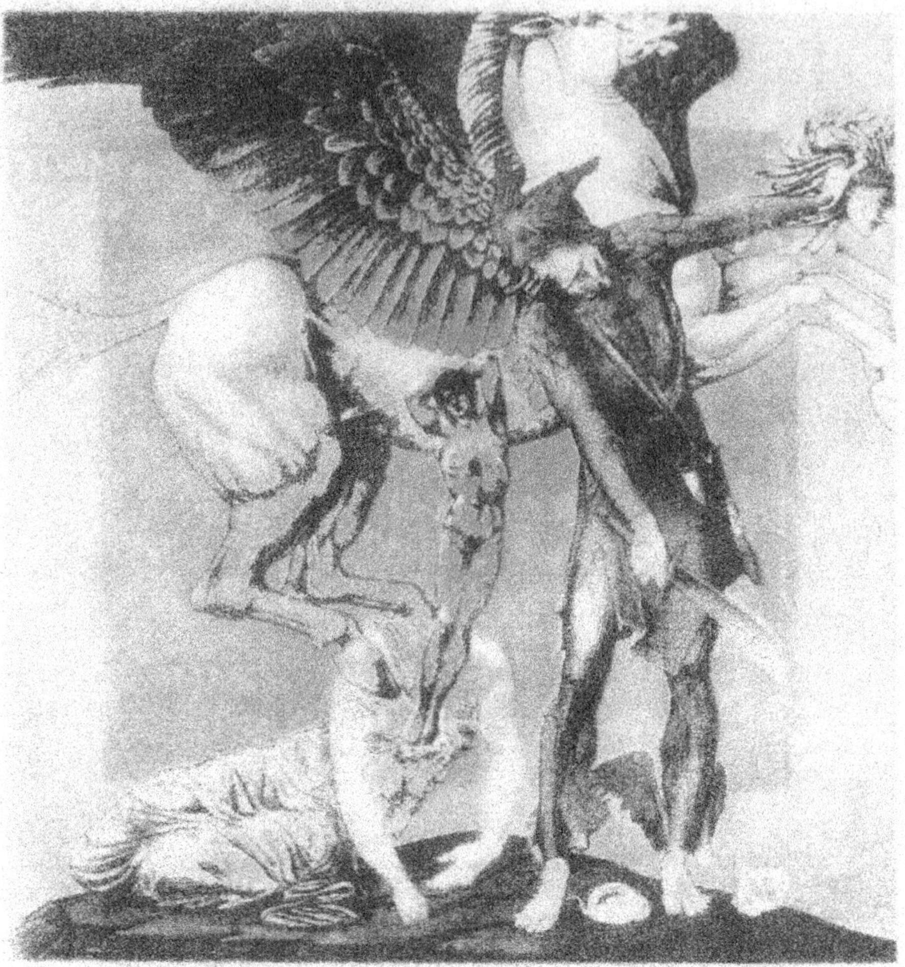

Fig. 6: Sir Edward Coley Brune-Jones, *The Death of Medusa I*, 1882, mixed media on paper, 1245 × 1169 mm. Southampton City Art Gallery, United Kingdom.

between 1960 and 1961,[7] the French philosopher of technology argued for the idea that certain technological processes had capacities to make play of cultural contents, that they had the capacity to inscribe themselves in the society not by their functions, their political contexts, or the speeches that they supported, but by their operativity itself, by their inherent technicality. The petrification process, the passage from the organic to the inorganic, the capacity not to be affected by death, but to be blocked in a state-stone, makes play of all these collective references. The term technophany itself

---

7 Gilbert Simondon, "Psychosociologie de la technicité (1960–61)," in *Sur la Technique*, ed. Gilbert Simondon (Paris: Presses Universitaire de France, 2014), 37.

makes direct reference to myths, because it was constructed in relation to Mircea Eliade's "hierophany."[8] Indeed, this historian and theoretician of alchemy, magic, and the sacred had established a term to speak about the way in which certain objects, in ancient cultures, could at the same time represent the sacred and make it exist in society, without losing their functionalities. An object would then have several existences: it would exist beyond functionality. But Eliade also stated a loss of this quality in modern times. According to him, objects were now considered in one dimension only, "not more than themselves,"[9] locked in their own existence of commodities. In reaction to this conclusion, Simondon speculated the existence of such relations in contemporary technical objects. He thus looked for links between technicity and culture, between operativity and symbolic dimensions—as a way to reunite these two regimes whose modes of existence have "dephased" according to his theory.[10]

Speculative Archaeology: Anthropocene and Diversity

The newly called "petrification" project thus took on a new dimension, not only through these relationships to the past, but also to the future (fig. 7). For petrification has always been a reference to death, to the destruction of life, but also to what endures, to what passes through annihilation. It is a pharmakon, in the sense of a poison and a cure, a fantasy and a dystopia intertwined. The stone statue, if it does not live, is also the means used in all ages to keep track of the living, to perdure, thus its use for important people, kings, emperors, and artists. The idea of petrifying humans, of finding technical devices to make bodies last beyond death, gave rise to numerous researches, notably in Italy and France in the nineteenth century.[11] Some even called it "Androlithe,"[12] as a new stone made from petrified human flesh.

But this question—our death and survival—is back in a very urgent fashion: the Anthropocene makes us consider the end of the living world. As such, the Anthropocene is calling for a petrified planet—a vision of horror, of the end of life, of the end of diversity, of a mass of reinforced concrete and grayish plastic replacing the wilderness of fauna and flora, material legacies and inherited techniques. The end of sub-

---

8 Mircea Eliade, *Le Sacré et le Profane* (Paris: Gallimard, 1965), 17.

9 Mircea Eliade, *Images et Symboles: Essais sur le symbolisme magico-religieux* (Paris: Gallimard, 1980 [1952]), 249.

10 Gilbert Simondon, *Du Mode d'Existence des Objets Techniques* (Paris: Aubier, 2012), 164–65.

11 Marta Licata, Chiara Rossetti, Chiara Tesi, Omar Larentis, Roberta Fusco, and Rosagemma Ciliberti, "To Save a Corpse from Decomposition – the Purpose of Petrification in the Second Half of the 19th Century," *Acta Medica Academica* 48, no. 3 (2019): 328–31.

12 Gian Marco Vidor, "Androlithe et pétrification des cadavres humains au XIXe siècle," *Frontières* 23, no. 1 (2010): 66–73, https://doi.org/10.7202/1004025ar.

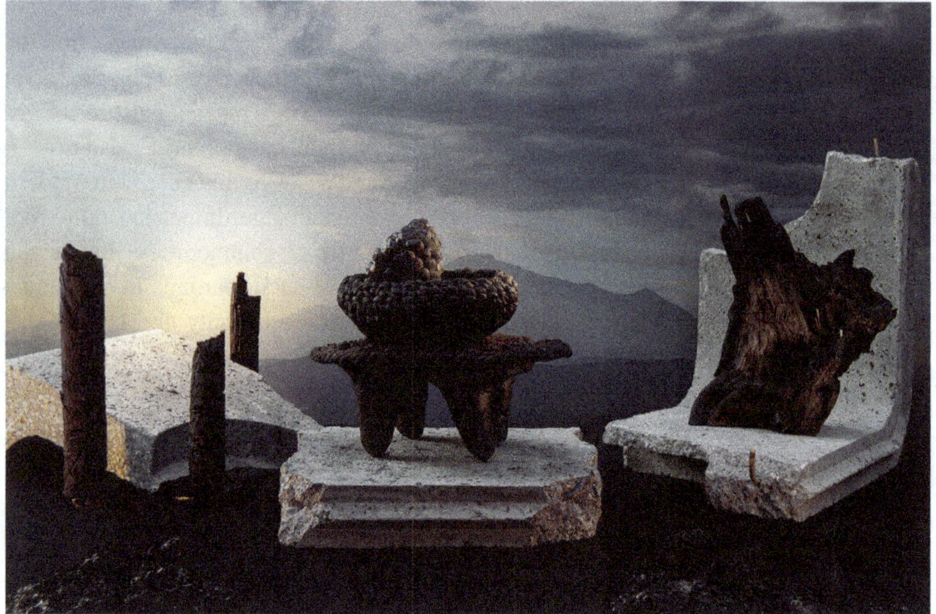

Fig. 7: Emile de Visscher and Ophélie Maurus, "Petrification: Material Transmutations and Speculative Archaeology," *able.journal* (2023), https://able-journal.org/petrification.

tle and diverse forms of life seems close, all of a sudden. What will be left? What will remain? What should remain?

To propose a process of transformation of fragile forms—paper, rope, fabrics, wools, cardboard foams—into stone, becomes then a means to question our future (fig. 8). A kind of speculative archaeology, the project asks us to choose what should remain, rather than to suffer the uncontrolled accumulation of waste. Can the Anthropocene layer be populated with diversity rather than industrial waste? Can we think of a joyful Anthropocene rather than just consider it a depressing fate? What will remain of the weak forms, the fragile techniques, the ephemeral materials, the know-how less visible than those of industry?

Conclusion: Democratic Enquiry

This issue should not remain in my hands. It must become a common object, a collective thing—a "res-publica." It is a question of inviting actuators of all kinds, designers, architects, craftsmen, researchers, to ask the question together and imagine our remains, our ruins, our heritage. To approach the question in functional, symbolic, aesthetic, political, geological, anthropological, mineralogical, archaeological terms, and to produce time capsules that build our collective visions of the future. Without providing definitive answers, these material experiments are ways to see beyond the end of the

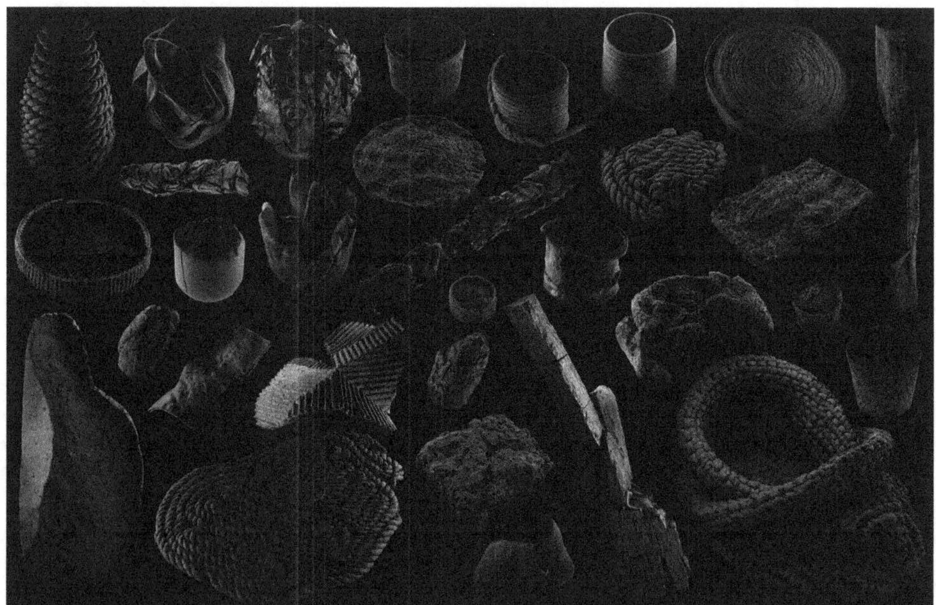

Fig. 8: Emile de Visscher and Ophélie Maurus, samples, presented in "Petrification: Material Transmutations and Speculative Archaeology," *able.journal* (2023), https://able-journal.org/en/petrification/.

world, without getting stuck in technological fixes for sustainable growth. By inviting Matters of Activity colleagues working on bacterial cellulose, or industrial designers, or textile artisans, the project becomes a place to meet and discuss the aftermath of the world: of what will remain—or what should remain—of our fragile existences.

Nina Samuel

Passivity Matters! Transience and Conservation Practices: Examples from the Eighteenth Century to Today

In 1902, a major architectural landmark in Venice, San Marco Basilica's bell tower, collapsed into a heap of rubble. Whether (and if so, how) to reconstruct the tower—which had been damaged, rebuilt, and added to several times throughout a history dating back to the ninth century—became a heated debate, revealing how the general public and experts viewed their relationship and duty to objects of cultural importance. This famous event brought to the fore questions about what the maintenance and restoration of an object should entail, both practically and conceptually; how to treat and re-purpose the materials of the object; and whether evidence of restoration and conservation processes should be made visible in the restored object.

In the following, I explore several competing concepts of conservation through a series of illustrative examples from European conservation practices of the last three centuries: the preservation of wet specimens, the use of wax moulages, climate-controlled spaces, and architectural restoration. These mini case studies demonstrate a range of approaches to the treatment of cultural objects, both technical and philosophical. They highlight differing paradigms of conservation: namely, whether conservation should aim to "passivate" the substance or material of a cultural object —to protect it and freeze it in a state of suspended animation through chemical or climatic means; to restore (or recreate) the "original" or "ideal" state of an object, even if only as a simulacrum or mimetic approximation of the object itself; or to view transformation, transience, and even decay of the object as integral and as processes worthy of preservation in their own right. A critical consideration of different concepts of conservation and restoration (both historical and contemporary) can help us think through our current moment's tendency toward "over-conservation." It can suggest alternative approaches to cultural memory—to the treatment, stewardship, and preservation of artifacts—and can help us chart a middle path between the extremes of idolatry and iconoclasm.

Introducing Specimens: Friction Points between Activity and Passivity

The conservation laboratory of the Berlin Museum of Medical History at the Charité accommodates various organic materials with the aim of keeping them materially stable for as long as possible.[1] To keep materials in such a passive state and slow down their natural transformation processes, conservators must have a comprehensive knowledge of the potential activity of the materials with which they work and, thus, must be experts in active and passive matter. In the following, I will examine the preconditions and ambivalences of passivation and its opposite, activation, and, second, their historical anchoring.

The collection of pathological-anatomical wet specimens at the Berlin Museum of Medical History was established by pathologist Rudolf Virchow and comprises around 10,000 objects.[2] One item in the collection is a sliced lung, whose structure and blackish discoloration are typical markers of tuberculosis (fig. 1). This specimen is undated, but it is known to be over one hundred years old. Its glass container was opened recently for restoration and research purposes. During conservation treatment, the specimen is stored inside a fume hood to protect humans from the vapors and gases emitted by the fluid preservative formaldehyde, which is classified as a carcinogen. This is a protective measure for both specimen and conservator: people must be protected from the vapors of the specimens, and the specimens must, in turn, be protected from people—or, more precisely, from the surrounding environment. Indeed, as soon as the glass containers are opened or are otherwise compromised, they become susceptible to the invasion of microorganisms and the development of mold. Leaking glass containers are one of the biggest problems in specimen storage, resulting in unintended molecular exchange processes and the gradual evaporation of the alcohol. For this reason, the preservative liquid must be regularly replaced. The "freezing" of an organic specimen through the use of preservative liquids requires not only regular maintenance but also enormous technical skill and specialized knowledge from the very beginning of the preservation process.

To better understand this process, it is essential to familiarize oneself with how organic specimens are created. First, organs are carefully dissected from a deceased body. Detaching the tissue from the adjacent body structures is particularly challenging. While the remains are left to the natural processes of organic decay or are cremated, the organs are subjected to an elaborate technical procedure of mechanical interventions and chemical manipulations by a trained preparator.[3] This fixation process

---

1 The article is based on a collaboration with the Human Remains Conservator and Museum Curator Navena Widulin, who generously opened the doors of the conservation laboratory for me, and whom I would like to thank very much. All factual information about the collection and conservation procedures that is not separately identified is obtained from communication with her.

2 In the history of knowledge techniques, anatomical specimens in natural history, medicine, and biology perhaps go back the furthest; see Hans-Jörg Rheinberger, "Epistemologica: Präparate," in *Dingwelten: Das Museum als Erkenntnisort*, ed. Anke te Heesen and Petra Lutz (Cologne: Böhlau, 2005), 67. We know of anatomical specimens from the seventeenth century on, for example, from Frederik Ruysch, one of the earliest collectors of anatomical specimens in The Hague.

3 See the description and images of this procedure in Thomas Schnalke and Isabel Atzl, "Magenschluchten und Darmrosetten: Zur Bildwerdung und Wirkmacht pathologischer Präparate," *Bildwelten des Wissens: Kunsthistorisches Jahrbuch für Bildkritik* 9, no. 1 (2012): 18–28.

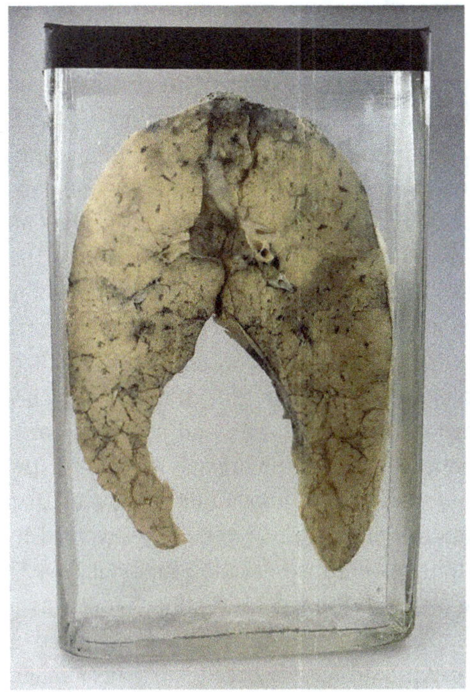

Fig. 1: Lung Specimen, Berlin Museum of Medical History at the Charité (BMM).

can take between a few days and up to several months to denature the proteins and delay the decomposition of the tissue. Although the fixative and fluid preservation processes cause a chemical alteration of the specimen and can lead to discoloration, shrinking, or swelling, these collections are able to last for hundreds of years under ideal conditions. In its natural and unprepared state, an organ would very quickly undergo a process known as autolysis, the destruction or self-digestion of a cell through the action of its enzymes. With the technology of fluid preservation, one could say, we come very close to a dream of material immortality—an almost perfect stillness of the organic substance.

In the case of the collection of the Berlin Museum of Medical History, there is an interesting tension between passivation and activation. Due to close cooperation with the pathology department of the Charité, tissue samples are regularly taken from the collection's specimens for histological examinations in order to gain new insights into old diseases. Also, the tissue of older specimens with sufficient volume is sometimes recut to refresh its color and structure. Both treatments (histological examination and recutting) activate a specimen. This implies that there is no way to passivate the substance itself in its status as an epistemic entity. On the contrary, the greatest possible deactivation of the material can even be identified as an irreducible condition of its ongoing epistemic activity.[4]

The collection also contains objects that are never "activated" in this sense or never (re)cut for research—namely, the specimens made by Virchow himself. Here, we encounter a completely different concept of the specimen as object, one that is similar to that of a work of art and that draws on the mythical charge of the original. Perhaps the integrity of his specimens would have pleased Virchow, particularly since he chose the concept of a "walk-in textbook" for the presentation of his collection at the opening of his Pathological Museum in 1899.[5] The production and editing process for this "textbook" took place in rooms of the museum that were not accessible to visitors. Photographs documenting the preparation and conservation laboratory—the so-called *Kittraum*—from the beginnings of the Pathological Museum in 1902, show bustling activity (fig. 2). The men, with concentrated facial expressions, mix and pour solutions, seal glass containers, and use sharp instruments to process the organs. The staged surroundings show specimens in different phases of preparation. Unlike today, at that time both pathologists and preparators made the preparations and were involved in the further development of preparation techniques and preservation methods.

Not evident from the photograph is the fact that their conservation liquid was a brand-new invention by Virchow's assistant, Carl Kaiserling, which had started a small revolution.[6] His novel mixtures of glycerin- and formol-based substances made specimens appear considerably truer to color, form, and structure. According to Virchow, this new invention made it possible to create specimens as "real images," in a way that was previously only known "from illustrations."[7] These would have the advantage of offering the visitor a "first-hand impression."[8] Virchow's claim was confirmed by twenty-first-century analyses of the history of scientific images, which have often described this method as an "image formation process" or even as "becoming images of themselves"—in other words, becoming images in which the represented and the means of representation are materially identical.[9]

---

4 This recalls the activation of certain collections in the Museum für Naturkunde, Berlin—for example, the collection of animal skins and hides. They are considered "biodiversity storage" and similarly used in research.

5 See Schnalke and Atzl, "Magenschluchten," 27–28. For a comprehensive discussion of Virchow, his Pathological Museum, and his collection of specimens, see Angela Matyssek, *Rudolf Virchow, das Pathologische Museum: Geschichte einer wissenschaftlichen Sammlung um 1900* (Darmstadt: Steinkopff, 2002).

6 Thomas Schnalke and Isabel Atzl, *Dem Leben auf der Spur im Berliner Medizinhistorischen Museum der Charité* (Munich: Prestel, 2010), 89–90.

7 Virchow, quoted in Schnalke and Atzl, *Dem Leben*, 90.

8 Virchow, quoted in Schnalke and Atzl, "Magenschluchten," 21.

9 See Hans-Jörg Rheinberger, "Präparate—'Bilder' ihrer selbst: Eine bildtheoretische Glosse," *Bildwelten des Wissens: Kunsthistorisches Jahrbuch für Bildkritik* 1, no. 2 (2003): 9–19; Johannes Grave, "Selbst-Darstellung: Das Präparat als Bild," *Kritische Berichte* 37, no. 4 (2009): 25–34; and Schnalke and Atzl, "Magenschluchten."

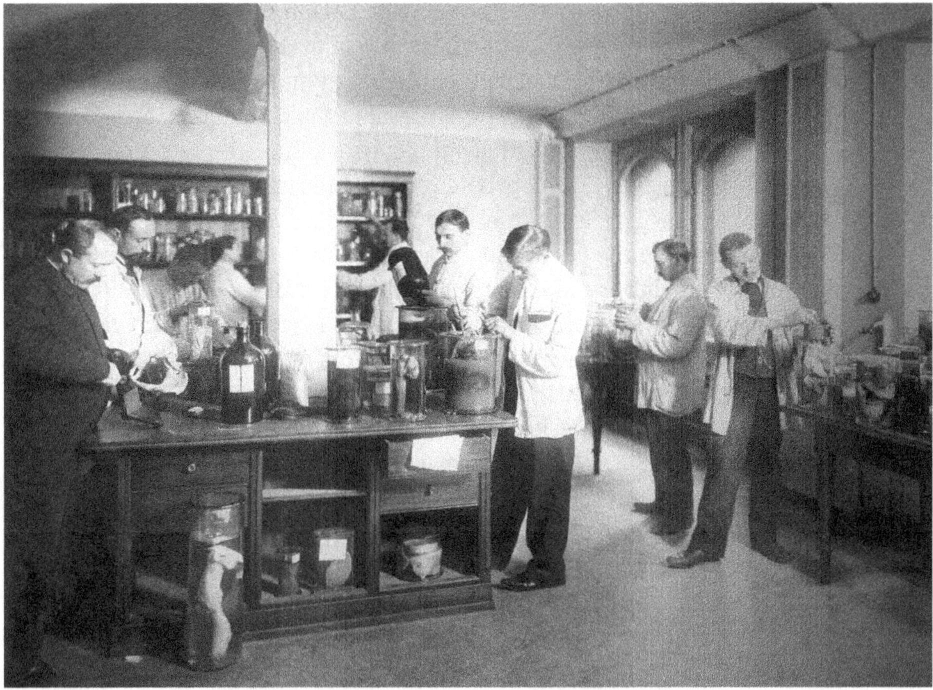

Fig. 2: *Kittraum* of Virchow's Pathological Museum, 1902 Berlin Museum of Medical History at the Charité (BMM).

The Restoration Boom and the Beginning of the Museum Age

Virchow's equating of specimens with textbook illustrations—the equating of the thing with its representation—points to an understanding of museum objects that is strongly influenced by the development of the paradigm of cultural preservation. This paradigm relies on successful passivation of the material, and, historically, it went hand in hand with the formation of the museum as an institution and the elaboration of approaches to building and maintaining collections in the eighteenth and nineteenth centuries. This parallel development can be seen in two illustrative examples: the professionalization of restoration and the history of exhibitions.

The development of the profession of painting restoration (between 1750 and 1830) occurred alongside the founding of large museums, beginning in the mid-eighteenth century. Although restoration methods had, to some extent, been passed down, starting from the Middle Ages, in art and recipe books (with such methods called "secrets"),[10] the quality and number of such transmitted texts began to change in the middle of the eighteenth century. By the end of that century, a modern understanding of restoration emerged.[11] In this context, "modern" means a preventive understanding of restoration, and thus the attempt to make interventions unnecessary through precautionary measures as well as the conviction that all procedures be based on the specific needs of the individual work.[12] In the late 1820s, the first handbooks were published, providing comprehensive insight into the restoration practices of that time.[13]

If one compares the history of the museum with the history of exhibitions, it becomes evident that the physical preservation of cultural assets was initially considered even more important than opening museums to visitors or lending works for exhibitions. In other words, the benefits of one were weighed against the risks of the other. In the nineteenth century, visitors were seen as a threat to the physical integrity of the cultural objects. In 1883, Johann Grässe, director of the Dresden Numismatic Collection, weighed the pros and cons of the plan to show art objects from public museums in exhibitions; as he wrote in the *Zeitschrift für Museologie und Antiquitätenkunde*, "Might the collection objects, especially in summer, with an unlimited admission of visitors, suffer from the dust and exhalations whirled up by those walking around?"[14] The human as breathing and sweating biological machine has long been considered an incalculable risk for the paradigm of conservation. Conservators must know not only about the intrinsic activity of objects and materials but also about the dangers posed by their proximity to visitors and by human interactions, which invariably stimulate them further. These are two different sides of a continuous transformation and material activation of museum objects that cannot be separated from each other. However, the idea of the museum as a shelter from contamination and decay, as a pure white room whose climate is continuously optimized for the perfect preservation of the objects it houses, persists to this day—despite numerous reforms and new, innovative, and experimental museum forms and formats.[15]

---

10 Cornelia Wagner, *Arbeitsweisen und Anschauungen der Gemälderestaurierung um 1800* (Munich: Callwey, 1988), 7.

11 Ibid., 15, 29.

12 Ibid., 15.

13 Ibid., 7. Here, however, it must be noted that the establishment of restoration science for different genres and objects did not occur homogeneously. As Wagner also writes, the restoration of antiquities, for example, was established earlier and more quickly than the restoration of paintings, while post-antique sculpture was considered for restoration only in the late twentieth century, and objects of decorative art not before the end of the twentieth century. Wagner, *Arbeitsweisen*, 9.

14 Johann Georg Theodor Grässe, "Ist es ratsam Kunstgegenstände aus öffentlichen Museen zu Ausstellungen abzugeben?," *Zeitschrift für Museologie und Antiquitätenkunde* 12 (1883): 89. Translation by the author.

15 It would be worthwhile to compare this idea to O'Doherty's analysis of the *white cube*; see Brian O'Doherty, *Inside the White Cube: The Ideology of the Gallery Space* (Berkeley: University of California Press, 2010). However, Doherty explicitly located his *white cube* in contrast to the outside world as po-

The rise of the conservation paradigm was also closely tied to mechanisms of value creation. The art market, which expanded as a result of social restructuring in the late eighteenth century, also contributed significantly to the upswing in the restoration sector.[16] An anonymous letter to the editor of a British journal from 1764 describes this development: "No picture bought at an auction, or elsewhere, must be sent home before it is sent to the picture cleaners, and the owners will take as much pride in letting you know who cleaned it, as who painted it."[17] Here, the restorer is equated with the artist in an astonishingly clear way, anticipating twentieth- and twenty-first-century debates on the role of restorers as editors of works of art.[18]

The Material Alchemy of Wax and the Knowledge of Conservators

All these factors led to the idea, beginning around 1900, that a material's natural activity had to be managed and controlled by the techniques of restoration and preservation. The conservation paradigm was also a driving force in the development of techniques in the medical field. Over the course of the nineteenth century, in some collections—in Florence, for example—specimens started to be replaced by another material that was considered superior from the point of view of conservation: wax.[19] Wax is the second most prominent material represented in the conservation laboratory at the Berlin Museum of Medical History. An examination of the ambivalences between strategies of passivating and activating and of the intrinsic "performance" of wax is useful for understanding the importance of wax in conservation, among other passivation strategies that date back to the late eighteenth century.[20]

On close inspection of a wax moulage, the most striking feature is its overwhelming capacity to hyperrealistically depict the human body and, in particular, the structure of the skin, right down to the smallest pores and surface structures that are barely visible to the human eye (fig. 3). This effect is often enhanced by additional elements,

---

litical reality. Here, by contrast, it is about the outside world as a biological influencing factor. In this shift lies a whole cultural history of the transformation of the museum as an autonomous space.

16 Wagner, *Arbeitsweisen*, 13.

17 Ibid., 13. The title of the journal, which published the anonymous letter, was *The Gentleman's Magazine*.

18 See Dušan Barok, Julia Noordegraaf, and Arjen P. de Vries, "From Collection Management to Content Management in Art Documentation: The Conservator as an Editor," *Studies in Conservation* 64, no. 8 (2019): 472–89, https://doi.org/10.1080/00393630.2019.1603921.

19 Sandra Mühlenberend, "Wachsmoulagen: Orte ihrer Etablierung," *Bildwelten des Wissens: Kunsthistorisches Jahrbuch für Bildkritik* 9, no. 1 (2012): 75–84, 78.

20 On the concept of the performance of materials see Bernadette Bensaude-Vincent, "Materials as Machines," in *Science in the Context of Application*, ed. Martin Carrier and Alfred Nordmann (Dordrecht: Springer, 2011), 101–11.

such as the incorporation of hair. The moulage achieves this mimetic proximity through a production process that relies on a direct imprint and thus on direct contact with the body. This process involves a number of steps: producing a plaster negative; coloring the melted wax in the corresponding skin tone; filtering the residues of resins and pigments through fine gauze; pouring the wax into the mold using a swivel technique; smoothing and hardening; and, finally, coloring the finished artifact with translucent oil paint.[21] This complex procedure is essential to achieving the artificial stillness of the material and to situating it in the intermediate realm between two states, thereby transcending dualisms—incidentally, this is one of the central properties attributed to wax, as often described by material theorists.[22]

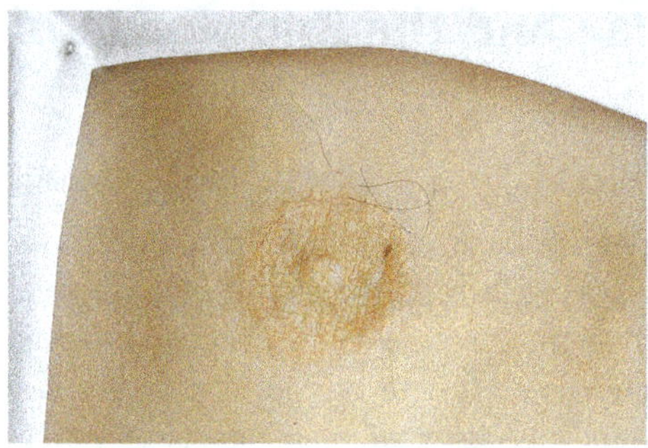

Fig. 3: Navena Widulin, skin detail of wax moulage, 2010.

One of the most important indicators of the inner activity of the material is contained in the production process of a wax moulage: the initial mixing of the substance. The lead preparator of the Museum of Medical History at the Charité, Navena Widulin, uses beeswax, calcium carbonate, and dammar resin, but the combination of ingredients varies considerably in each historical moulage, posing an additional challenge for conservation—or, as Widulin puts it, "Wax is never the same."[23] In her material experi-

---

21 For more details on the production process, see Navena Widulin, "Der Blick auf die Haut – Die Heidelberger Scharlachmoulage und die Fertigung des klinischen Wachsabdrucks," in *Spiegel der Wirklichkeit: Anatomische und Dermatologische Modelle in der Heidelberger Anatomie*, ed. Sara Doll and Navena Widulin (Berlin: Springer, 2019), 159–63.

22 "Wax is solid but can be easily melted. . . . It can be sculpted, modeled or cast, so the traditional hierarchies of the fine arts are unknown to him from the outset. It can be worked with the hands or with all kinds of tools; it can be opaque or transparent, matte or polished, smooth or sticky; its consistency can be modified infinitely by adding various resins. It is a fragile and perishable material, but it is usually used—precisely because of its textural richness—to make objects that are meant to last. The plasticity (or performance) of wax would thus consist primarily in this range of physically ambivalent properties." Georges Didi-Huberman, "Die Ordnung des Materials: Plastizität, Unbehagen, Nachleben," in *Die Ordnung des Materials: Vorträge aus dem Warburg-Haus* 3, ed. Wolfgang Kemp et al. (Berlin: Akademie-Verlag, 1999), 10–11. Translation by the author.

23 Navena Widulin (Human Remains Conservator and Museum Curator, Berlin Museum of Medical History at the Charité), in discussion with the author, October 6, 2020. Besides beeswax, paraffin or stearin are used far more frequently today. There is a multitude of possible waxes, both organic and synthetic.

ments in her laboratory, Widulin investigates the chemical and physical response of wax to its environment (humidity, climate conditions, pollution, pressure, chemical reactions to adjacent materials such as wood, lacquer, textiles, etc.). This method is in the historical tradition of using the ingredients of wax as indicators to monitor and influence its material response to the surrounding environment. Since at least the Renaissance, various wax recipes have been known to us from the writings of artists and art theorists, who used a range of wax ingredients to determine wax's response to various climatic conditions and mechanical impacts—for example, the degree of its malleability and durability.[24] One could say that—through extensive experiments and passed-down knowledge—the material has been individually "coded" to show different intrinsic activities and to fulfill different functions. The specific ingredients of the wax used could therefore be described as an operational mechanism, in the sense of wax being an active material.[25]

The "alchemical" use of the inherent material activity of wax is rooted in its very constitution. Starting from its origin as a product of bees' metabolism, it is one of the most complex organic substances produced, and it is characterized by continuous transformation processes. Wax moulages, for instance, can be melted down as often as desired without causing damage to the material. Yet, this "fixed instability" of wax also makes it one of the most fragile materials in the museum: "It deteriorates, passes away, vanishes more easily than others."[26] Unlike water, wax always remains wax in all its different states and conditions.[27] It is this "alchemical" constitution of wax that generates a very different kind of activation of the material, done at the margins of laboratory practice and unnoticed by the public.

To grasp the significance of this kind of activation, one must understand the direct precursor of the wax moulage: the anatomical wax model, of which Florence became the center of production in the eighteenth century.[28] In contrast to the moulage, which depicted pathologies, wax models showed a healthy and idealized human body. One famous type was the so-called anatomical Venus. The sexualization and passivation of

---

24 Vernon J. Murrell, "Some Aspects of the Conservation of Wax Models," *Studies in Conservation* 16, no. 3 (1971): 95–109.

25 See the description of material as a "set of operational mechanisms" in Wolfgang Schäffner, "Active Matter," in *23 Manifeste zu Bildakt und Verkörperung*, ed. Marion Lauschke and Pablo Schneider (Berlin: De Gruyter, 2018), 6.

26 Georges Didi-Huberman, "Viscosities and Survivals: Art History Put to the Test by the Material," in *Ephemeral Bodies: Wax Sculpture and the Human Figure* (Los Angeles: Getty Research Institute, 2008), 155.

27 This has been called a "paradox of consistency"; see Didi-Huberman, "Die Ordnung," 13. This evaluation is consistent with other theoretical reflections on the wax, such as its "amorphous and polymorphous potential"; see Monika Wagner, Dietmar Rübel, and Sebastian Hackenschmidt, ed., *Lexikon des künstlerischen Materials: Werkstoffe der modernen Kunst von Abfall bis Zinn* (Munich: Beck, 2002), 231.

28 Mühlenberend, "Wachsmoulagen," 78.

the female subject associated with it have been extensively analyzed in the literature.[29] The figure lies supine, often with a languid gaze—a symbol of passive surrender. It can be actively opened for the study of anatomy and its organs can be removed. As it happens, Navena Widulin's laboratory is populated by another Venus, a countermodel to this Florentine version of feminine passivity: a large number of casts of a small replica of the Venus of Willendorf, dated around 30,000 BC and often identified as a mother goddess (fig. 4). They stand defiantly, seemingly ready to march; even though their arms are not visible, they appear confident and combative. Without individual facial features, they seem to symbolize empowered femininity itself, whose physicality and fertility are proudly shown, not passively exposed. Begun as a personal project for the purposes of both documentation and entertainment, the Venuses are cast from the wax residues of each new moulage that Widulin produces, thus capturing the different skin tones that had been mixed. Widulin's Venuses are an impressive testament both to the importance of museum laboratory practice beyond public visibility and to the fact that the power of transience, decomposition, and change can also be a subversive force.[30]

Besides this potential of the material to transform and to provoke, it is the capacity of wax to store or record that characterizes its cultural history: from mummification to death masks, from "memory material" (*Gedächtnisstoff*) to the first writing tablets, wax has been considered a medium of preservation since antiquity.[31] "Wax never forgets," as Widulin puts it, referring to the fact that even the smallest pressure or temperature fluctuations are stored in the deepest layers of the material, and sometimes even dust fuses with it.[32] This in-between state of activity and passivity is also noticeable when shaping a piece of wax with one's hands. Wax records the unintentional, unconscious movements of the hand of the person shaping it, and even the imprints of the fingers. For Georges Didi-Huberman, this represents a liveliness beyond passivity.[33]

---

29 See Melissa Bailar, "Uncanny Anatomies/Figures of Wax," *Journal of the Midwest Modern Language Association* 49, no. 2 (2016): 29–53; Corinna Wagner, "Replicating Venus: Art, Anatomy, Wax Models, and Automata," *19: Interdisciplinary Studies in the Long Nineteenth Century* 24 (2017), https://doi.org/10.16995/ntn.783.

30 On the subversive and epistemic power of museum remains in a broader context, see Nina Samuel and Felix Sattler, ed., *Bildwelten des Wissens: Kunsthistorisches Jahrbuch für Bildkritik*, vol. 18, *Museale Reste* (Berlin, De Gruyter, 2022).

31 Didi-Huberman, "Die Ordnung," 5. On wax and memory, see also Pliny's report on the "imagines"—portrait busts and death masks made of wax, which record the remembrance of ancestors. In the presence of the lifelike depiction, the absence of the deceased is transcended. Wax effigies have been associated with the cult of the dead since antiquity (see Wagner et al., *Lexikon*, 233). On wax as a writing medium, see Andrea Jördens, Michael R. Ott, and Rodney Ast, "Wachs," in *Materiale Textkulturen: Konzepte – Materialien – Praktiken*, ed. Thomas Meier, Michael R. Ott, and Rebecca Sauer (Berlin: De Gruyter, 2015).

32 Widulin, discussion.

33 He calls this the "power of the material" of wax, which is related to its "viscosity." Didi-Huberman, "Die Ordnung," 13.

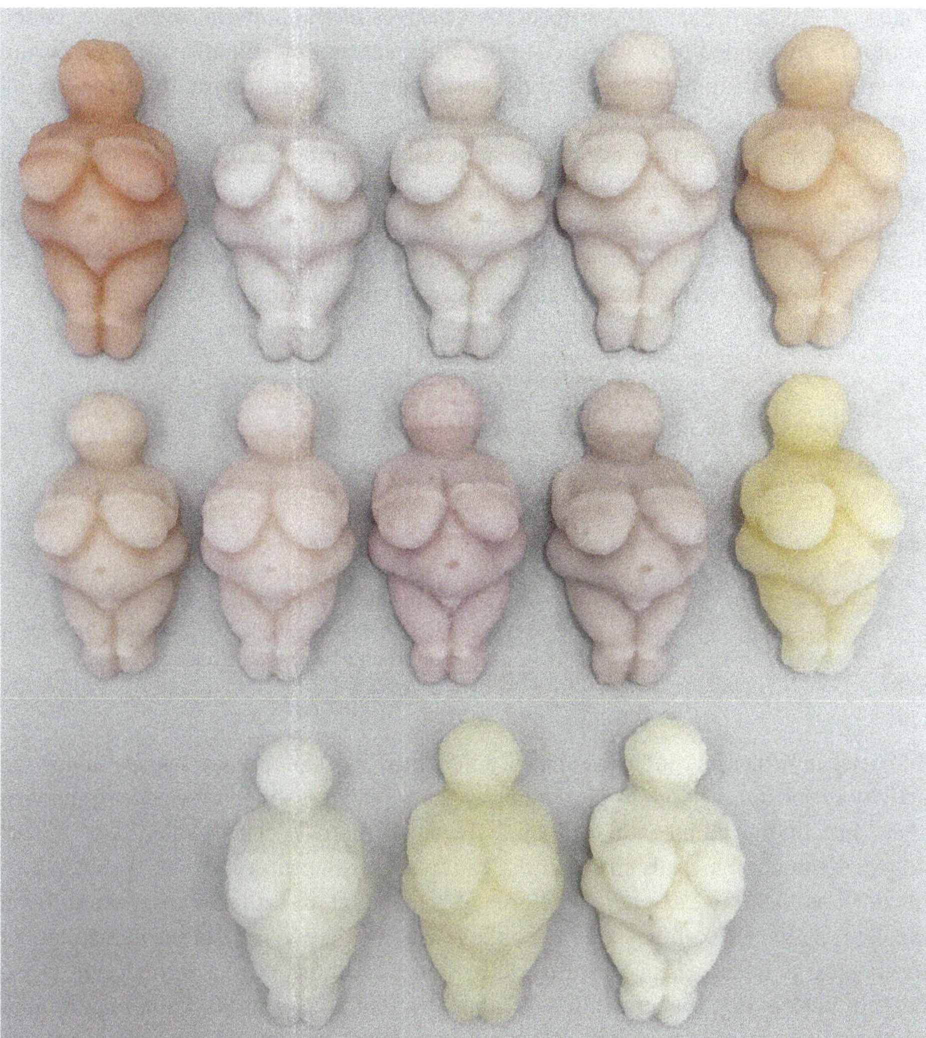

Fig. 4: Navena Widulin, wax copies of Venus of Willendorf, ongoing.

If one considers museum objects and their unintentional material activities, this intermediate state between transformation and preservation also proves to be a practical problem.[34] This is evident, for example, in the phenomenon of "efflorescence" or "blooming" of wax: an organic, internal process (or "auto-activity") of the material that is accelerated by its environment (fig. 5). Chemical reactions are triggered within the wax's constituents, and fatty acids come to the surface and form a white crystalline

---

34 Monika Wagner has concisely summarized these two aspects: in wax, "metamorphosis merges with conservation." Wagner et al., *Lexikon*, 235.

coating.[35] This is a normal transformation process of the material over time, which cannot be suspended or stopped but can be accelerated by climatic fluctuations and humidity.[36] Hence, impermanence and transience can lead to growth. However, in conservation, this auto-activity poses a major challenge because it can distort the appearance of the documented disease and raise questions about how to distinguish between the evolution of the material and the pathological condition. But wax reacts extremely sensitively not only to climatic fluctuations; in the history of conservation, moisture has been identified as one of wax's archenemies. This can be seen in the prehistory of today's struggle against the effects of water in conservation.

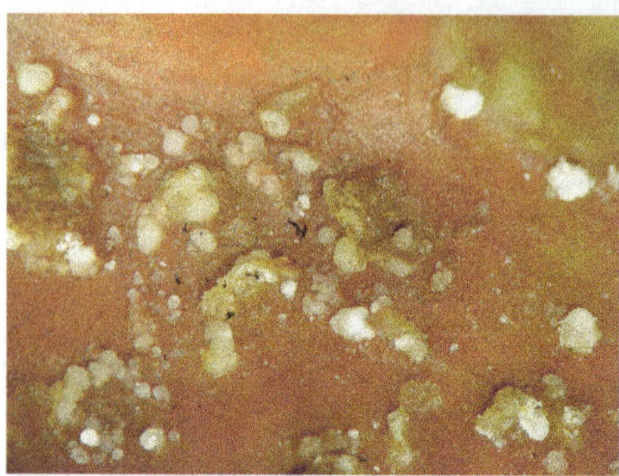

Fig. 5: Microscopic photography of wax surface with efflorescence and crystalline coating induced by climatic fluctuations, 1955. Moulage from the collection of Deutsches Hygiene Museum Dresden.

The science of conservation was heavily informed by scientific knowledge about the activity of substances and, in particular, by the rapidly growing knowledge of modern chemistry in the nineteenth century.[37] However, there is clear evidence of expert knowledge about materials and their activation already a century earlier. Water, in particular, was identified as one of the greatest challenges in artifact conservation from the late eighteenth century on, coinciding with the growing professionalization of the field. It is unsurprising that this discussion was initiated by a commission of experts in Venice—a place known for its fluctuating climatic conditions and high humidity. From their detailed report in 1777, it can be deduced that the experts' advice on conservation was based on a comprehensive knowledge of the activity of materials, especially of wood and dyes.[38] Moreover, the first-known discussions about creating

---

35 Widulin, discussion.

36 See Widulin, "Der Blick," 151; Patrick Dietemann, Ursula Baumer, and Christoph Herm, "Wachse und Wachsmoulagen: Materialien, Eigenschaften, Alterung," in *Körper in Wachs: Moulagen in Forschung und Restaurierung*, ed. Johanna Lang et al. (Dresden: Sandstein Verlag, 2010): 61–81, 76. For a different perspective on the influence of humidity on wax, see Alicia Sánchez Ortiz and Sandra Micó Boró, "Preventive Conservation Strategies for Wax Bodies in Scientific University Collections," *Conservation Science in Cultural Heritage* (2012): 219.

37 Wagner, *Arbeitsweisen*, 25.

38 "Picture supports made of canvas or wood work under the influence of fluctuating humidity, salts damage some pigments, and the adhesive power of the binder weakens when exposed to humidity,

a microclimate for storing works of art can also be dated to around the same time. Art-technological treatises of the time, for instance, recommend achieving this by paneling or wallpapering rooms to regulate humidity.[39] Thus, as early as the end of the eighteenth century, the foundation was laid for the conservation paradigm of dryness as the supreme condition for the stagnation of material activity.[40]

The demand for artificial cold was soon added to the paradigm of dryness in order to successfully passivate the material—a development that gained momentum as industrialization progressed in the second half of the nineteenth century: from the first patent for mechanical cooling, in 1851, in the United States to the founding of the first big firms for the large-scale production of artificial cold in Europe, around 1900.[41] The first air-conditioning systems installed in British museums in the 1920s were the result of an well-established debate about the ideal "museum climate"—a debate that was conducted at a high level in specialist journals already decades before.[42]

From the examples of the preparation of specimens and of the use of wax in conservation, we can conclude the following: fluids with different chemical compositions play an important role in passivation technologies in museums. For the creation of specimens, alcohol, as a highly manipulative liquid, ensures that the natural process of the decomposition of human organs is suspended; moulages, meanwhile, must be protected from humidity in order to slow down their material transformation processes. This realization by conservators marked the beginning of a long process that led to the establishment of a completely new concept of indoor air, in the museum and else-

---

which also endangers the adhesion of the painting layers." Wagner, *Arbeitsweisen*, 14. Translation by the author.

39 M. Mauclerc, *Traité des couleurs et vernis* (Paris: Ruault: L'auteur, 1773), 116; Wagner, *Arbeitsweisen*, 14.

40 This finding is also confirmed by research on textile restoration in the nineteenth century: "Contrary to possible general assumptions, many current conservation and restoration practices have their roots in the nineteenth century, or earlier. The negative effects of moisture and humid indoor climates on artifacts were known. … Some materials for maintenance and preventive conservation used more than a century ago, such as water repellents, have remained almost the same." Maria Brunskog and Johanna Nielsson, "Restoration of Flat Textiles: Ideological Framework, Ideas and Treatment Methods in Sweden before 1900," in *Conservation in the Nineteenth Century*, ed. Isabelle Brajer (London: Archetype, 2013), 178.

41 Kostas Gavroglu, "Historiographical Issues in the History of the Cold," in *History of Artificial Cold, Scientific, Technological and Cultural Issues*, ed. Kostas Gavroglu (Dordrecht: Springer, 2014), 3.

42 For the UK context, see J. P. Brown, and William B. Rose, "Humidity and Moisture in Historic Buildings: The Origins of Building and Object Conservation," *APT Bulletin: The Journal of Preservation Technology* 27, no. 3 (1996): 12–23, 14. In Germany, this development started only in 1955; see Martina Griesser-Stermscheg, *Tabu Depot: das Museumsdepot in Geschichte und Gegenwart* (Vienna: Böhlau, 2013), 86–88. See also Mattias Legnér, "On the Early History of Museum Environment Control Control, Nationalmuseum and Gripsholm Castle in Sweden, c. 1866–1932," *Studies in Conservation* 56 (2011): 125–37; Andrea Luciani, "Historical Climates and Conservation Environments: Historical Perspectives on Climate Control Strategies within Museums and Heritage Buildings." PhD diss. (Politecnico di Milano, 2013), https://www.academia.edu/3505879/Historical_climates_and_conservation_environments_Historical_perspectives_on_climate_control_strategies_within_museums_and_heritage_buildings.

where: the idea of "synthetic air."[43] Temperature control and manufactured microclimates have played a central role in perfecting the passivation of materials in museum interiors.

The Paradigm of Conservation and Architectural Discourse:
From Broken Wax to Cultural Heritage

Drying and cooling are important control parameters for preventive conservation, which has a purely precautionary character. Just as central to daily museum work, however, are issues arising from active conservation and restoration, which involve either substance-preserving or reconstructive measures.[44] In the case of significant damage to or destruction of an object, what is to be done is decided on an individual basis. That very different approaches can be taken is evident from comparing the treatment of two moulages from the collection of the Berlin Museum of Medical History at the Charité, both of which had been largely destroyed (figs. 6/7). Figure 6 shows a woman with an eye tumor, from around 1909. We know from the records that the patient, "an old woman from the infirmary [Siechenanstalt]," refused surgery and instead covered her face with a damp cloth.[45] Thanks to the fortunate find of an illustration in the *Atlas der äußeren Augenkrankheiten* (1909) that shows the moulage before its destruction, it was decided, in 2007, to restore it completely, in such a way that the traces of the reconstruction would not be visible—a rather unusual decision in view of today's *Code of Ethics for Museums* (fig. 8).[46] This was accomplished by one of the most renowned experts in wax moulages, Elfriede Walther of the Hygiene-Museum Dresden. She glued the existing wax fragments together, added missing parts, applied new paint, and inserted a new linen binding. The moulage in figure 7, however, could not be assigned a medical diagnosis and is undated. In this case, experts decided not to reconstruct it, other than to stabilize it, leaving it in this condition and thus respecting its deteriorated state.[47]

---

43 Wulf Böer, "Synthetic Air," *Future Anterior: Journal of Historic Preservation, History, Theory, and Criticism* 13, no. 2 (2016): 77–101.

44 On the priority of active and preventive conservation over restoration, see Ute Hack, "Wachsmoulagen: Restauratorische Grundsätze," in *Körper in Wachs: Moulagen in Forschung und Restaurierung*, ed. Johanna Lang et al. (Dresden: Sandstein Verlag, 2010); see especially pages 43–44.

45 This description is from the label text for the moulage at the Berlin Museum of Medical History at the Charité. Widulin, discussion.

46 International Council of Museums, *Code of Ethics for Museums* (Paris: 2017), 15; also available online at https://icom.museum/wp-content/uploads/2018/07/ICOM-code-En-web.pdf.

47 Widulin, discussion.

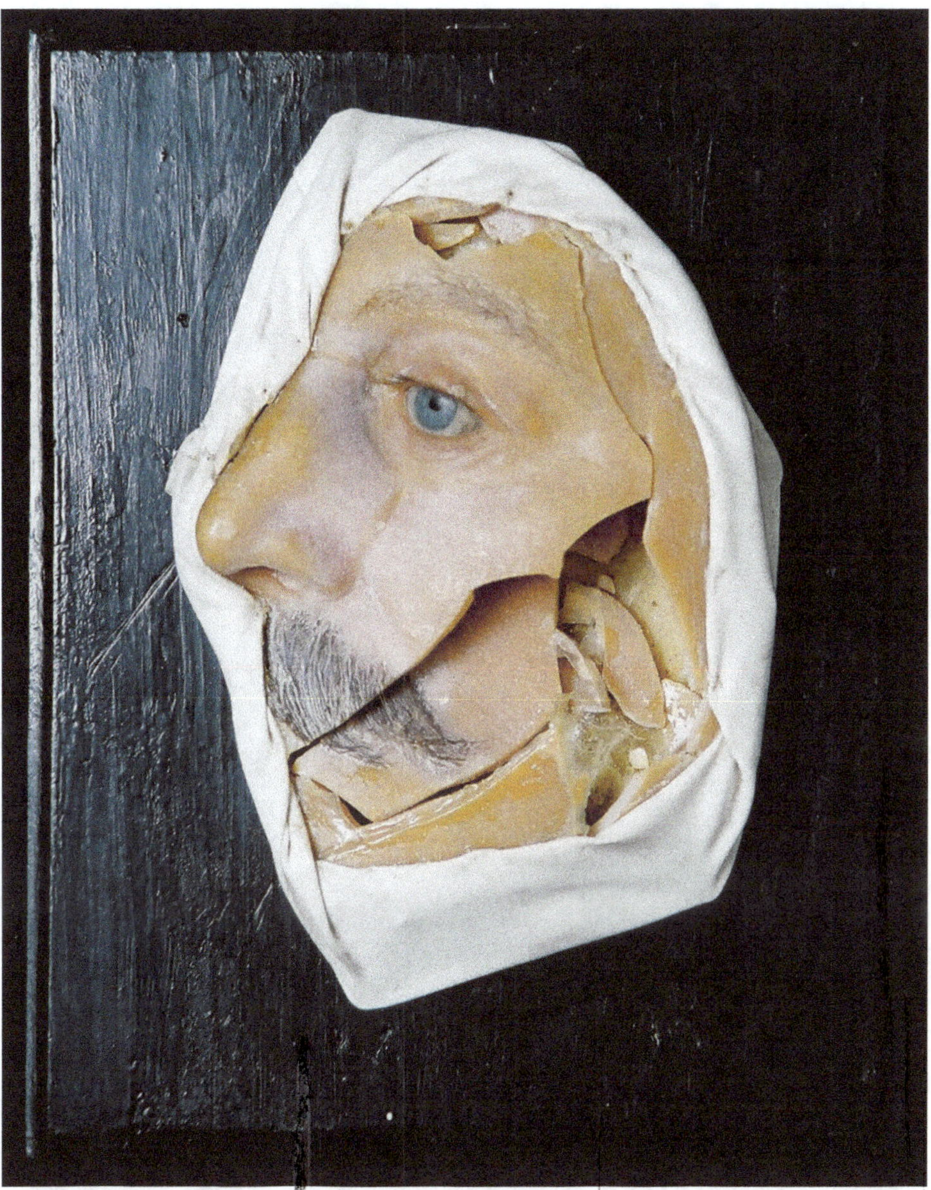

Fig. 6: Destroyed moulage, ca. 1919, Berlin Museum of Medical History at the Charité (BMM).

In these two decisions, two different concepts of an object (and thus its treatment) crystallize—both of which emerged in the history of conservation over the course of the nineteenth century. The first concept holds that an object was "complete" at a clearly defined point in time in the past, and thus this completeness should be achieved again, even after total destruction, or the completeness of its past state should be frozen. This can be called the path of reconstruction and the imitation of a known or imagined

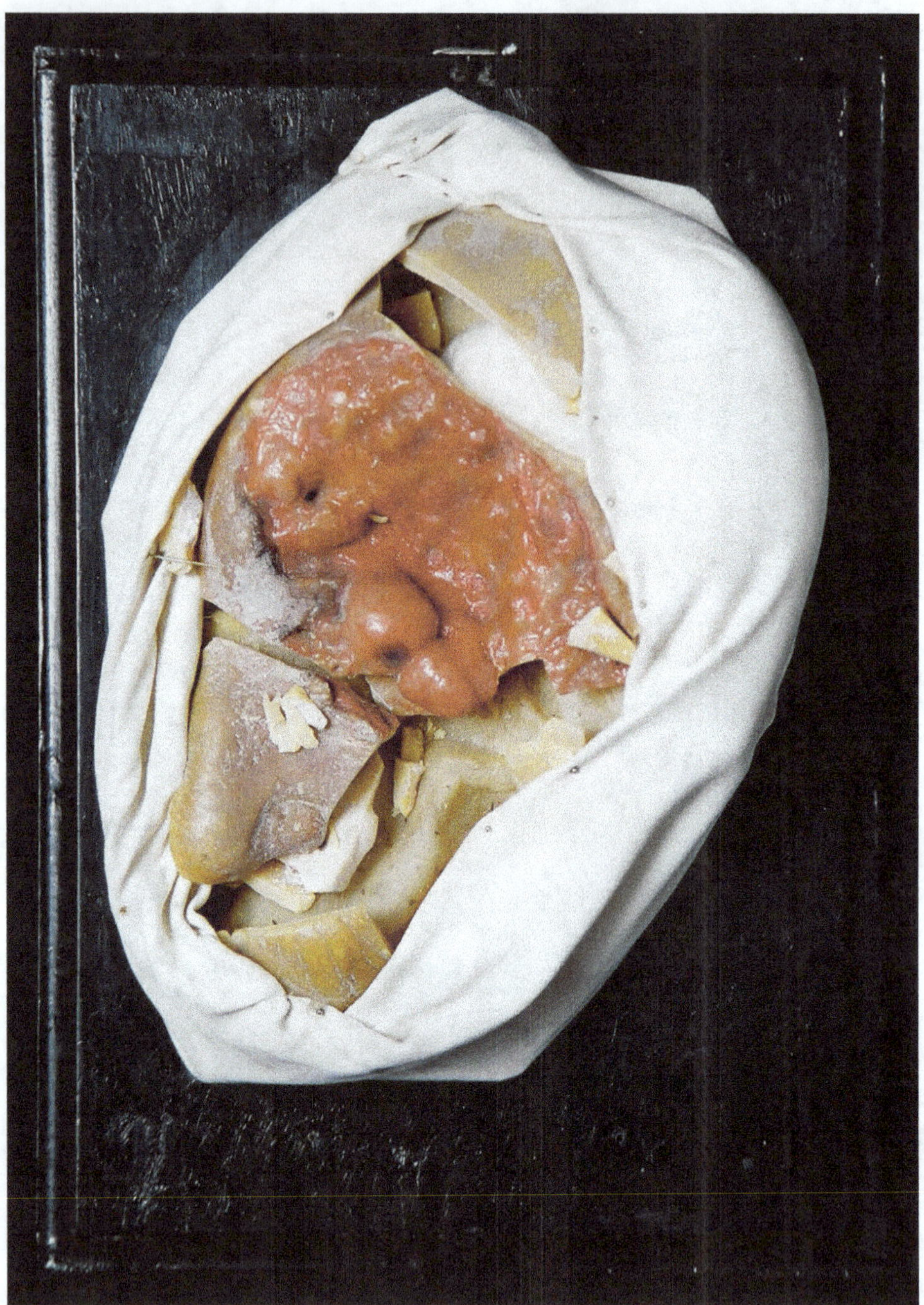

Fig. 7: Destroyed moulage, *Carcinoma epitheliale palpebrarum* (reconstructed findings), ca. 1910, Berlin Museum of Medical History at the Charité (BMM).

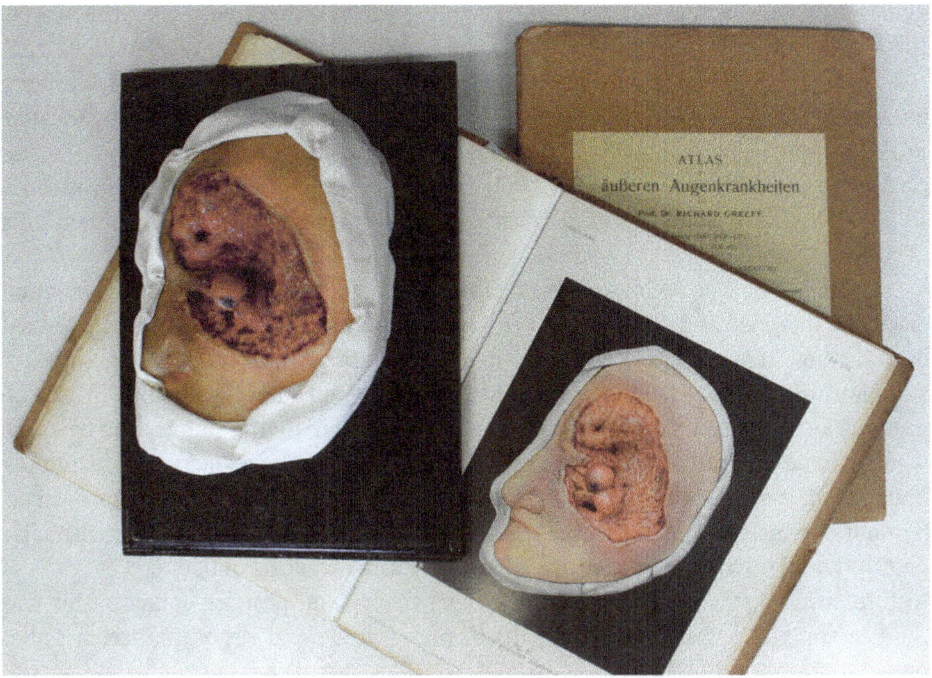

Fig. 8: Restored moulage by Elfriede Walter (1919–2018) based on an illustration in the Atlas of External Eye Diseases, Dr Richard Greeff, 1909, Fig. 21. Berlin Museum of Medical History at the Charité (BMM).

style. The second concept of an object views it from the standpoint of the course of time: all the effects of material change are seen as worthy of preservation, considered "complete" in their states of transience, impermanence, and change. This can be called the path of maintenance or of freezing the status quo—to preserve the physical side of the object, its historical substance. This second notion of the object was particularly influential in the theories of conservation in the twentieth century and led to an understanding of materiality as information. This involved the principle that "every restoration . . . implies a modification of cultural heritage and means a loss of information"[48] —and hence restoration should remain an exception.

Perhaps the best-known and most influential clash of these two concepts of the object—and, by extension, two different ways of grappling with the past and with remembrance—occurred over theories of the preservation of cultural heritage in the nineteenth and early twentieth centuries, as exemplified by the controversy between two of the most prominent architectural theorists of their time, Eugène-Emmanuel Viollet-le-Duc and John Ruskin.[49] When San Marco Basilica's bell tower collapsed in

---

48 Katrin Janis, *Restaurierungsethik im Kontext von Wissenschaft und Praxis* (Munich: Martin Meidenbauer, 2005), 139. Translation by the author.

49 See David Spurr, *Architecture and Modern Literature* (Ann Arbor: University of Michigan Press, 2012), 142–61.

Venice in 1902, the contrast between these two concepts became more pronounced, sparking a public debate that continued the previously established dialectic but that reduced it to mere polar oppositions. Although both Viollet-le-Duc and Ruskin were no longer alive at the time of this specific debate (the former died in 1879, the latter in 1900), their ideas remained vital. Those subscribing to Viollet-le-Duc's theory advocated the need to restore such a symbolic feature of the city's skyline, whereas those siding with Ruskin's opposed its reconstruction because this action would amount to a falsification of history.[50]

Viollet-le-Duc was famous for his definition of restoration as "reconstruction of an originally intended and perfect state, as it had possibly never existed before."[51] He had demonstrated this approach, beginning in the 1830s, through his numerous creative restorations of medieval buildings in France.[52] Ruskin spoke out against this in his collection *The Seven Lamps of Architecture* (1849): "Neither by the public, nor by those who have the care of public monuments, is the true meaning of the word *restoration* understood. It means the most total destruction which a building can suffer: a destruction out of which no remnants can be gathered: a destruction accompanied with false description of the thing destroyed."[53] Ruskin protested what he understood as a loss of truthfulness, which he thought Viollet-le-Duc's theory and practice entailed. This was the beginning of a discourse on "truth" in preservation and restoration science that persisted until the 1980s.[54]

The school of thought inspired by Viollet-le-Duc won in the case of the bell tower, whose reconstruction was finished in 1912.[55] In the long run, however, it was Ruskin's idea of prioritizing conservation over reconstruction that prevailed as the guiding prin-

---

50 Quoted in Natalia Escobar Castrillón, "Introduction," *OBL/QUE* 1 (2016): 4.

51 Violett-le-Duc in 1866, quoted in Gerald Unterberger, "Restaurierung – Restauration: Eine moderne Begriffsdifferenzierung und die ursprüngliche Bedeutung eines Wortes im mythisch-kultischen Kontext," *Muttersprache: Vierteljahresschrift für deutsche Sprache* 122, no. 3 (2012): 203–14, 204. Peter Kurmann has pointed out the ambivalences and contradictions of this definition; see Peter Kurmann, "Viollet-le-Duc und die ironisierte Kunstgeschichte," *Geschichte der Restaurierung in Europa* 2 (1993): 54–56.

52 The attitude of imitation and creation embodied by Viollet-le-Duc was not only discussed in the preservation of historical monuments. This development can also be clearly traced in the history of the restoration of mural paintings. Again, the change occurred around 1900: conservation before restoration was now the guiding principle. See Markus Santner, "Erhaltung, nicht Restaurierung – Der Paradigmenwechsel um 1900," in *Bild versus Substanz: Die Restaurierung mittelalterlicher Wandmalerei im Spannungsfeld zwischen Theorie und Praxis (1850–1970)* (Vienna: Böhlau, 2016).

53 John Ruskin, *The Complete Works of John Ruskin*, ed. E. T. Cook and Alexander Wedderburn, vol. 7, *The Seven Lamps of Architecture* (New York: Longmans, Green, and Co, 1903), 242.

54 Natascha Bäschlin, *Fragile Werte: Diskurs und Praxis der Restaurierungswissenschaften 1913–2014* (Bielefeld: transcript, 2020), 86.

55 This case continues to shape how we think about conservation even now, according to Natalia Castrillón: "The idea that history could be studied as empirical science and that historic truth and authenticity could be retrieved through rigorous restorations." Castrillón, "Introduction," 4.

ciple in the protection of cultural assets.[56] Systematized and further developed by Georg Dehio and Alois Riegl, it became a foundation of the influential Venice Charter for the Conservation and Restoration of Monuments and Sites in 1964.[57] It was in this context that the still-valid principles of preservation and acceptance of changes—caused by natural aging, minimal intervention, reversibility, and making the restoration process visible—were formed.[58] One of the key consequences of this development was the demand by the International Council of Museums, in their 1986 Code of Professional Ethics, that museum collections only include works of art that the museum is also able to preserve.[59] The prevailing paradigm of conservation and perfect passivation also affected collection development and strategies. The possibility of objects' material preservation thus acted as a filter for which objects were allowed to enter museums at all.

Three main elements of a history of material passivation can be identified in the examples provided above, and all of them share a legacy stretching from the late eighteenth to early twentieth century: first, the correlation with the beginning of the age of the modern museum; second, the increase in discussions about water as a threat to conservation and the role of climatic influences and intrinsic material activities for conservation strategies (of a material such as wax); and, third, the introduction of a new discourse in the field of architectural preservation through Ruskin, from which the principles of contemporary conservation have developed. The various approaches to conservation and preservation have contributed to a historical moment that now makes it possible, from both academic and practical perspectives, to reflect on what preservation actually means: either reconstructing the past or maintaining the status quo. From today's perspective, it seems safe to add that these historical developments and constellations have significantly contributed to our current situation: living in an age that, in the words of archaeologist Cornelius Holtorf, is "obsessed with preservation."[60] Urgent questions arise for the present.

---

56 The paradigm shift of "conservation before restoration" began around 1900. See, for example, Santner, "Erhaltung" (see note 52). Santner refers mainly to historical wall painting. The development of the debate in architecture should be distinguished from debates over cultural assets (see Castrillón, "Introduction," 50]). However, all these debates have their origins in the discourses on the preservation of monuments (or heritage preservation), which was influential in the nineteenth century (see Janis, *Restaurierungsethik*, 19).

57 Janis, *Restaurierungsethik*, 18–24.

58 See Hack, "Wachsmoulagen." At this point, a new understanding of the preservation of "authenticity" was emerging, which encompassed the entire materiality of a work of art (e.g., in the case of paintings, preservation included the frame, the reverse side, etc.); see Bäschlin, *Fragile Werte*, 86–96.

59 Bäschlin, *Fragile Werte*, 85. It was not until around 1998 that the debate began about whether this requirement was outdated.

60 Cornelius Holtorf, "The Heritage of Heritage," *Heritage & Society* 5, no. 2 (2012): 153–74.

Notes on the Future of Conservation

To paraphrase Rem Koolhaas's proclamation in his 2004 manifesto, while preservation was an invention of modernity, we currently find ourselves in a moment in which preservation appears to be "overtaking us."[61] He illustrates this development with two diagrams (fig. 9). Since the end of the eighteenth century, the length of time between the construction of a building and it being classified as worthy of preservation has steadily decreased; in fact, buildings are now often classified as worthy of preservation even before their completion.[62] This development has long since reached museums where, for example, in the British Museum, only 1 percent of preserved items are on display, and the rest are stored in cool and dry places.[63] In addition to this danger of *over-conservation*, there is also the other extreme in our relationship with the past and cultural heritage, which might be called *under-conservation*. Consider, for example, the natural history displays of Charles Willson Peale in the United States (fig. 10). In his displays from around 1800, his interest in taxidermy and zoological specimens merged with his enthusiasm for two-dimensional pictorial spaces: he placed taxidermy animals in the foreground, against a landscape painting in the background, in order to bring the habitat of the animals to life in a new way.[64] Unfortunately, in the late twentieth century, all his displays were destroyed—through neglect and a lack of awareness of their historical value—and, with them, the opportunity to study the displays as precursors to the dioramas that became popular in nineteenth-century natural history museums.[65]

The question of how to engage with the paradigm of conservation is crucial for the present because it determines our notion of remembrance. An example from around 1800 underlines this: when Alexander von Humboldt visited Latin America, two fundamentally different concepts of memory culture collided. He was disturbed when he learned how old monuments were used to construct new buildings, and he set about ending this practice, which he saw as sacrilege, in order to preserve the monuments.[66]

---

61 Rem Koolhaas, "Preservation Is Overtaking Us," *Future Anterior, Journal of Historic Preservation: History, Theory, and Criticism* 1, no. 2 (2004): 1–3. See also Rem Koolhaas and Jorge Otero-Pailos, *Preservation Is Overtaking Us* (New York: Columbia University Press, 2014).

62 Koolhaas, "Preservation," 2.

63 Griesser-Stermscheg, *Tabu Depot*, 90. On the "expansive evolution" of the concept of objects which are now judged to be worthy of preservation, see Salvador Muñoz Viñas, "Contemporary Theory of Conservation," supplement, *Studies in Conservation* 47, S1 (2002): 27–28.

64 I thank Claudia Blümle for mentioning Peale to me. See Claudia Blümle, "Hier – Schau hin: Überzeugungsfiguren im Bild," in *Überzeugen: Szenarien von Zeugenschaft und ihre Akteure*, ed. Matthias Däumer, Aurélia Kalisky, and Heike Schlie (Paderborn: Wilhelm Fink, 2017), 167–93, especially 181–87.

65 See Stephen Christopher Quinn, *Windows on Nature: The Great Habitat Dioramas of the American Museum of Natural History* (New York: Abrams, 2006), 13–14.

66 I thank Wolfgang Schäffner for this reference. For more details on Alexander von Humboldt's journeys, see Vera M. Kutzinski, Ottmar Ette, and Laura Dassow Walls, ed., *Alexander von Humboldt and the America* (Berlin: Verlag Walter Frey, 2012).

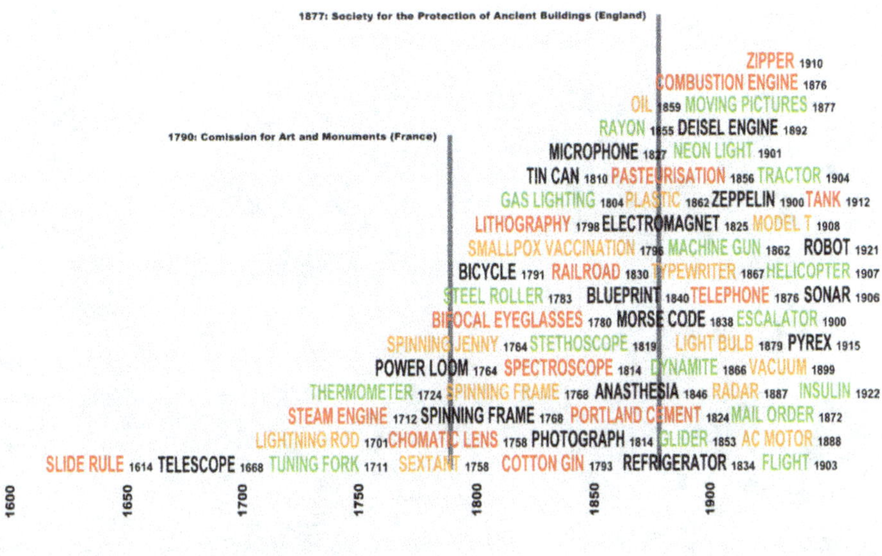
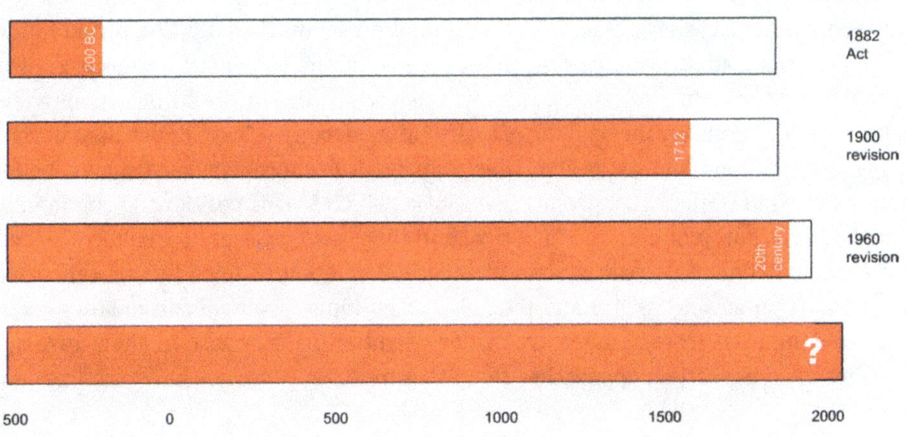

Fig. 9: Rem Koolhaas, *Preservation Is Overtaking Us* (New York: Columbia University Press, 2014), 45.

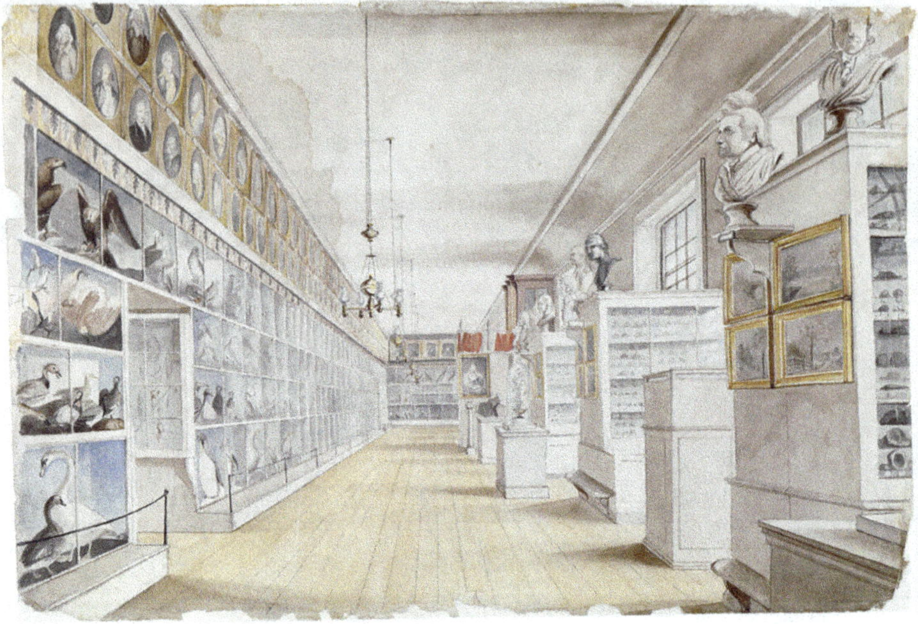

Fig. 10: Charles Willson Peale and Titian Ramsay Peale, *The Long Room, Interior of Front Room in Peale's Museum*, 1822, watercolor over graphite pencil on paper. Detroit Institute of Arts.

The central question is thus as follows: Do we incorporate objects and make something new out of them, or do we freeze and immobilize them?

There is a third possible answer to this question, which stems, once again, from architecture. Starting in the 1950s, new constellations between conservation and restoration emerged. A groundbreaking early example of this development is Carlo Scarpa's transformation of Castelvecchio in Verona into a museum (fig. 11).[67] The preservation of the historical substance triggered innovative solutions in the field of architecture, transforming the entire site into a display; meanwhile, the medieval objects on view acquired a new visibility through the *dispositif* of modernity—Scarpa's use of materials and forms and their contrast to the older historical substance. Figure 11 shows the famous equestrian statue of Cangrande della Scala, which Scarpa positioned prominently against the backdrop of both the historical and the recently restored architecture, and which he embedded in a path system through the site that enabled the viewer to see the statue from various angles and thus also in multiple historical constellations.

These impulses from architecture can be used to further reflect on alternative approaches to conservation, approaches that focus more on processes and practices—or,

---

67 See Manuel J. Martín-Hernández, "Architecture from Architecture: Encounters between Conservation and Restoration," *Future Anterior: Journal of Historic Preservation, History, Theory, and Criticism* 4, no. 2 (2007): 62–69.

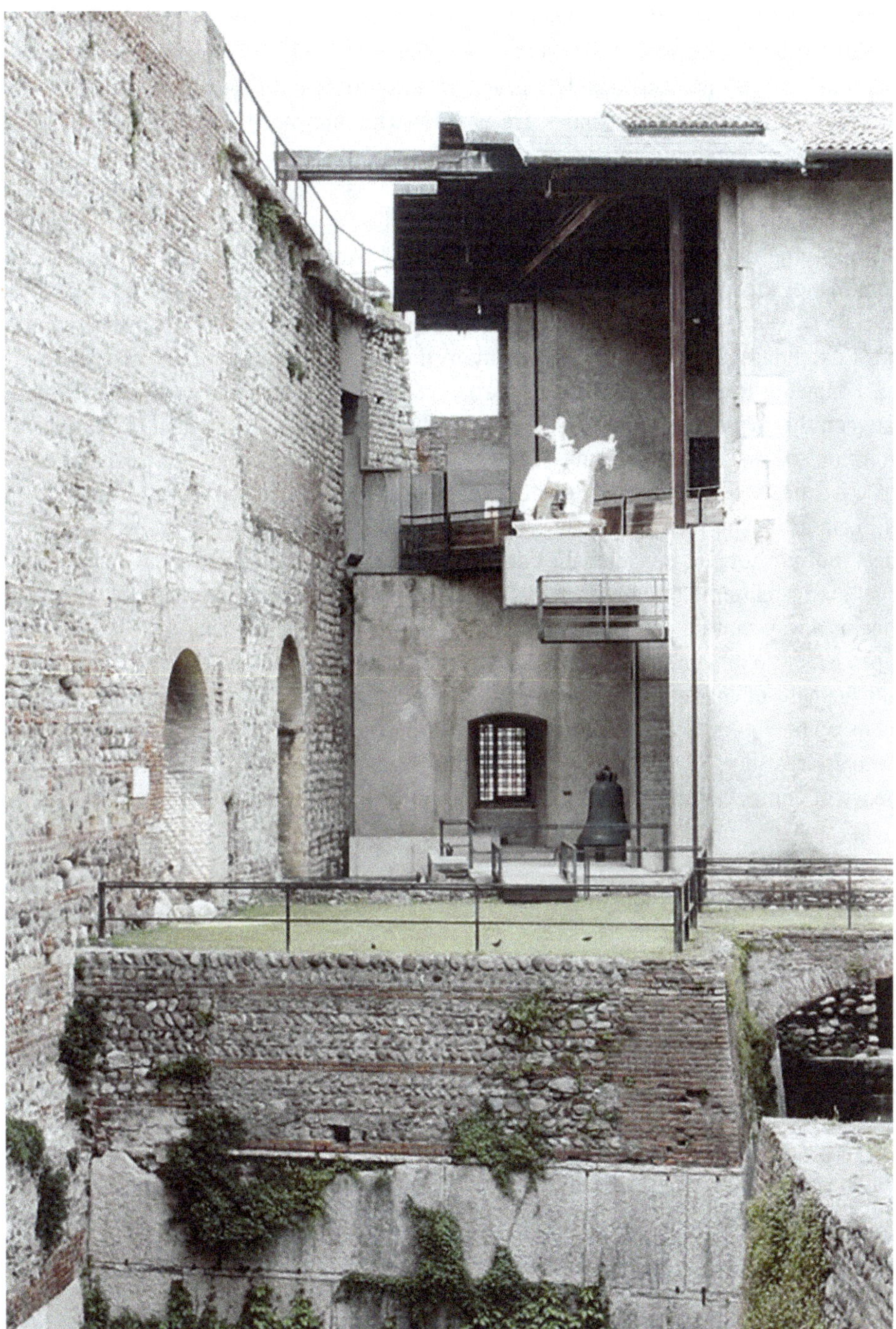

Fig. 11: Carlo Scarpa (architect), Castelvecchio Museum, Verona, 1956 – 74.

as cultural heritage researcher David Lowenthal summarized: "Preferring fragments to wholes; written, painted or mental images to physical objects; and processes to material entities."[68] If one also considers practices and processes as worthy of preservation, different conservation approaches are possible. One approach is based on repetition instead of perpetuation and is exemplified by the well-known Ise Grand Shrine in Japan. Cyclical destruction and reconstruction are used to conserve both practices of worship and practices of the architectural construction associated with it, as opposed to preserving the actual substance of the shrine.[69] Another approach can be derived from one of Ruskin's terms: "voicefulness."[70] If voices of the past are stored in the objects, they cannot be made audible by material preservation, but rather through an ethics of extended usage, as discussed today in the field of ethnographic conservation.[71]

Following the thinking of Bruno Latour, what is needed is an intermediate path between the two extremes of idolatry and iconoclasm—that is, between the uncritical attention to images (the attitude of over-conservation) on the one hand, and the destructive neglect of images (the attitude of under-conservation) on the other.[72] To use Latour's term, this third way would be one of "iconophilia," in the sense of "respect, but not for the image itself, but for the movement of the image . . . for the passage, the transition from one pictorial form to another."[73] Applied to conservation science in a way that takes into account its developments over the last two decades, this approach of no longer focusing exclusively on protection could be described as "careful management of change."[74] As much as this challenges core conservation values, it also opens up new perspectives, one of the most important being the "shift from protection towards creation . . . and embracing communities' associations."[75] The contemporary theory of conservation, which is critically reflecting on the preservation paradigm, is

---

68 David Lowenthal, "Material Preservation and its Alternatives," *Perspecta* 25 (1989): 66–77, 77.

69 Lowenthal, "Material Preservation," 73.

70 Bäschlin, *Fragile Werte*, 88.

71 Because "the objects themselves are not important, what matters is what the objects represent." Miriam Clavir, "Preserving the Physical Object in Changing Cultural Contexts," in *The International Handbook of Museum Studies: Museum Transformations*, ed. Sharon Macdonald and Helen Rees Leahy (West Sussex: John Wiley and Sons, 2015), 389.

72 Bruno Latour, "Wie wird man ikonophil in Kunst, Wissenschaft und Religion?" *Zeitschrift für Ästhetik und Allgemeine Kunstwissenschaft* 57, no. 1 (2012): 19–44. It must be noted here that the attitude of under-conservation cannot be equated across the board with the iconoclastic destruction of images or objects, as is known from the history of art; see Dario Gamboni, *The Destruction of Art: Iconoclasm and Vandalism since the French Revolution* (London: Reaktion Books, 1997). That said, indifference and historical oblivion, which can be described as attitudes driving under-conservation, can also lead to material destruction, as in the example of Charles Willson Peale's displays. I thank Claudia Blümle for the reference to Latour.

73 "We should pay respect to the series of transformations for which each individual image is only a provisional one," Latour, "Wie wird man," 21. Translation by the author.

74 See Bäschlin, *Fragile Werte*, 140.

75 Ioannis Poulios, "Moving beyond a Values-Based Approach to Heritage Conservation," *Conservation and Management of Archaeological Sites* 12, no. 2 (2010): 170–85, 182.

being called upon to incorporate this question into the development of new strategies for the present.[76] This involves, as the research of Hanna Hölling has shown, "recognizing larger goals related to the intangible: the transmission of tradition, memory, skill, technique, and tacit knowledge."[77]

As we have seen from Scarpa, each conservation method also produces new visibilities. Yet this raises the question of what we in fact want to make visible, and how. More specifically, we might ask: What kind of new visibility can be created that frames "processes of decay … as … productive,"[78] and how can this influence contemporary artifact design? And, to take this further, what might happen if the activity or performance of the material were not understood as a disturbance or dysfunction but were instead viewed in a positive light?

To touch upon this question briefly, it is worth returning to one of the materials discussed above—wax—and its uses in applied sciences that deal with processes of decay, as opposed to its use in conservation. It is precisely the aforementioned intrinsic activity of wax that has recently been researched in the field of transient electronics. Tailored wax-based materials are currently being coded in such a way that they can operate in a stable fashion over extended periods of time in aqueous environments before dissolving completely. For example, researchers studied the functional use of a partially biodegradable wireless LED (fig. 12). There are highly promising applications of these waxes, such as in biomedical implants that are able to self-destruct after a predetermined lifespan, thereby avoiding surgical extraction and the waste management of electronics.[79] One could call this a "coded decay of the material."[80]

While this would not be a sustainable approach for museum objects such as specimens or wax moulages, since they would lose their function as educational material, it opens up perspectives beyond questions of museum conservation. For the future, it remains to be seen whether part of our ecological responsibility will be to primarily produce artifacts that dissolve without residue on a fixed date. If materials were truly understood in terms of their "performance" rather than merely as an accumulation of

---

76 On the change in preservation science in progress, see Bäschlin, *Fragile Werte*, 141. See also Jorge Otero-Pailos, Erik Langdalen, Thordis Arrhenius, ed., *Experimental Preservation* (Zurich: Lars Müller, 2016).

77 Hanna Hölling, Francesca G. Bewer, and Katharina Ammann, *The Explicit Material: Inquiries on the Intersection of Curatorial and Conservation Cultures* (Leiden, Boston: Brill, 2019), 4. See also Hanna Hölling, "The Technique of Conservation: On Realms of Theory and Cultures of Practice," *Journal of the Institute of Conservation* 40, no. 2 (2017): 87–96.

78 Caitlin DeSilvey, *Curated Decay: Heritage Beyond Saving* (Minneapolis: University of Minnesota Press, 2017), 5.

79 Sang Min Won et al., "Natural Wax for Transient Electronics," *Advanced Functional Materials* 28, no. 32 (2018): 1801819, https://doi.org/10.1002/adfm.201801819.

80 Another historical example of the intrinsic activity of wax could be one of the oldest molding processes for metal and glass casting known since antiquity: the lost-wax casting in which the wax form is lost (i.e., melts) in order to obtain a metal sculpture. During the process, both the model and the mold are destroyed.

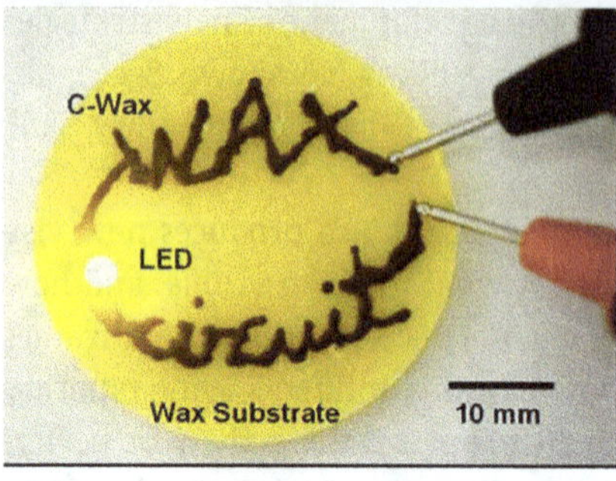

Fig. 12: Sang Min Won et al., "Natural Wax for Transient Electronics," *Advanced Functional Materials* 28, no. 32 (2018): 1.

properties,[81] the age of cooling and drying, which this article has explored, could soon give way to a new era in which the various states and process qualities of materials are not fought, mitigated, or suspended, but are made more productive.

---

81 Bensaude-Vincent, "Materials as Machines," 20.

Material Energies

Horst Bredekamp

The Material Metaphysics of Felt

The Hat

A man, with his upper body exposed, is gazing at the viewer with an intensity that is hard to forget (fig. 1). The features of the man portrayed seem familiar, but it is difficult to put a name to his face, especially as his pointed ears also lend his visage a Mephistophelian air. The reason for this mixture of familiarity and uncertainty is that this head lacks that prop with which it seems to have grown together (fig. 2). It was taken by the photographer Charles Wilp, a friend of Joseph Beuys, on a working vacation in Kenya in 1975.[1] The characteristic hat, as part of the artist's identity, represented a kind of trademark, without which his appearance seemed incomplete and unfamiliar.

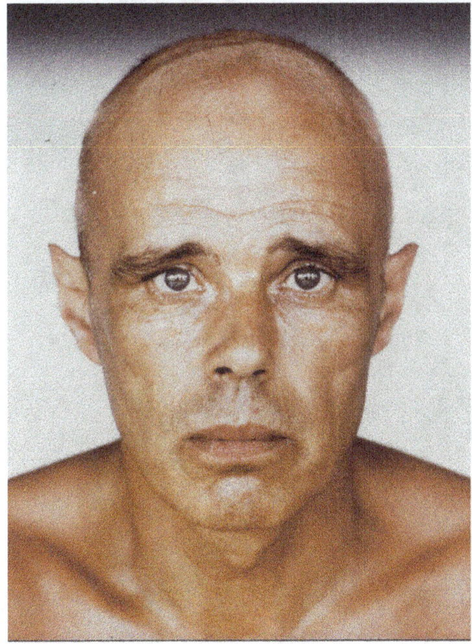

Fig. 1: Joseph Beuys, photograph by Charles Wilp, Kenya, 1975, published by Qumran Verlag, Frankfurt, Untitled, from the portfolio "Nature Experiences in Africa."

---

1 See Franz-Joachim Verspohl, "Joseph Beuys – Das ist erst einmal dieser Hut," *Kritische Berichte* 14, no. 4 (1986): 77.

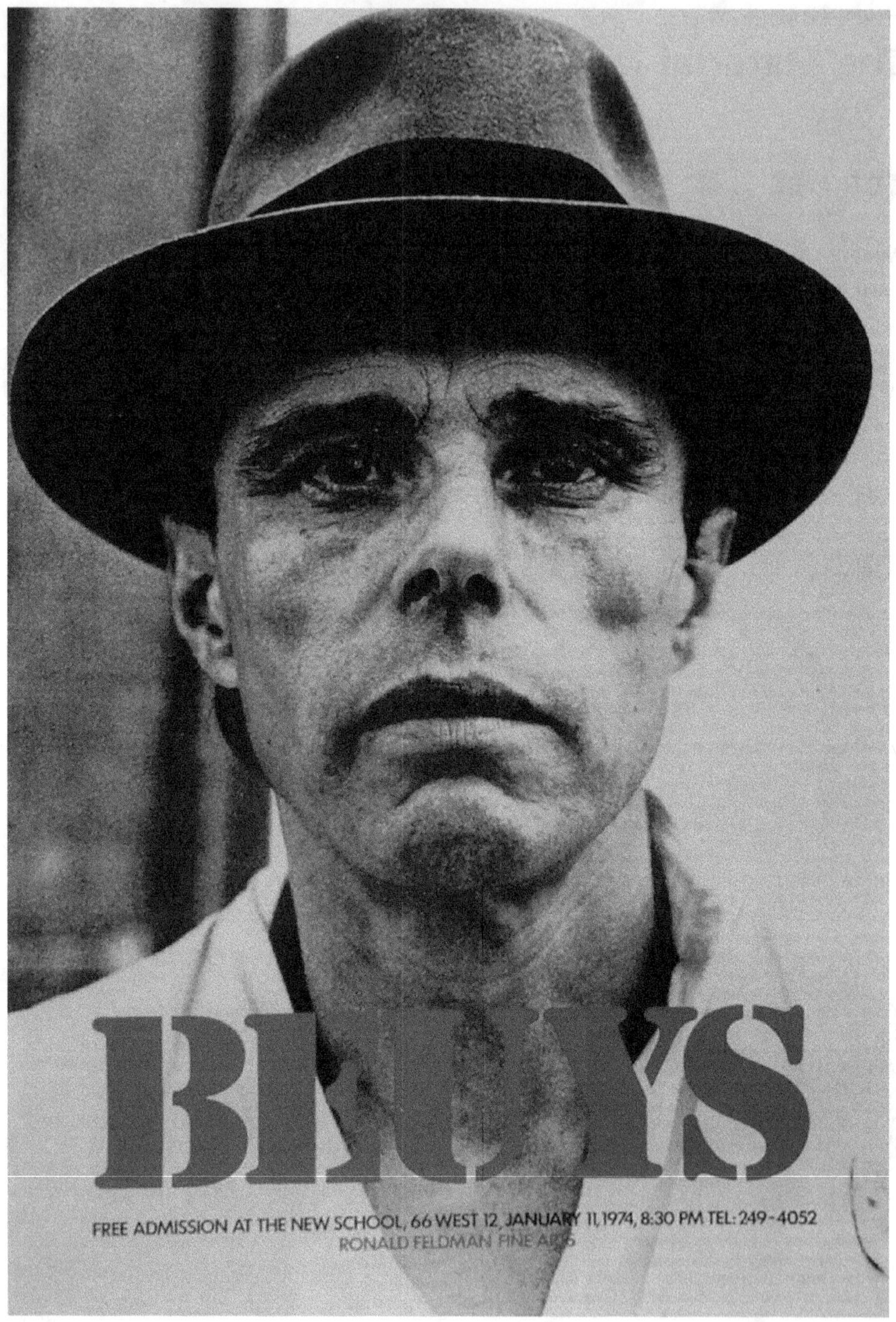

Fig. 2: Joseph Beuys poster for the US lecture tour *Energy Plan for the Western Man*, organized by Ronald Feldman, offset print, 1974.

In 1980, Andy Warhol produced a portrait of Beuys, in which by inverting the colors he emphasized the hat and, here especially, the hat band as the artist's insignia.[2] This is probably the most accurate portrait of Beuys ever made. The hat is irrevocably associated with the artist. He almost never took it off, not even in moments of extreme challenge. I experienced an enigmatic confirmation of this *habitus* myself as a young student on February 22, 1969, at the Akademie der Künste (Academy of Arts) in Berlin, during an hours-long political brawl. Firehoses were activated, flooding the stage so that those present cowered or moved away. Beuys, by contrast, never removed his hat (fig. 3).[3] On a superficial level, keeping his hat on would have been connected to protection from the water; like a compact umbrella, something a hat always is, and the reason head coverings were invented. An additional meaning resulted automatically, however, associating this mechanical shield from water with an aura of inviolability.

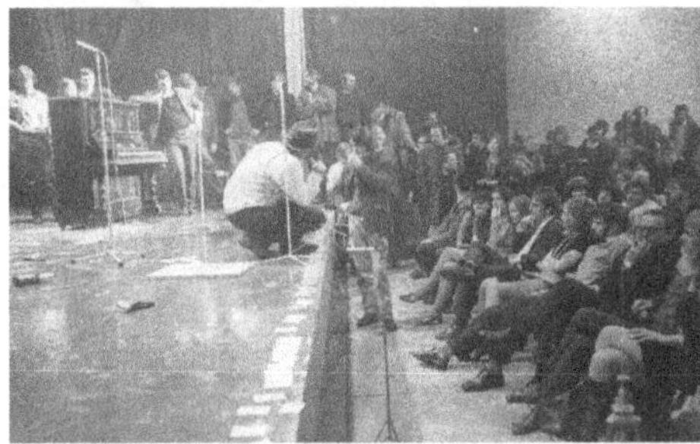

Fig. 3: Josef Beuys, *Wasserstrahl* performance, February 27, 1969.

Felt

That was precisely what was intended. The hat is made of firm felt, whose thick weave works like a wall, defending against outside influences. That is because the hairs and remnants of fat from the skin get tangled up and combine to form a weave so tight that it does not let water penetrate, yet remains soft and elastic and, above all, conveys an animal warmth. This material has been given the name "felt." It is a combination of materials taken from dead animals and therefore conventionally regarded as equally material per se. Thanks to its quality of providing warmth, however, felt preserves

---

2 See Andy Warhol's *Diamond Dust Joseph Beuys* from 1980, in exh. cat. *Andy Warhol: Rétrospective* (Paris: Centre Pompidou, 1990), 329; Verspohl, "Joseph Beuys," 81, fig. 3; Michael Groblewski, "'… eine Art Ikonographie im Bilde': Joseph Beuys – von der Kunstfigur zur Kultfigur?" in *Kultfigur und Mythenbildung: Das Bild vom Künstler und sein Werk in der zeitgenössischen Kunst*, ed. Michael Groblewski and Oskar Bethmann (Berlin: Akademie Verlag, 1993), 43–44, fig. 4.
3 See Horst Bredekamp, "Beuys als Mitstreiter der Form," in *Joseph Beuys: Parallelprozesse: Archäologie einer künstlerischen Praxis*, ed. Ulrich Müller (Munich: Hirmer, 2012), 23–41.

the living quality of its origin, and for that reason an aura of vitality is attributed to it. This may go back to the prehistory of humankind.[4]

The Italian historian of science and founder of an important publishing house Leo Olschki unfolded a concise history of this pseudo-alive material and its overarching semantics from Mongolia to Europe, playing a special role in Dante's *Divine Comedy* in a talk delivered in 1947, which became a small but highly influential book published in 1949.[5] The Arte Povera movement was very much inspired by this. The combination of materiality and the metaphysical semantics directly associated with felt can be seen as a motto for the question of the material's own activity, which in our cluster is being raised with the concept of "active matter." This formulation has three components—matter, space, and image—and all three are associated with felt. For that reason, this material can be understood as a ferment for various areas of the cluster and here, of course, especially for the section on weaving and that on symbolic material.

Artists of all eras have regarded the fact that materials take on a life of their own when they are shaped as a prerequisite for their own activity. Joseph Beuys was possessed—if not obsessed—by this assumption like few others, and felt played a special role here too. For him, this determination of the activity of the formed material had a particular and, as he emphasized again and again, almost existentially dimensioned significance. In World War II, he was the radio operator of a bomber that crashed while flying over the Crimea in March 1944. Beuys survived, with severe head injuries. According to his account of this event, Beuys would not have survived if Tatars had not rescued him, brought him into a tent, and taken care of him; one of the vital components of his healing process was said to have been wrapping him in felt blankets. The story cannot have happened in this manner; it should be ranked with the series of myths about artists that are no less relevant than the events they are transcending.[6] Beuys connected the warmth of the felt to an energy that benefitted and crucially promoted his healing. With this story, and thanks to this material his felt hat was for Beuys a proof of his identity. Felt, which is made from the hair of sheep, hares, and all animals with thick fur, is strikingly transferred as a material into the design of works of art.[7]

Fourteen years after that event, Beuys began, with the so-called *Eurasier* (*Eurasian*) of 1958, to shape this context into works of art (fig. 4).[8] That work consists of a

---

4 See Monika Wagner, Dietmar Rübel, and Sebastian Hackenschmidt, ed., *Lexikon des künstlerischen Materials: Werkstoffe der modernen Kunst von Abfall bis Zinn* (Munich: C. H. Beck, 2002), 97–101.

5 See Leonardo Olschki, *The Myth of Felt* (Berkeley/Los Angeles: University of California Press, 1949).

6 See Philip Ursprung, *Joseph Beuys: Kunst Kapital Revolution* (Munich: C. H. Beck, 2021), 15–23; in general, Werner Busch, *Die Künstler Anekdote: 1760–1960: Künstlerleben und Bildinterpretation* (Munich: C. H. Beck, 2020).

7 See Monika Wagner, *Das Material der Kunst: Eine andere Geschichte der Moderne* (Munich: C. H. Beck, 2001), 214–17.

8 See Klaus-Dieter Pohl, "Eurasier 1958," in *Joseph Beuys: Eurasienstab*, ed. Birgit Stöckmann (Göttingen: Steidl, 2005), 10–11.

thick felt mat whose dark shade is clearly distinguished from a looming form wrapped in gauze but nevertheless established a connection to the felt floor via its material. A small bent metal staff rises up out of this figure and takes the form of a shepherd's crook. Without question, it is a kind of self-portrait, as Beuys put it in a taped interview from 1972: "I can still remember that for years I behaved like a shepherd. I went around with a staff, a sort of 'Eurasian staff,' which later appeared in my works, and I always had an imaginary herd gathered around me. I was really a shepherd who explained everything that happened in the vicinity. I felt very comfortable in this role."[9] The gauze as a trace of a protection for a wound, the shepherd's or bishop's crook, the felt as the basis for a warming and healing materiality, to which according to his own words he owed his survival, and finally the anthropological linking of cultures that were separated by an impenetrable wall during the Cold War[10]—all this has taken form in this self-portrait installation, which surrounds the almost breathing materiality with the space of its unfolding and with an aura that has become an image.

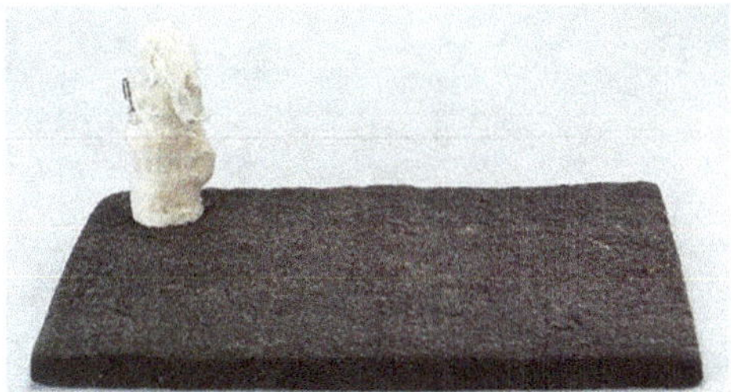

Fig. 4: Joseph Beuys, *Eurasier*, 1958, felt, gauze, metal, 14 × 33.5 × 16.5 cm. Hessisches Landesmuseum Darmstadt, in Joseph Beuys, *Eurasienstab*, no. V of the series "Joseph Beuys Medien-Archiv," ed. Nationalgalerie im Hamburger Bahnhof, Museum für Gegenwart Berlin, Staatliche Museen zu Berlin (Göttingen: Steidl, 2005), 10.

Seven years later, Beuys expanded this iconology of materials into a large installation titled *Ein Mal 90 Grad Filzwinkel* (One Ninety-Degree Felt Angle) and *Zwei 90 Grad Filzwinkel* (Two Ninety-Degree Felt Angles), clearly taking up the thread of El Lissitzky's *Proun Room*, that presented a dynamic installation of a chamber in which a large felt mat lying on tree trunks is the central object.[11] Here, the entire room is connected to felt props, so that the iconology of this material has now become the main person of an entity that determines the surroundings.

Around the same time, in one of his first and most impressive Fluxus actions, *Der Chef: Fluxus Gesang* (The Chief: Fluxus Singing), Beuys associated felt with himself in

---

9 See Götz Adriani, Winfried Konnertz, and Karin Thomas, *Joseph Beuys, Life and Works* (Woodbury, NY: Barron's Educational Series, 1979), 12.

10 See Ursprung, *Joseph Beuys*, 100–5.

11 See Birgit Stöckmann, ed., *Eurasienstab*, 23–24.

such a way that he assimilated his own physiognomy.[12] In this presentation Beuys tried to make it understandable how seemingly dead objects and beings preserve a quasi-electrical energy that can be measurably accessed via the material. He rolled himself up completely in a felt blanket so that his figure could only be surmised from the bulge from his shoulders.[13] His feet were turned toward the viewers, and his head was connected by a rope lying on the axis of his body to a microphone into which he made guttural animal sounds, which were supposed to be identified with the sounds of stag. Draped in the axis of this orientation were two dead hares, whose pelts defined the still unworked, natural, organic field, which was then reshaped and semantically prepared in the felt. For Beuys, following the tradition of European iconography, hares were symbols of vitality and fertility.

Seen from the line of the hares, their form transitioned directly into that of the artist, who was lying on the floor completely wrapped in felt and remained in this position of absolute motionlessness, merely making noises for sixteen to twenty-four hours, in order to embody, on the one hand, the stability of lifelessness and, on the other, the vitality of the dead (fig. 5).[14]

The Critique of Felt

This was continued in the form of the action *I Like America and America Likes Me* in New York in 1974, which was one of the most impressive, most cryptic, and perhaps also most problematic actions the artist ever carried out.[15] Wrapped in felt, he had an ambulance pick him up at the airport in New York and bring him to the René Block Gallery on East Broadway without ever coming into contact with the urban space of New York and Manhattan.[16]

The goal was the encounter with a creature in the gallery that in the supposed mindset of the American settlers represented the lowest, ugliest, and shiftiest animal: the coyote. For three days in a row, Beuys, wrapped in felt and fitted out with a triangle, a flashlight, and a walking stick, entered a space separated by bars to spend the entire day with the animal (fig. 6). It had been difficult to bring the coyote to New York, especially as the animal in question was considered aggressive and could be tamed by its keepers only with difficulty.

In Beuys's case, however, it was the felt that clearly introduced a harmony into the encounter, so that while the meeting was not without risk, at no point did it require intervention by outsiders. It resulted in harmony and synergy between the movements

---

12 See Uwe M. Schneede, *Joseph Beuys: Die Aktionen. Kommentiertes Werkverzeichnis mit fotografischen Dokumentationen* (Ostfildern-Ruit: Hatje Cantz, 1993), 68–79.
13 See Ibid., 76.
14 See Ibid., 78.
15 See Ibid., 330–53.
16 See Ibid., 342.

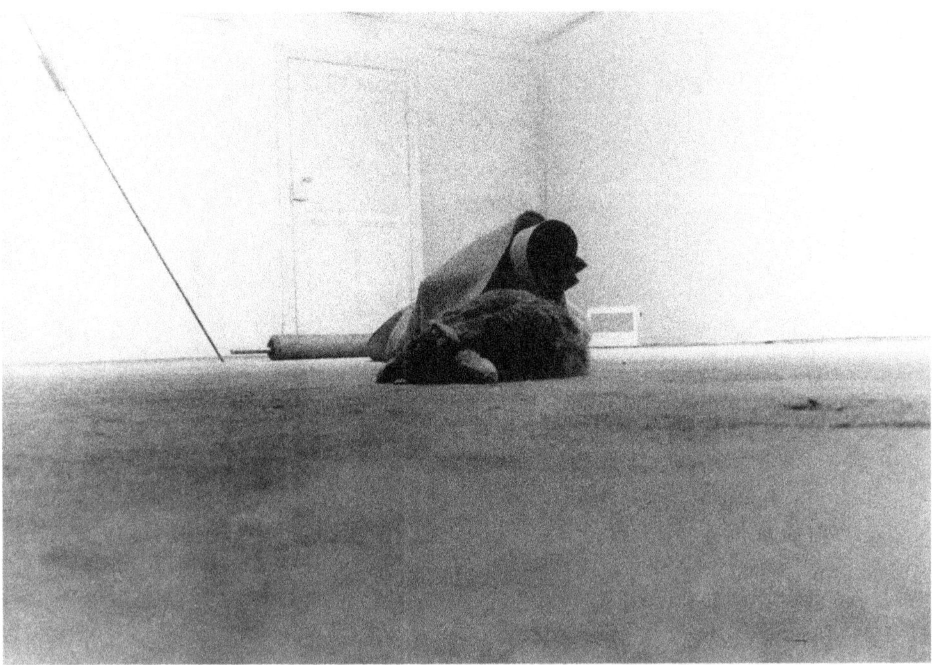

Fig. 5: Joseph Beuys, *Der Chef / The Chief: Fluxus Gesang*, August 30, 1964, Billedhuggersalen Charlotten-
borg, Copenhagen and December 1, 1964, René Block Gallery, Berlin, in ed. Uwe M. Schneede, *Joseph
Beuys: Die Aktionen: Kommentiertes Werkverzeichnis mit fotografischen Dokumentationen* (Ostfildern-Ruit: Verlag Gerd Hatje, 1994), 78.

of the felt man and the coyote (fig. 7). When the former was still, the animal was also
prepared to calm down; when the felt moved, the coyote reacted as well, sniffing the
material, making it his by urinating on it, and tearing it to pieces and making a kind of
cave with it.

In a unique—and perhaps revealing—way, the coyote viewed the pieces of felt that
had been released or torn away as his property, on which he rested and even slept,
whereby the straw that had been brought into the gallery space especially for him
did not interest him (fig. 8). He only ever went over to the straw when Beuys had
lain down on it.

This action at the Block Gallery was politically symbolic in that a European protected himself with felt and arrived as if through a tunnel at a gallery where an ostracized
animal of the American prairie awaited him. The animal's reaction could not be predicted, but in essence it behaved exactly as the artist had hoped. By way of felt he established an empathetic relationship with the animal that ultimately resulted in a kind
of friendship. On the third day, the coyote was already expecting his roommate to enter
that morning.

Through felt and the calmness and self-confidence of his movements, Beuys transformed himself into some form of projection of a Native American who was trying to

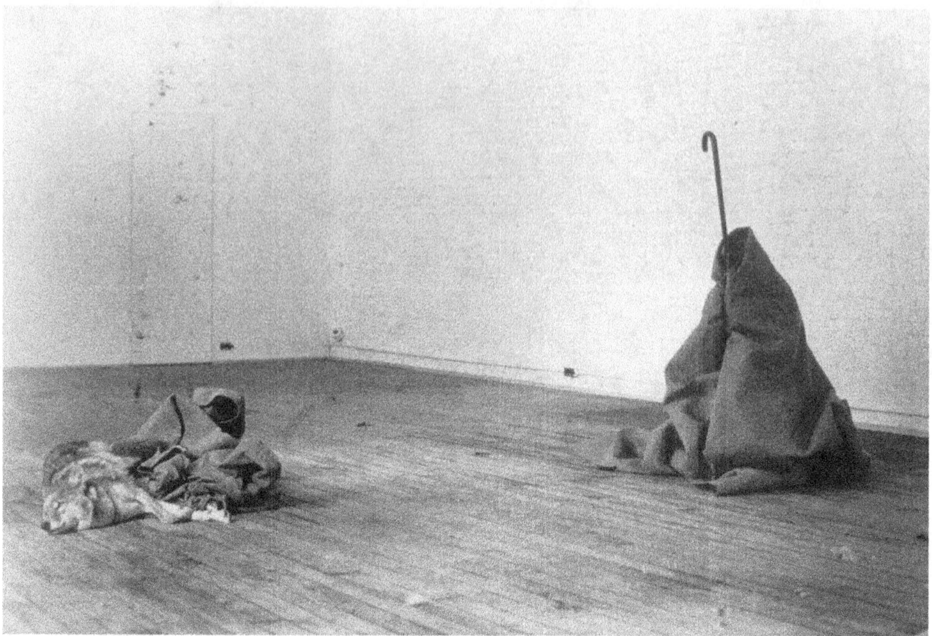

Fig. 6: Joseph Beuys, *I Like America and America Likes Me*, May 23 – 25, 1974, René Block Gallery, New York, in *Joseph Beuys: Die Aktionen*, 343.

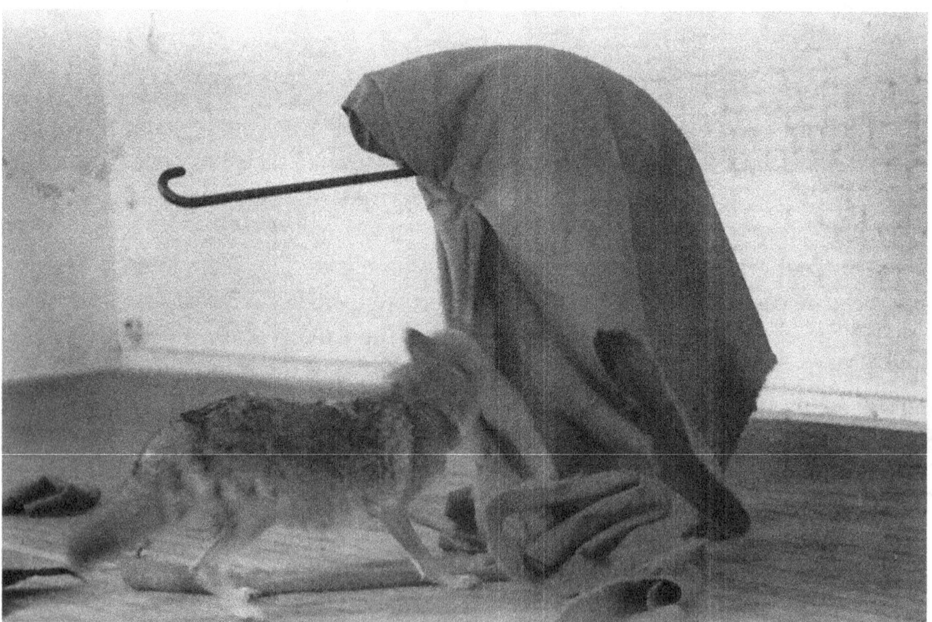

Fig. 7: Joseph Beuys, *I Like America and America Likes Me*, May 23 – 25, 1974, René Block Gallery, New York, in *Joseph Beuys: Die Aktionen*, 344.

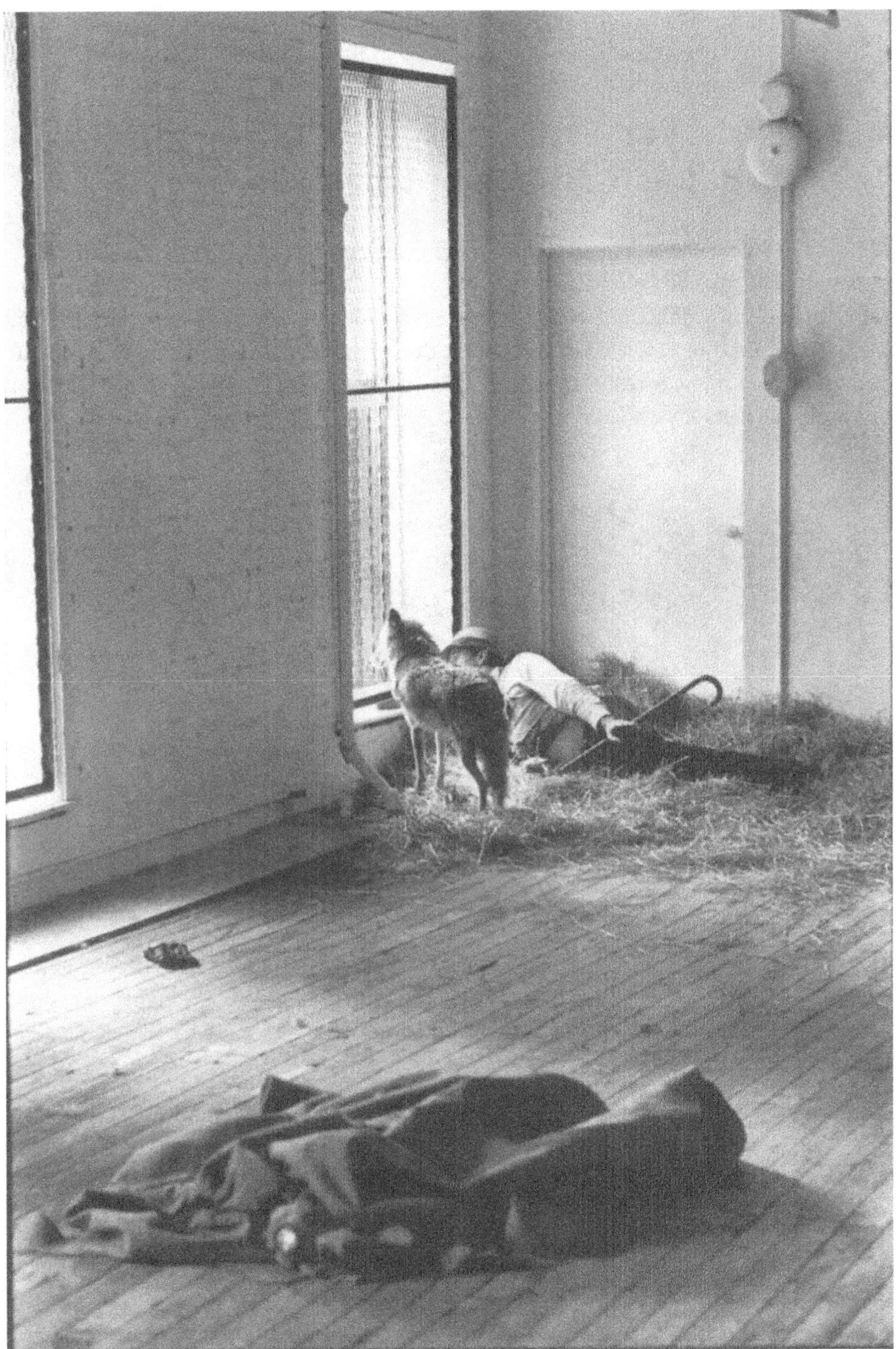

Fig. 8: Joseph Beuys, *I Like America and America Likes Me*, May 23–25, 1974, René Block Gallery, in *Joseph Beuys: Die Aktionen*, 351.

live in harmony with nature. The space played a crucial role as an actor in that the white cube was reversed into the landscape of a prairie in which human being and animal met via matter without fighting.

The Felt Hat as Talisman of Liberty

All these characteristics remained preserved in Beuys's felt hat. Concentrated in it was a spatial, physical, and cultural autonomy that made him immune to authorities, as in a conversation with politician Heinz Kühn after a debate in the state chancellery in Düsseldorf.[17] In this way, the hat and with it the felt had become a symbol of independence and freedom as a result of the protection that the material had to offer in both physical and metaphysical ways.

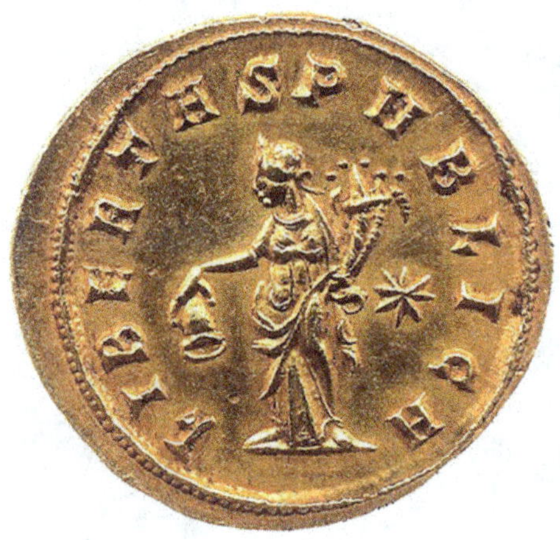

Fig. 9: Aureus of Marcus Aurelius Julianus, 284–285 CE. British Museum, Reg. Nr. 1864, 1128.294.

This side of felt has its own history, which is part of the foundations of the political iconology of Europe. Its origins have not been clarified, but it is reasonable to assume that the materiality of felt—that is, the combination of individual hairs tangled together to produce a new, inseparable whole—stood for the community formed not by one individual but by the collective working together. As a materiality that does not permit individual strands to be emphasized and in which every single fiber vouches for autonomy within the whole, felt was considered the fabric of a form of society in which no single individual can rule the whole. For that reason, a felt hat, the pileus, became the sign of the freedom of the Roman citizen of antiquity. Already in the second century BCE, a denarius was minted on which the personification of liberty holds out the pileus as a sign of a free community (fig. 9). Over the centuries, this felt cap

---

17 See Charles Wilp, *Heinz Kühn und Joseph Beuys,* 1975, photograph, in Verspohl, "Joseph Beuys," 84, fig. 7.

signifying freedom was retained on the coins of the Roman empire; the pileus was still regarded as a sign of the freedom of the Roman citizen, and so in the legal act in which a slave was granted the rights of a citizen, he was handed the pileus.

This act took on a combative component when Marcus Junius Brutus and his group of coconspirators murdered Julius Caesar on March 15, 44 BCE, by stabbing him twenty-three times. He had his deed legitimized by a portrait medallion, on the obverse of which the pileus of freedom is framed by two daggers (fig. 10). Ever since, felt in combination with daggers has been considered the icon of tyrannicide and of the struggle against autocracy.

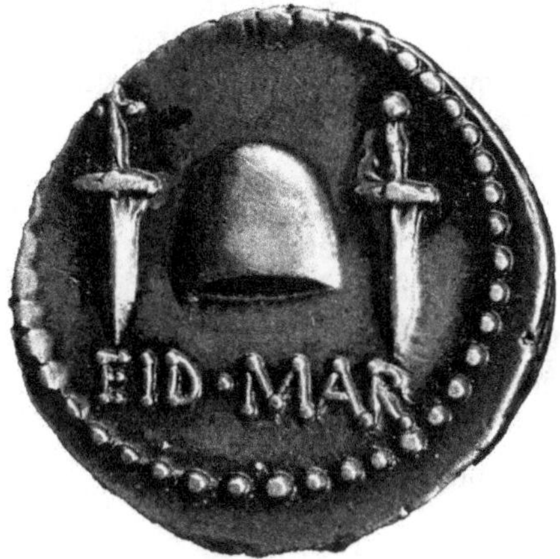

Fig. 10: Denar of Marcus Junius Brutus, 43–42 CE. British Museum, Reg. Nr. 1860, 0328.124.

In the sixteenth century, this motif became topical again when the Republic of Florence was transformed into an absolutist form of government. One of the first and probably also most incompetent protagonists of that process was Alessandro de' Medici, who was stabbed to death as a tyrant in 1537 by Lorenzino de' Medici, a republican-minded member of the family. The new Brutus had a medal of honor coined, in parallel with the ancient denarius of Marcus Junius Brutus, with the felt hat of liberty flanked by two daggers pointed downward (fig. 11). As the iconological handbook of Cesare Ripa demonstrates, the felt hat of freedom has existed in the world ever since—both in the form of a cap and as a broad-brimmed hat, never again to disappear (fig. 12).

It was joined by a third motif of the felt hat in the form of the Phrygian cap. An alternative tradition associated the act of liberating a slave not with the pileus but with the felt cap with its peak drawn back as worn by the Phrygians. This symbolic form of felt is probably the basis of the most successful political iconology because the protagonists of the French Revolution identified themselves with this symbol. A sign of liberty and overthrow, along with bringing down all previous authorities and

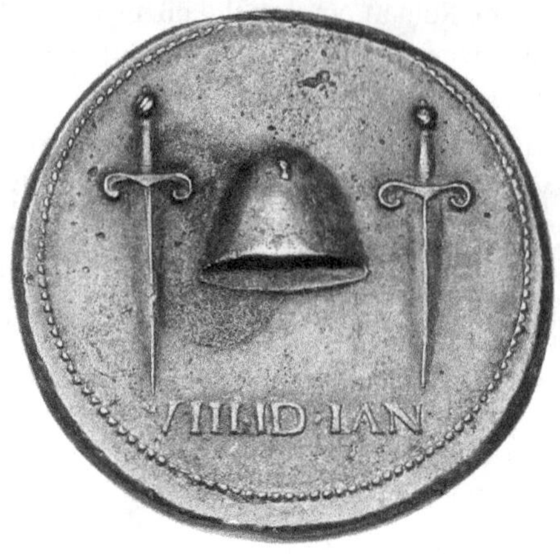

Fig. 11: Giovanni Cavino (attributed), bronze medal of Lorenzino de' Medici, 1537, in Horst Bredekamp, *Michelangelo* (Berlin: Verlag Klaus Wagenbach, 2021), 558, fig. 31.

the signs that represented them, within a few years the Phrygian cap became the emblem of the Reign of Terror.[18]

But this felt hat of liberty remained an icon of the positive destiny of the revolution. As such it was depicted in combination with the rebellious "Marianne" personifying the entire country with the spring cap in the mythical painting by Delacroix—perhaps the most successful image in all political iconology.[19]

The Felt Hat of Artists

In the 1540s, Michelangelo Buonarroti was one of the Republican-minded Florentines living in exile in Rome after the autocracy of the Medici had been established in Florence. At the beginning of that decade, he created, in an obvious allusion to the assassin Alessandro de' Medici, a marble bust of Brutus, in which the assassin was characterized in a mysterious way as both a hero and a violent criminal. Michelangelo was a republican, yet he rejected introducing freedom with an act of murder.[20]

But this only made him identify more strongly with the felt hat of freedom. He reportedly confessed to the Portuguese artist Francisco de Holanda: "Sometimes, I may tell you, my important duties have given me so much license that when, as I am talking to the Pope, I put this old felt hat nonchalantly on my head, and talk to him very frankly."[21] The Portuguese seasoned these words with a rather clumsy portrait of Michelan-

---

18 See "Libertà," in Cesare Ripa, *Iconologica*, 1611; Würtenberger, *Symbole der Freiheit*, 145, fig. 74.
19 See Nicos Hadjinicolaou, *Die Freiheit führt das Volk von Eugène Delacroix: Sinn und Gegensinn* (Dresden: Verlag der Kunst, 1991).
20 See Bredekamp, *Michelangelo*, 559–65.
21 See Francisco de Holanda, "Three Dialogues," in Charles Holroyd, *Michael Angelo Buonarroti*, 2nd ed. (London: Duckworth, 1911), 229–80, esp. 237.

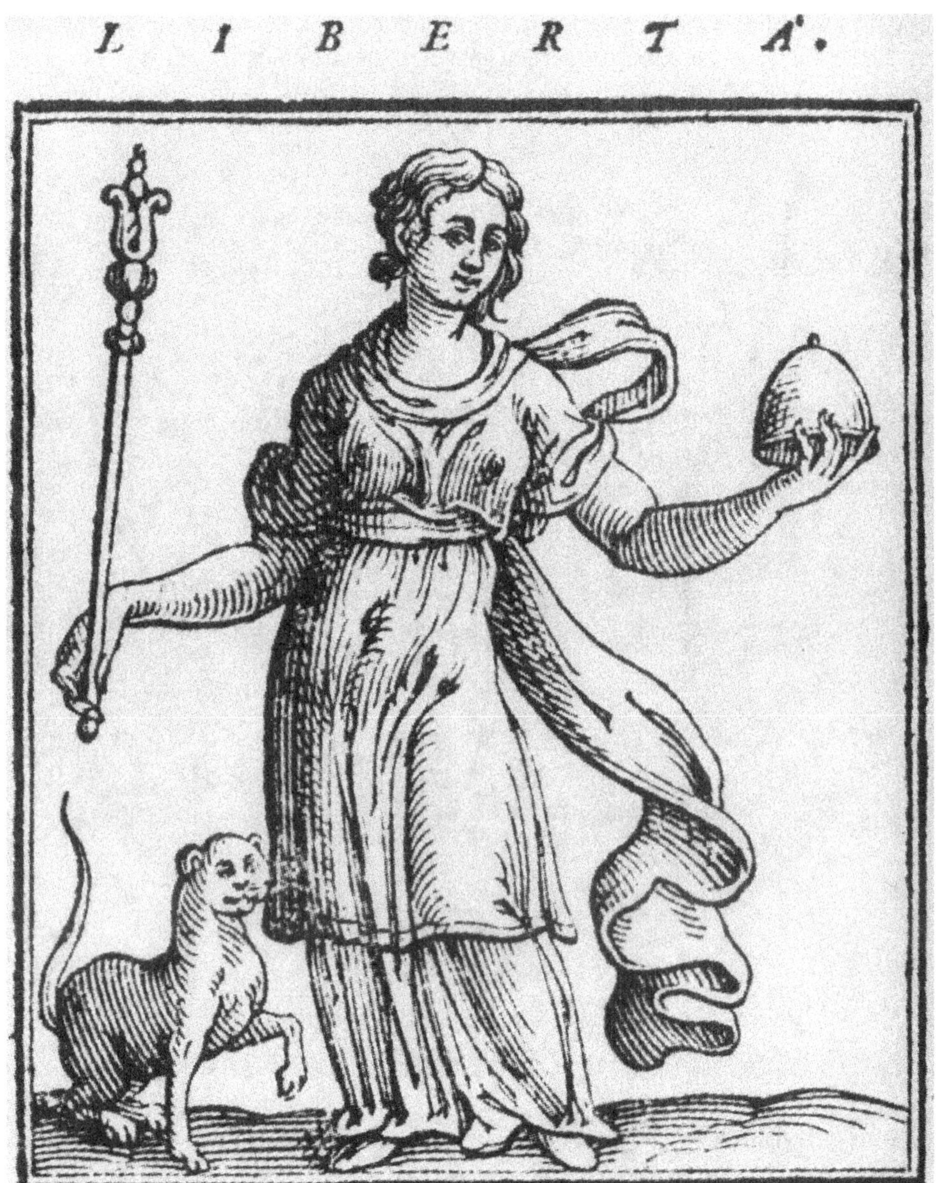

Fig. 12: Libertà, in Cesare Ripa, *Iconologica*, 1611. Bayerische Staatsbibliothek München, 4 L.eleg.m, in Thomas Würtenberger, *Symbole der Freiheit: Zu den Wurzeln westlicher politischer Kultur* (Vienna et al.: Böhlau Verlag, 2017), 48, fig. 9.

gelo, in which the felt hat testifying to freedom dominates the scene (fig. 13).[22] Before the highest authority of his age, Michelangelo used the felt hat to embody his independence.

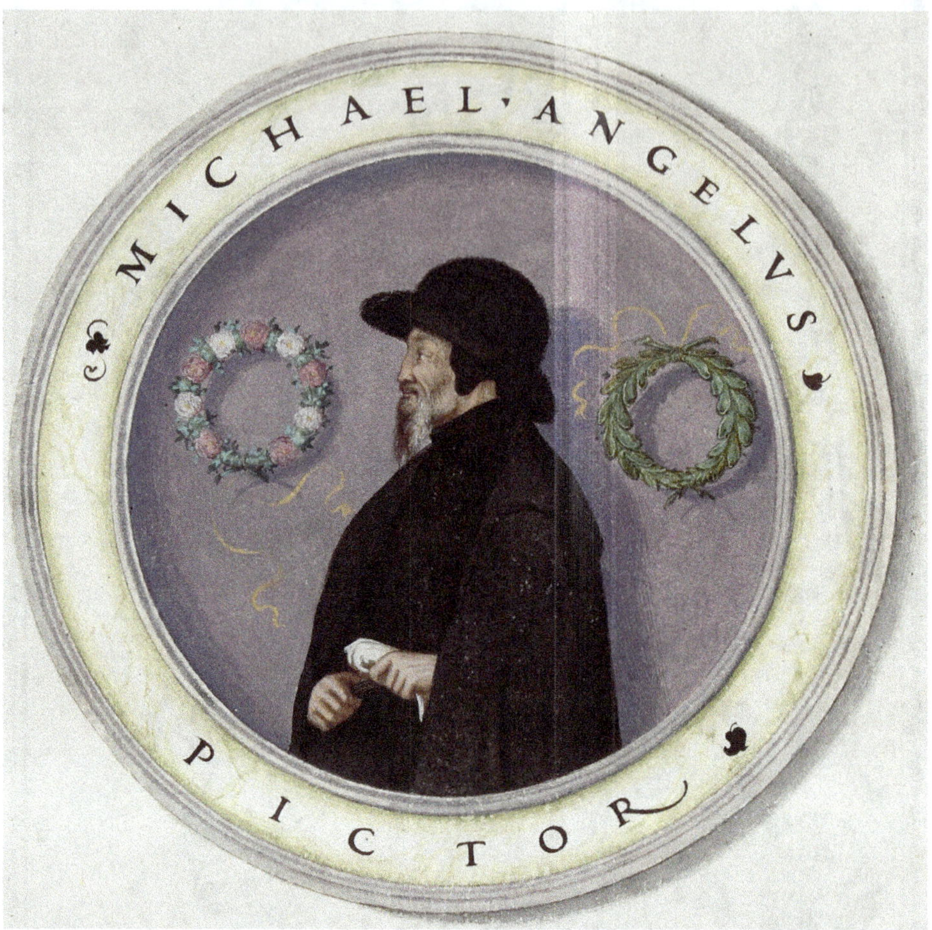

Fig. 13: Francisco de Hollanda, Michelangelo, miniature portrait, engraving, in Franciso de Hollanda, Os Desenhos da Antiqualhas, ca. 1540, Library of the Monastery of San Lorenzo de El Escorial Biblioteca del Real Monasterio de San Lorenzo, Ms. 28-I-20.

Michelangelo also claimed for himself the original forms of the pileus, the cap. One of his most impressive portraits is a drawing that his fellow artist Fra Bartolomeo made of him (fig. 14).[23] The already aged face is clearly characterized by the pressed in box-

---

22 See Francisco de Holanda, *Album dos Desenhos das Antigualhas*, ed. José da Felicidade Aves (Lisbon: Livros Horizonte, 1989), 7v (24); Verspohl, "Joseph Beuys," 84, fig. 6.
23 See Bredekamp, *Michelangelo*, 685, fig. 2.

er's nose, and the felt cap of the liberation of the slave and the right of the free citizen looks merely like a streak of light, so that its color stands out particularly well. The Beuys scholar Franz Joachim Verspohl, to whose study I owe much, has impressively shown that the bust that the sculptor Walther Brüx made of Beuys in 1947, in obvious allusion to that portrayal of Michelangelo, presents Joseph Beuys wearing the pileus, which as a felt hat combined protection, life, and liberty (fig. 15).

The comparison to the felt-armed Michelangelo can be extended to Beuys's Manhattan action. In Michelangelo's lifetime, his fellow artist Leone Leoni had created an extraordinary medallion on which he characterized his friend as a shepherd holding out his shepherd's crook.[24] It is equally to be taken as a blind man's staff, because the man portrayed, completely confident in his actions and views, no longer needs external orientation. He is accompanied by a dog, the symbol of absolute fidelity, who is sniffing out the surroundings for him. On his head, Michelangelo wears the felt hat as symbol of freedom in the version of the Phrygian cap.

When Beuys, transforming himself into a felt persona, and once again acting as a shepherd, tried to establish a similar relationship with a canine creature, he was not just following a momentary insight but rather uniting in himself the entire iconography of felt as an emblem of freedom, of protection, of warmth, and of life (fig. 6). If there is one symbolic fabric that deserves the name of a living articulation of the material, it is felt. The issue of justifying one's own material, spatial, and visual activity is rendered particularly powerful by this material. A constructivist perspective would indubitably try to demonstrate that the felt's attributed vitalist semantic variations were mere projections. Yet, a perspective of material and image activity is hopefully able to approach it as a phenomenon resonance between the symbolic production of an active material, the form, the artist, and the user and viewer. Perhaps the question of the *agens* cannot be answered satisfactorily. But the existence of this open question fuels the mystery that determines our design and perception of the world. In that spirit, I hope that I have shown that material iconology has at its disposal an essential aspect that constitutes this cluster in the very best sense of the noun "felt."

---

24 See Horst Bredekamp, "Im Zeichen der Freiheit. Zwei Bildnisse Michelangelos," in *Contactzone: Ein Prinzip "der guten Nachbarschaft,"* ed. Sarah Hübscher and Christopher Kreutchen (Dortmund: Kettler, 2021), 366–70.

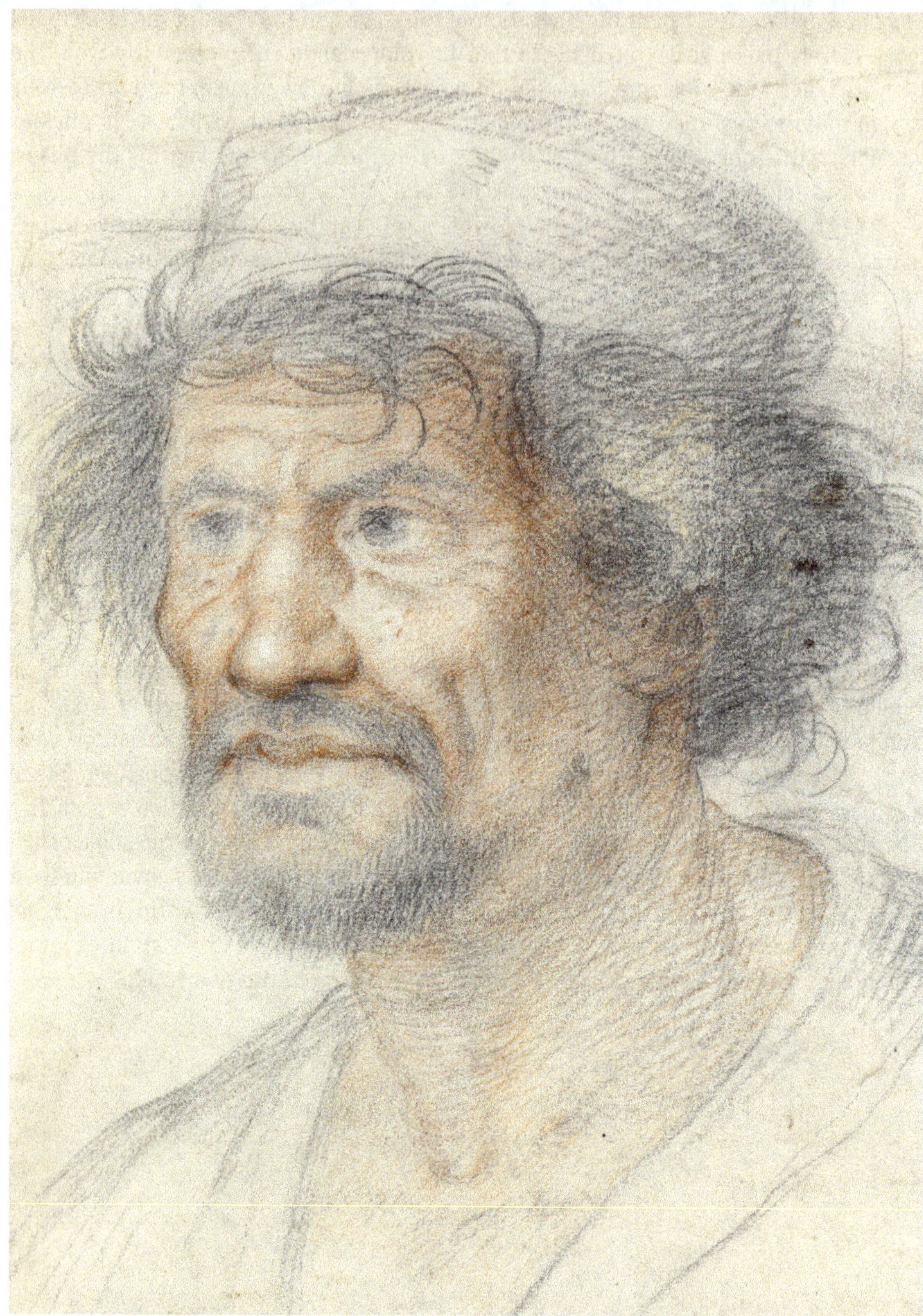

Fig.14: Fra Bartolomeo (Bartolomeo-Domenico di Paolo del Fattorino, Baccio della Porta), portrait of Michelangelo Buonarroti, circa 1516–1517. Collection Museum Boijmans Van Beuningen, Rotterdam.

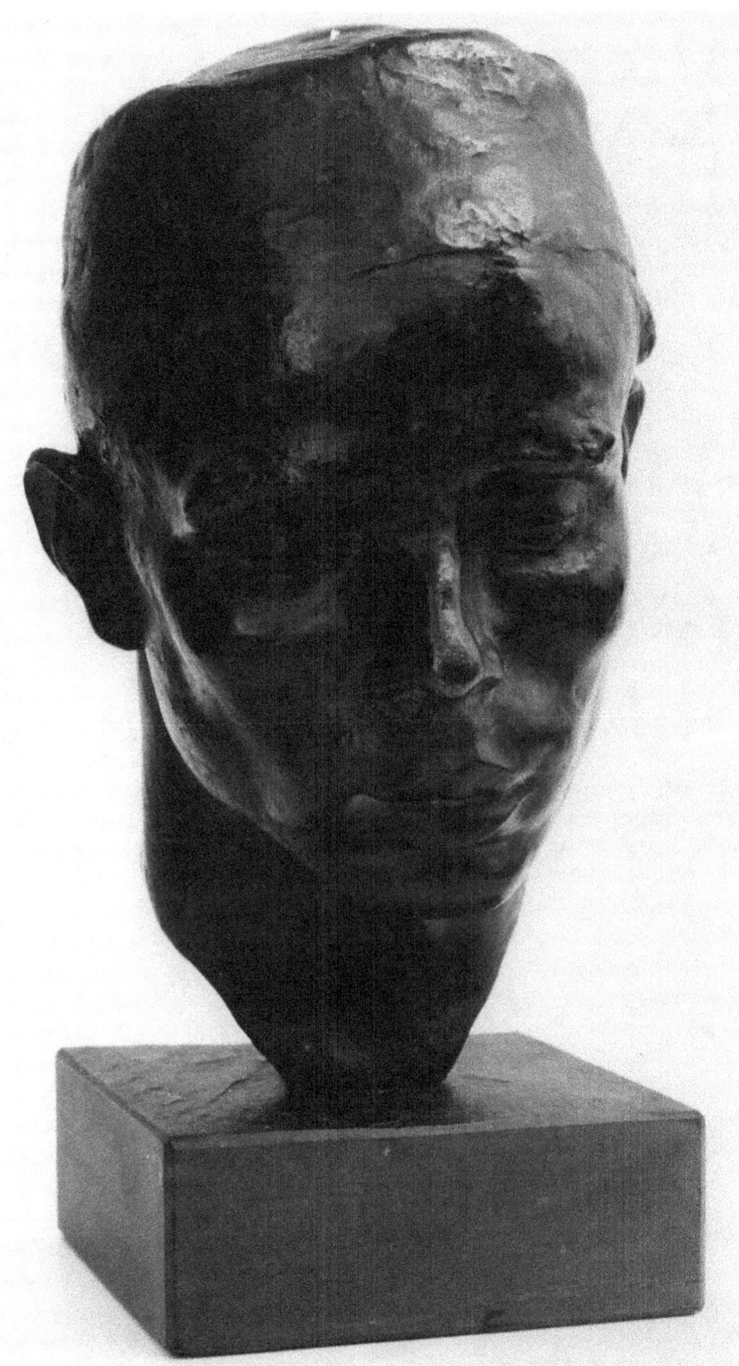

Fig. 15: Walther Brüx, Joseph Beuys, 1947, bronze, Museum Kurhaus Kleve, in Verspohl, "Joseph Beuys,"
fig. 4.

Jürgen P. Rabe

Matters of Free Energy and a Tesseract

Introduction

Life on earth depends critically on the energy coming in from the sun, as well as on the materials existing in our biosphere. The latter may be considered a very large and open physical system with visco-elastic materials and many mobile molecules, including the essential and abundant water in our atmosphere and the oceans.

Matters of activity in such a physical system can be ascribed to its *free energy, F,* which means, that part of the *inner energy U* of the system that can potentially be converted into physical work. In 1882, Hermann von Helmholtz contributed to the Königlich Preußische Akademie der Wissenschaften and coined the term *freie Energie*, with similar work carried out around that time by James Clark Maxwell and Josiah Willard Gibbs in England. This stimulated major research into the various forms of energy, including mechanical, electrical, chemical, or thermal energy, and their interconversions. Moreover, it coincided with major technological developments, which revolutionized various industries, particularly by exploiting electricity and chemistry.

The difference between the *inner* and the *free energy*, meaning the remaining energy *U minus F,* finally ends up in *thermal energy,* producing global warming in the long run, increasing the system's temperature T times its entropy S. Remarkably, the temperature allows to quantify precisely the *average* thermal energy of a single particle in the system, $k_B T$, with k_B being the Boltzmann constant. On the other hand, the entropy S is a measure of the disorder, i.e., a lack of precise information about a large system. More exactly, it characterizes the number of possible microscopic states of the individual particles of a system that comply with its macroscopic state.

Coming back to the particularly valuable free energy, the activities of nature on earth are largely determined by our environment, particularly temperature and humidity, but also light and carbon dioxide. As a consequence, there are circadian and annual cycles. For example, the optimum temperature for elongation growth in tulip petals differs for the upper and lower side of the mesophyll by 10° Celsius, resulting in the blossoms' opening in the morning and closure in the evening.[1] On the other hand, materials such as papers based on partially hydrophilic fibers may keep their shape at low humidity levels but swell and therefore buckle at elevated ones.

In the following, we illustrate how the free energy of water molecules may be used to act precisely and exclusively at inner interfaces of materials. As an example, we focus on graphene, a two-dimensional material, on top of a cleaved natural layered

---

1 W. G. van Doorn and C. Kamdee, "Flower Opening and Closure: An Update," *Journal of Experimental Botany,* 65 (2014), 5749–57.

crystal, mica. The internal interface between these two materials is employed to pull in selectively only water from a humid environment, thereby demonstrating its activity for *filtering molecules*. The intervening water also acts as a knife, *cutting* the tightly bonded materials at the interface. Moreover, wetting the interface with different molecules such as water and ethanol allows to control the formation of nanostructured two-dimensional molecular systems at internal interfaces. This controls the mechanical properties of the system: water *lubricates* better than ethanol, and—much to our surprise—heavy water with deuterium replacing the hydrogen lubricates an order of magnitude better than normal hydrogenated water.

The underlying activities for all these effects—cutting, filtering, and lubrication—are due to many-body effects, and therefore due to temperature and thermal energy, as well as entropy. This allows for very precise predictions of the behavior of large systems such as materials, the human body, or the Earth's atmosphere, without the need to know all details: a few degrees Kelvin—out of about 300 K on earth—determines whether an individual human or even mankind may survive. The concept of temperature allows predicting quite precisely what happens with a complex system such as a material, a biological organism, or an ecological system, even without precise knowledge of the microscopic details.

We therefore decided to expand our previously investigated metamodel for the established fundamental theories of physics for few-body systems, the Cube of Physics[2] into a four-dimensional Tesseract, the Hypercube of Physics for large many-body systems. Working practically with a model in 4D space is challenging, and we aim to accomplish this with a movie of a rotating 3D perspective of the 4D object in time.

Activities of Interfaces: Filtering and Cutting

Our work on filtering and cutting with interfaces started with an accidental observation about a decade ago.[3] Nikolai Severin from our group had deposited single layers of graphene, a two-dimensional version of graphite, onto a freshly cleaved piece of mica. Graphite and mica are naturally occurring layered minerals and were obtained as macroscopic pieces that can be easily cleaved, providing atomically smooth surfaces over macroscopic areas, up to many square centimeters. Much to our delight—and to some surprise—the deposited graphene pieces were atomically flat over large areas, which could be examined with optical microscopy and on the atomic scale with atomic force microscopy, where a scanning tip, tapping at high eigenmodes, can reliably image

2 Cube of Physics, https://cube-of-physics.org/.
3 Nikolai Severin, Philipp Lange, Igor M. Sokolov, and Jürgen P. Rabe, "Reversible Dewetting of a Molecularly Thin Fluid Water Film in a Soft Graphene-Mica Slit Pore," *Nano Letters* 12 (2012): 774–79, https://doi.org/10.1021/nl2037358.

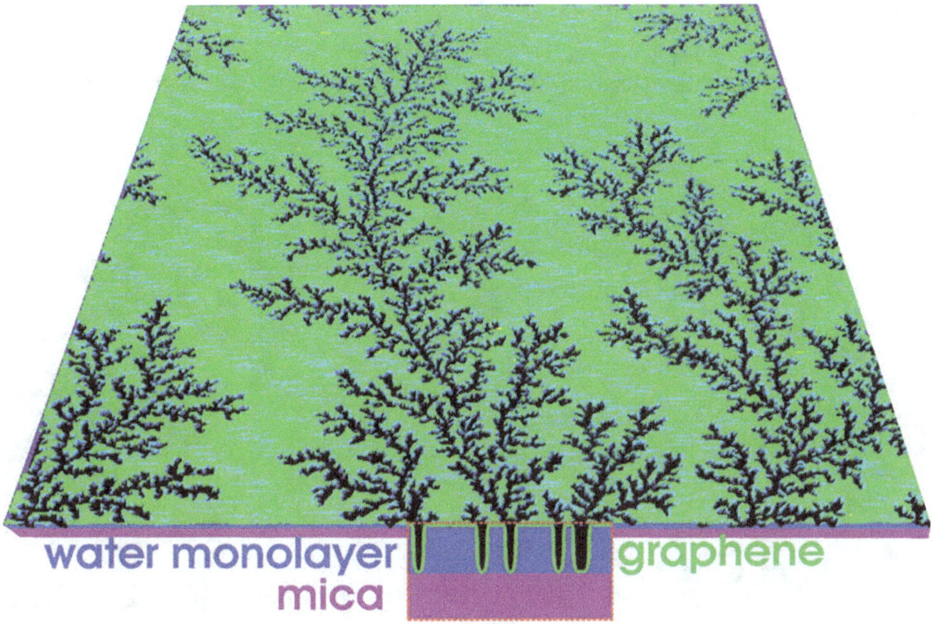

Fig. 1: Fractals in ultrathin films of water at the interface between mica and graphene in dry air during winter. The lateral and vertical dimensions are on scale of a few microns and less than a nanometer, respectively. Nikolai Severin, Philipp Lange, Igor M. Sokolov, and Jürgen P. Rabe, "Reversible Dewetting of a Molecularly Thin Fluid Water Film in a Soft Graphene-mica Slit Pore," *Nano Letters* 12 (2012): 774–779, doi:10.1021/nl2037358.

surface structures with atomic precision.[4] This effect went on during the summer, but when the winter came, some fractals were observed that we had not seen before. It did not take long until we realized that the flat graphene in summer may not have been in direct contact with the mica, but instead deposited on a very thin layer of water absorbed from the humid air onto the hydrophilic mica surface. During the deposition of the graphene onto the mica surface, almost all absorbed water molecules were removed, except a tightly bound first layer. In winter, however, when the ambient humidity was lower, some of the water molecules escaped the graphene-mica interface at the graphene edges, and this caused holes in the two-dimensional water layer, which became larger with time. Interestingly, the growth proceeded through fractal shapes (fig. 1), which allowed for systematic investigation and thereby for a better understanding of the underlying growth process.

Similarly, other hydrophilic molecules, such as ethanol, can wet and de-wet the interface, and mixtures between the two of them lead to interesting new phases at the

---

4 Nikolai Severin et al., "Atomic resolution with high-eigenmode tapping mode atomic force microscopy," Physical Review Research 4 (2022), 023149. https://doi.org/10.1103/PhysRevResearch.4.023149.

interface between mica and graphene (fig. 2).[5] [6] Other 2D materials such as MoS_2 and similar transition metal dichalcogenides have been employed,[7] providing a toolbox for shaping surfaces and interfaces of 2D materials on mica with intercalating water and ethanol.[8]

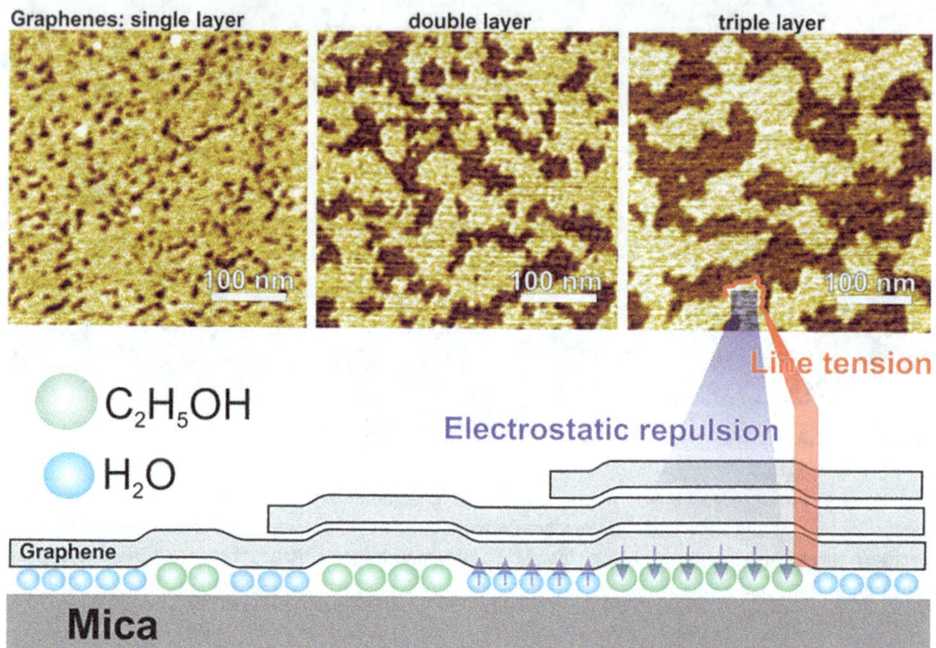

Fig. 2: Water and ethanol mix well in bulk liquids. At the interface between mica and graphene they phase separate on the nanoscale. Nikolai Severin, Jonas Gienger, Vitalij Scenev, Philipp Lange, Igor M. Sokolov, and Jürgen P. Rabe, "Nanophase Separation in Monomolecularly Thin Water-ethanol Films Controlled by Graphene," *Nano Letters* 15 (2015): 1171–1176, doi:10.1021/nl5042484.

5 Nikolai Severin, Igor M. Sokolov, and Jürgen P. Rabe, "Dynamics of Ethanol and Water Mixtures Observed in a Self-Adjusting Molecularly Thin Slit Pore," *Langmuir* 30 (2014): 3455–59, https://doi.org/10.1021/la404818a.

6 Nikolai Severin et al., "Nanophase Separation in Monomolecularly Thin Water-Ethanol Films Controlled by Graphene," *Nano Letters* 15 (2015): 1171–76, https://doi.org/10.1021/nl5042484.

7 Abdul Rauf et al., "Non-Monotonous Wetting of Graphene-Mica and MoS2-Mica Interfaces with a Molecular Water Layer," *Langmuir* 34 (2018): 15228–37, https://doi.org/10.1021/acs.langmuir.8b03182.

8 Abdul Rauf et al., "Shaping Surfaces and Interfaces of 2D Materials on Mica with Intercalating Water and Ethanol," *Molecular Physics* 119 (2021): e1947534, https://doi.org/10.1080/00268976.2021.1947534.

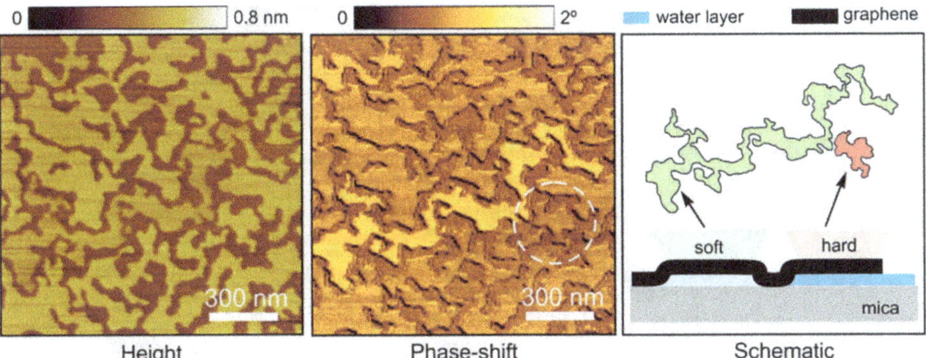

Fig. 3: Non-monotonous wetting of interfaces between mica and MoS$_2$ with water. Nikolai Severin, Jonas Gienger, Vitalij Scenev, Philipp Lange, Igor M. Sokolov, and Jürgen P. Rabe, "Nanophase Separation in Monomolecularly Thin Water-ethanol Films Controlled by Graphene," *Nano Letters* 15 (2015): 1171–1176, doi:10.1021/nl5042484.

Interfacial Lubrication

The discovery of the easy control of ultraflat but dissimilar interfaces with small molecules, provided at atmospheric pressures and temperatures, raised the question: how to use this to control matters of materials' activities? We focused on two key interfacial materials properties: lubrication and electronic properties.

In order to investigate lubrication, we considered the graphene and MoS$_2$ on a mica sample as a beam, which can be bent (fig. 4) by putting it on two knife-edges and pushing on the ends of the slab with two screws. From the geometry of a thick mica slab and a monolayer of graphene or MoS$_2$ on top, one expects that the 2D materials would stretch upon bending the whole piece. If a 2D material sticks firmly to the mica, its bond lengths should be stretched, and if it slides, they should not. In order to monitor the bond lengths, we used Raman spectroscopy and photoluminescence, respectively, which are highly sensitive in these particular cases.[9]

Fig. 5 displays the result: for the *dry* interface without any intervening water (fig. 4B), one finds a stick-slip behavior, i. e., no sliding, while for the *hydrated* interface the 2D material is initially stretched upon a quick bend and then relaxes on a time scale of several minutes. This characterizes the lubrication due to the monolayer of water, which made its way from the ambient into the interface.

---

9 Hu Lin et al., "Influence of Interface Hydration on Sliding of Graphene and Molybdenum-Disulfide Single-Layers," *Journal of Colloid and Interface Science* 540 (2019): 142–47, https://doi.org/10.1016/j.jcis.2018.12.089.

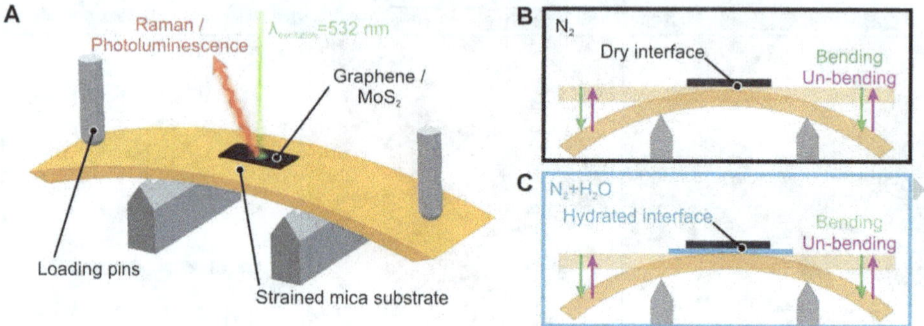

Fig. 4: Experimental set-up for the investigation of interfacial lubrication. Hu Lin, Abdul Rauf, Nikolai Severin, Igor M. Sokolov, and Jürgen P. Rabe, "Influence of Interface Hydration on Sliding of Graphene and Molybdenum-disulfide Single-layers," *Journal of Colloid and Interface Science* 540 (2019): 142–147, doi:10.1016/j.jcis.2018.12.089.

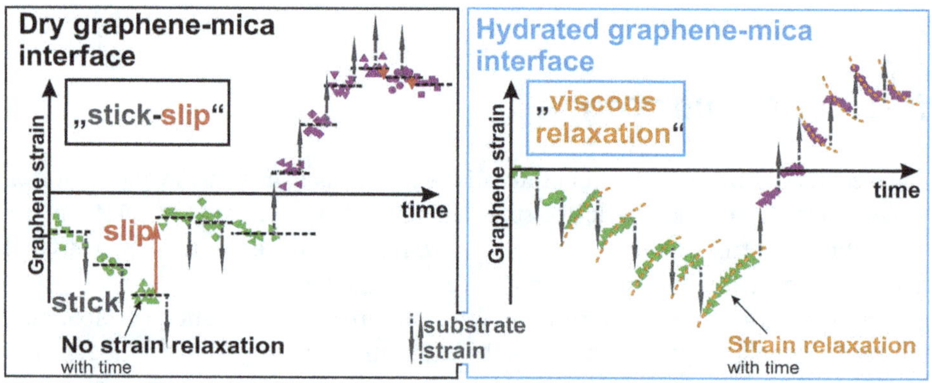

Fig. 5: Influence of interface hydration on sliding of graphene and molybdenum-disulfide single layers. Hu Lin, Abdul Rauf, Nikolai Severin, Igor M. Sokolov, and Jürgen P. Rabe, "Influence of Interface Hydration on Sliding of Graphene and Molybdenum-disulfide Single-layers," *Journal of Colloid and Interface Science* 540 (2019): 142–147, doi:10.1016/j.jcis.2018.12.089.

Of course, the lubricating activities depend on both the interfaces and the lubricants. The described setup allows for the investigation of the lubricating activities of a broad range of materials, both with regard to the solids and the lubricants.

Thin Film Electronics

When Herbert Kroemer, in his Nobel Prize lecture in 2000, coined the provocative phrase "the interface is the device," he referred to the revolutionary development of information technologies based on thin film semiconductors. Today, understanding and controlling charge transfer through molecular nanostructures at interfaces is still of paramount importance, particularly for electronic devices, and contact electrification as well as bioelectronics.

We investigated the influence of intercalation and exchange of molecularly thin layers of small molecules (water, ethanol, 2-propanol, and acetone) on charge transfer at the well-defined interface between an insulator (muscovite mica) and a conductor (graphene). Raman spectroscopy has been used to probe the charge carriers in graphene. While a molecular layer of water blocks charge transfer between mica and graphene, a layer of the organic molecules allows for it. The exchange of molecular water layers with ethanol layers switches the charge transfer very efficiently from off to on and back. We proposed a charge transfer model between occupied mica trap states and electronic states of graphene, offset by the electrostatic potentials produced by the molecular dipole layers, as supported by molecular dynamics simulations. Our work demonstrates how the intercalation of molecules of volatile liquids can reversibly affect charge transfer at interfaces. This implies its strong impact on the function of hybrid inorganic–organic electronic devices in different ambients and potential applications, including sensors and actuators.

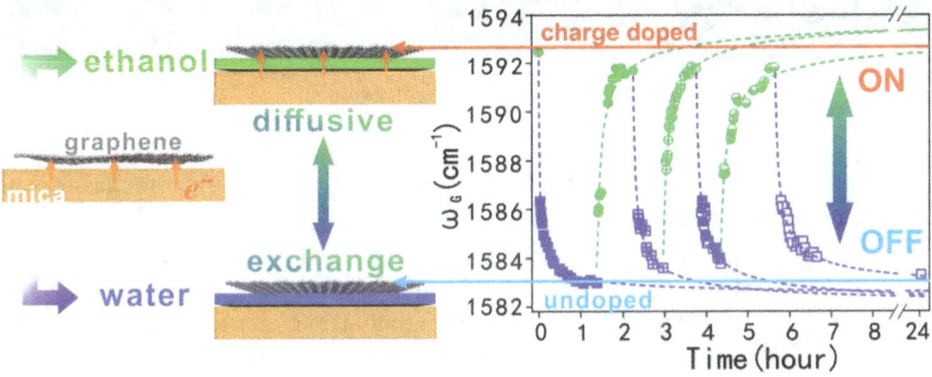

Fig. 6: Molecular layers switch interfacial charge transfer ON/OFF. Hu Lin, José D. Cojal González, Nikolai Severin, Igor M. Sokolov, and Jürgen P. Rabe, "Reversible Switching of Charge Transfer at the Graphene-mica Interface with Intercalating Molecules," *ACS Nano* 14 (2020): 11594–11604, doi:10.1021/acsnano.0c04144.

The Cube of Physics and the Hypercube

Towards the end of the twentieth century, the standard model of physics has been shown to encompass the current knowledge of physics. It describes all known elementary particles and the interactions between them. It even predicted phenomena such as the Higgs-Boson, which was discovered only decades later, when giant particle accelerators had become available. Nevertheless, the model exhibits a major deficiency: gravity does not fit into the picture. For more than fifty years, physics did not manage to integrate gravity into the theoretical quantum-mechanical framework. As a consequence of this theoretical deficiency, events such as the big bang or the interior of black holes can still not be explained convincingly.

The Cube of Physics (fig. 7) is a spatial model of physics. The idea to establish a map of physics based on the natural constants G, c, and h goes back to the Russian physicist Matwei Bronstein in 1933.

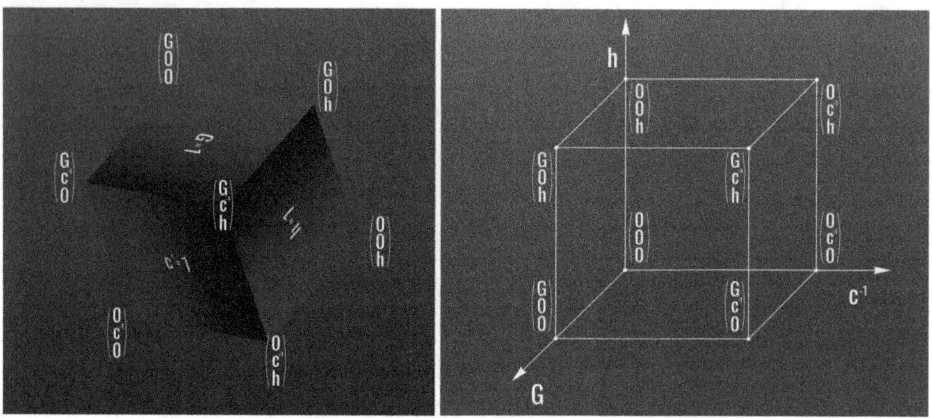

Fig. 7: The Cube of Physics 2022, https://cube-of-physics.org.

The gravitational constant G determines the force between two separate masses. Implicitly, this was measured for the first time at the end of the eighteenth century by determining the density of the earth. Also, the bending of space-time in general relativity is connected to G. The speed of light c is the natural constant known for the longest time. Its value is fixed exactly in the meantime. Firstly, it was determined in the second half of the seventeenth century based on astronomical observations. All electromagnetic and gravitational waves travel at speed c. In quantum mechanics, Planck's quantum of action h describes the ratio of energy and frequency of a photon. Max Planck discovered it at the end of the nineteenth century.

However, the Cube of Physics does not include Boltzmann's constant k_B, which Max Planck had introduced explicitly into physics in his seminal lecture on the temperature-dependent radiation of a black body.[10] Indeed, the three-dimensional Cube of Physics does not assign any distinct space to thermodynamics and statistical physics, and this calls for a novel metamodel of physics with four axes: the analog of the Cube of Physics in four-dimensional space, i.e., a tesseract (fig. 8), or the Hypercube of Physics.

---

10 Max Planck, "Zur Theorie des Gesetzes der Energieverteilung im Normalspectrum," *Verhandlungen der Deutschen Physikalischen Gesellschaft* 2 (1900): 237–45.

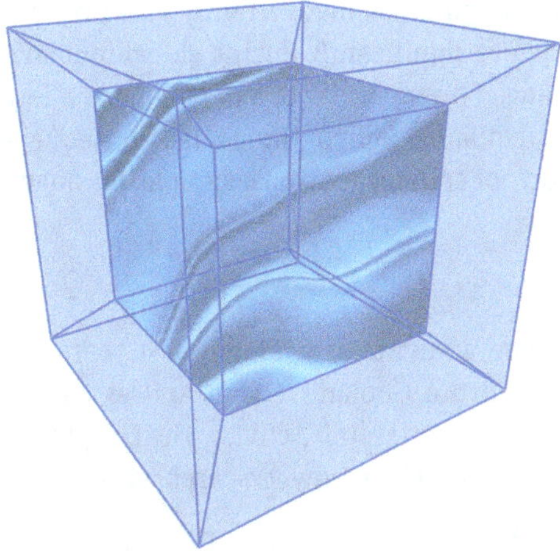

Fig. 8: 2D image of the 4D hypercube (tesseract).

Mathematically, a tesseract is the 4D analog to the 3D cube and the 2D square. However, while we know how to display a 3D object in a 2D plane by employing a perspective view, we are not used to displaying a 4D object in a 2D plane. One can achieve this upon employing time: a movie displaying sequential images of perspective drawings of the 4D Cube allows one to experience the structure and the symmetries of the tesseract.[11]

Concluding Comments

In large physical systems, such as the materials in our biosphere, *matters of activity* can be ascribed to their *free energy*, i. e., that part of the *inner energy U* of the system, which potentially can be converted into physical work. Temperature is a single key parameter, which can control its activity independent of the exact system size. Internal interfaces of materials are typical structural elements, which permit one to employ the free energy of, for instance, water in our atmosphere to drive processes such as filtering and cutting of materials: the water may act as a *sensing knife* for heterogeneous materials, being first filtered from the ambient and then used to cut internal interfaces of the materials. Moreover, water may act there as a lubricant, or it may work as the active element of an electronic device.

The key to all these processes is the temperature of the system, which explains why we should take good care of it. Our biosphere has established a rather stable average annual temperature over thousands of years, with only moderate fluctuations dur-

---

11 "Tesserakt," Wikipedia, https://de.wikipedia.org/wiki/Tesserakt, and https://de.wikipedia.org/wiki/Tesserakt#/media/Datei:Tesseract.gif. The "snapshot" in fig. 8 may be irritating, as it suggests an inequivalence of the eight 3D cubes, notabene a smaller inner and a larger outer cube, as well as six quadratic pyramids. The "Tesserakt" video, however, displays their equivalence.

ing ice ages and warm times. However, physical processes inadvertently also produce thermal energy, which has heated our atmosphere significantly during global industrialization: the average annual temperature changed on the order of a percent, i.e., a few degrees, since the late nineteenth century. As globalization continues, the consequences may become dramatic rather soon, if we do not counteract—for example, by smartly employing free and thermal energy.

Acknowledgments:

This work is based on the interaction with many individuals over the past years, including particularly José D. Cojal González, M. Fardin Gholami, Carlos-Andres Palma, Christian Kassung, and Matthias Staudacher from Matters of Activity, Hu Lin, Abdul Rauf, Bita Rezania, Nikolai Severin, and Stefan Kirstein of my research group, and Igor Sokolov in the Department of Physics.

Wolfgang Schäffner

Material Energy Information: Toward an Analog Code

According to the classical notion of codes, symbolic elements like signs are usually regarded as immaterial elements and therefore act according to special logical or mathematical rules, which can be treated completely separately from the physical reality they can represent.[1] Addressing the idea of an analog code goes beyond this classical idea of alphanumeric code based on discrete symbolic elements such as letters and numbers that has been implemented on earth. The analog can also transcend the fundamental antagonism between technical operations and nature, which has been the basis for modern engineering and, above all, for its most recent version as digital technology. The strategy of implementing symbolic operations that determine how things work and what they are transforms physical reality into a passive carrier of human intentions. This strategy and its corresponding features have been vastly executed and inserted into our physical and social reality for at least 200 years and on an unprecedented scale since the 1950s, thus producing a tremendous impact on nature and earth. The dominant *modus* of implementation has to be considered as an essential cause of the Anthropocene crisis.

In contrast to this logic of implementation related with so-called "immaterial" symbolic or digital operations, I will present in the following a new relationship between material, energy, and information, where materials are no longer seen as passive carriers, where the symbolic is not at all immaterial activity, and where the symbolic is intimately related to energy. Conceiving this constellation relies on the decisive shift toward the idea of active materials as an embracing quality of both the symbolic and the material.[2] This shift is fundamental since the dichotomy of activity and its inert material basis is still an unquestioned foundation of our modern culture. Above all, since the nineteenth century, modern technology is based on passive materials—such as iron and steel, concrete and silicon—and has excluded the materials' own activity (e.g., wood) as a failure, defect, or dysfunction. Even digital hardware is based on a neutral material carrier for the execution of the externally programmed activity.

---

1 For the notion of code see: Friedrich Kittler, "Code (or How You Can Write Something Differently)," in *Software Studies: A Lexicon*, ed. Matthew Fuller (Cambridge, MA: MIT Press, 2008).
2 Peter Fratzl et al., ed., *Active Materials* (Berlin: De Gruyter, 2021).

A New Workbench

Against this painful heritage of "spirit," "pure symbolic operations", and "immateri-
alities" separated from their material basis, this decisive shift from the material's passivation to the usage of its own activity responds at the same time to the present-day
anthropocenic urgency—deeply linked to the destructive impact of modern technology.
This strategy will have enormous consequences: it allows, first of all, to think and conceive the idea of a fusion of the symbolic and the material by experimenting with the
materials and designing the corresponding new artifacts that blur the boundary between culture and nature. Here lies the necessary and pivotal challenge: only then
will our present century "have been"—if such a future perfect will exist for mankind—and if so, it will have been the beginning of a merging of the symbolic and
the material, of code and matter, of nature and technology. It will no longer be guided
by digitally implementing alphanumeric code into physical reality and the ruling of nature; it will no longer be cybernetic by separating the information flow from the working machine.

This small but far-reaching shift toward active matter faces a triple challenge: firstly, to examine the fundamental structures of the inherent activity of matter in nature,
and secondly, to make evident the extent to which this activity was and still is ignored
within most of our technologies and human cultural activities, and thirdly, to transform these new insights into a new idea about matter and its relationship to energy
and information which finally has to provide the conditions of possibility for different
design strategies. All this means, in an essential way, resetting our modern cultures.
Moving the switch from passive matter to active materials therefore responds to this
triple challenge by integrating three fundamental perspectives: experimental practice
based on biology, physics and materials science, historical and critical epistemology of
the humanities, and practice-based design-driven projecting, which need to be combined as mutually completing features within an integrative approach of a large variety of disciplines (fig. 1).

To do so, one needs a workbench where the still isolated features, materials, and
objects that are necessary for developing such a transversal and integrating perspective can be brought together. This kind of table establishes a space of encounter for materials and structures: wood, woven tissue, felt structures, biofilm layers, plywood,
sheets of paper, etc. This collection seems to be as surrealistic as the dissection table
in Lautréamont's famous poem "Les Chants de Maldoror," where a sewing machine
and an umbrella meet.[3] For André Breton, this ensemble exemplified the "convulsive

---

3 Breton talks about "the fortuitous meeting of two distant realities on an inappropriate plane (this is
said as a paraphrase and a generalization of Lautréamont's famous phrase: 'As beautiful as the fortuitous meeting of a sewing machine and an umbrella on an operating table')." André Breton, "Surrealist
Situation of the Object [1935]," in *Manifestoes of Surrealism* (Ann Arbor: University of Michigan Press,
1969), 275.

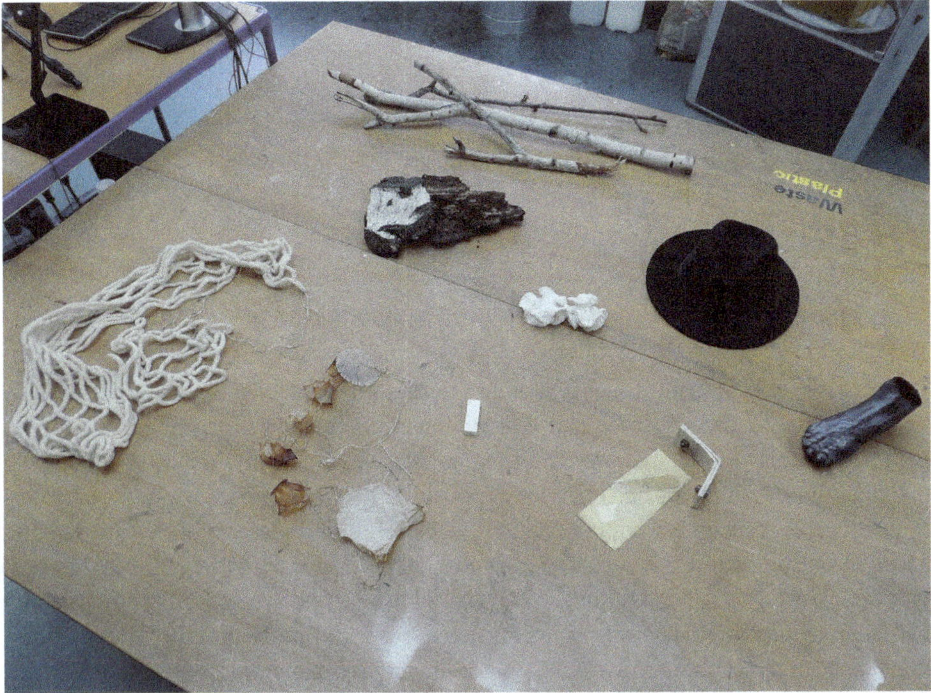

Fig. 1: Operating table.

beauty" of surrealism,[4] and in *The Order of Things* Michel Foucault took it as a challenge that strongly attacks our modes of thinking.[5] In our case, it represents a completely changed mode of action: it is a real workbench displaying a "convulsive activity" of materials, where you actually have to roll up your sleeves and work with your hands.

Beyond the Graphic Surface

After centuries of transforming physical realities into symbolic operations flattened on paper, we have arrived at a situation where this strategy can and needs to be changed in a fundamental sense (fig. 2).

---

4 "La beauté sera CONVULSIVE OU ne sera pas" (Beauty will be convulsive, or it will not be at all). André Breton, *Nadja* (Paris: Editions Gallimard, 1998), 161. See also: Raymond Spiteri, "Convulsive Beauty: Surrealism as Aesthetic Revolution," in *Aesthetic Revolutions and the Twentieth-Century Avant-Garde Movements*, ed. Aleš Erjavec (Durham, NC: Duke University Press, 2015).

5 Michel Foucault, *The Order of Things: An Archeology of the Human Sciences* (New York: Vintage Books, 1994), XVI.

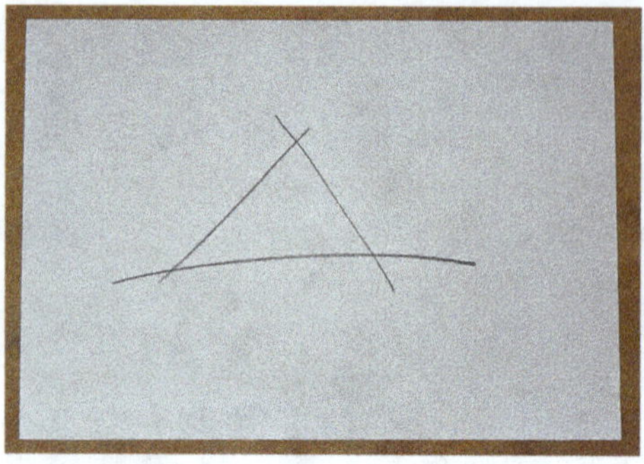

Fig. 2: Sheet of paper with text.

The classical way of projecting everything into the 2D space of texts and images has a lot of advantages, one of which is the reduction of complexity in the mode of processing information. Graphic surfaces have had a decisive impact on all cultures that used this medium over long periods of time, from stone surfaces to papyrus, wax, paper, or silicon. It was regarded as some sort of external mind and memory, or, vice versa, the mind was also imagined as a surface. Graphic surfaces contain at least three different types of operations: 2D operations, when the whole surface is used in its simultaneity by plane geometry, 1D operations, when there is a focus on sequential, linear operations, and, perhaps the strangest, 0D operations, when points are taken into account as unextended elements, incisions, perforations, or fundamental absences. This is in its essence a flat geometrical world that allows the performance of drawing, writing, and calculating as three fundamental symbolic cultural procedures that occur on that graphic surface but are related by its symbolic character to absent physical realities that could be represented by them.

Symbolic operations were regarded as extracted from the physical world or the "book of nature" to the graphic surface, fostering the idea that there exists a separated and isolated realm of the symbolic. René Descartes identified it as a space of thinking at the scale of a mathematical point that can represent the physical world without being physical.[6] This dichotomy of the symbolic and the physical is represented if not produced by the graphic surface, and above all by the elements of the alphanumeric code. It seemed as if the world of symbolic codes even controls the physical realm: this is the case in all our technology seen as an implementation of our ideas and codes into our material environment.

This strategy becomes even more obvious in digital technology that fully inherited this dichotomy. Alan Turing still envisioned the computer as a paper machine: "A man provided with paper, pencil, and rubber, and subject to strict discipline, is in effect a

---

6 René Descartes, *Meditations on First Philosophy: With Selections from the Objections and Replies*, ed. John Cottingham (Cambridge, UK: Cambridge University Press, 1996), 37.

universal machine."[7] This machine exists only on paper,[8] but also transforms the classical sheet of a graphic surface into a long tape, the interface between the physical world and the symbolic alphanumeric operations becoming a flat and neutral physical surface. Symbolic procedures require a passive carrier where materiality does not intervene actively.

Graphic surfaces are a medium to process and store information, and they also make transmission possible. If we look at the history of symbolic operations, it is above all transmission that has fundamentally changed our alphanumeric code: from postal systems for transmitting graphic surfaces in terms of letters to telegraphic systems, an important shift to electrification and sequentialization took place. This included the fundamental transformation of code from images, words, and letters to the endless chains of dots and dashes, 0 and 1. Under the conditions of electric transmission, the alphanumeric code maintained and even radicalized its fundamental separation from the material. This focus on transmission essentially shaped our idea of code and information.

Today, however, we have arrived at a moment when we can overcome this fundamental dichotomy of the symbolic and the material and move out of this passive surface, or paper machines. Already in 1959, in his famous lecture There is Plenty of Room at the Bottom, Richard Feynman announced this move from the graphic surface to the interior of the material, when he speculated about the possible micro-storage of the *Encyclopedia Britannica* on the head of a pin as a minute surface. Feynman also proposed to leave the graphic surface for using the 3D interior spatial structure of the material for processing information.[9] This transformation is still a challenge today: moving beyond the fundamental separation of the symbolic from the material, the Cluster of Excellence "Matters of Activity" (MoA) focuses on material activity as a new mode of symbolic material operation, of relating information, energy, and material. We step out of the flatness of paper and classical switching circuits, and we do not follow the shift to electrification: rather, we have to switch from passive graphic surfaces to active material interfaces.

This shift can easily be experienced on the flatness of an office desk with a classical sheet of paper that is used as a passive and blank surface of writing (fig. 3).

When the environment is very humid, paper is no longer an obedient passive, neutral, and flat carrier for drawing and writing—it starts to act itself. The direct interaction with the environment intervenes: under these circumstances, paper becomes a disturbance, transforming the sheet into something dysfunctional.

---

7 Alan Turing, "Intelligent machinery [1948]," in *The Essential Turing: Seminal Writings in Computing, Logic, Philosophy, Artificial Intelligence, and Artificial Life: Plus The Secrets of Enigma*, ed. B. Jack Copeland (Oxford: Clarendon Press, 2004).

8 Sybille Krämer, *Symbolische Maschinen: Die Idee der Formalisierung in geschichtlichem Abriß* (Darmstadt: Wissenschaftliche Buchgesellschaft, 1988), 171.

9 Richard Feynman, "There is Plenty of Room at the Bottom: An Invitation to a New Field of Physics," *Caltech Engineering and Science* 23, no. 5 (1960): 22–36.

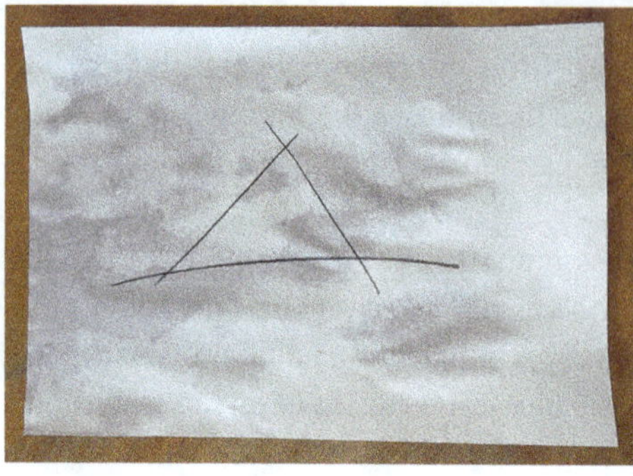

Fig. 3: Same sheet wrinkled by humidity.

The history of texts has suffered these processes: when one talks about Euclidean geometry, it refers to a text written more than 2,300 years ago, a physical text that is no longer existent. There is a fragment from 100 AD that is the only part of the oldest still existing material version of Euclid's famous *Elements* (fig. 4).

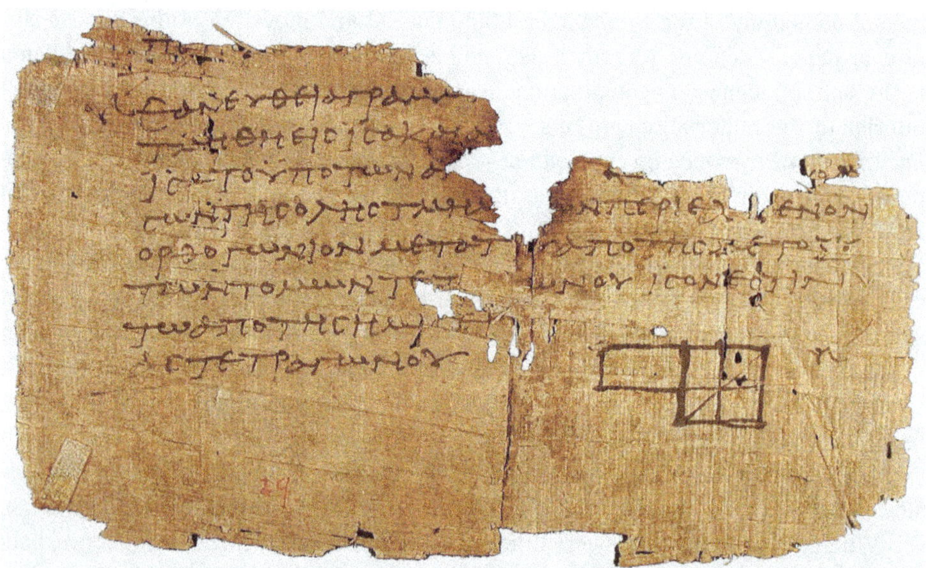

Fig. 4: Papyrus fragment of Euclid's *Elements*.

Geometry obviously did not survive only on graphic surfaces but was transmitted by practices and instruments. The example of the papyrus shows something very clearly: to generate a stable symbolic memory on the flatness of graphic surfaces, the activity of the carrier is silenced and suppressed. Long-term storage was and still is a great challenge within a world of active matter, since the intrinsic activity of the graphic support always produced severe alterations and decay.[10] Above all, historical objects to be kept as cultural heritage in museums—which is a rather modern mode of constructing history—shows the huge effort that is necessary for storing them as stable memory.[11]

In the case of paper, the material base of cellulose is a vegetal fiber structure produced by extracting fibers from their original structure in plant tissues that is destroyed in the production process and transformed into a rather passive, relatively dry and thin, felt-like structure that is also quite flexible. By adding water, an immediate, uncontrollable activity occurs.[12]

The disturbance of the classical scenario of flat symbolic operations leads back to the objects on the table, where the sheet of paper is not part of a book or a library, but a material 3D object among a whole series of already-mentioned active materials. The activity of the cellulose fibers that was silenced in the sheet of paper has now become the object of concern: with active materials we enter a new field, where materials are no longer seen as passive, where the symbolic is not at all immaterial, both different modes of active processes intimately related to energy.

External Code

Material activity, which is the object of our research, goes beyond implementing alphanumeric code into physical reality; it will no longer be cybernetic by separating the flow of information from the working machine. The classical relationship of material and symbolic operation that has been demonstrated by the sheet of paper can easily be explained with the technical example of widely used robotic arms.

A robotic arm is a piece of material mechanics. There are controllable joints connecting stiff elements to a kinetic chain that defines their possible movements. But only the double input of external energy and code for producing the controllable movements makes the passive material arm move. This structure follows the classical three-part model of machines separating motor, transmission or gearbox, and working unit as it was coined at the Paris École Polytechnique in the early-nineteenth century.[13]

The robot is also shaped by the historical development of electrical machinery and the corresponding passive materials. The industrialized and standardized design of artifacts, combined with electrification and digitization emphasized the necessity that

---

10 Elena Verticchio et al., "Climate-Induced Risk for the Preservation of Paper Collections: Comparative Study among Three Historic Libraries in Italy," *Building and Environment* 206 (2021): 1–16, https://doi.org/10.1016/j.buildenv.2021.108394.

11 See Peter Miller and Soon Kai Poh, ed., *Conserving Active Matter* (Chicago: University of Chicago Press, 2022).

12 Hannelore Derluyn et.al., "Hygroscopic Behavior of Paper and Books," *Journal of Building Physics* 31, no. 1 (2007): 9–21, https://doi.org/10.1177/1744259107079143.

13 Jean Victor Poncelet, *Traité de mécanique industrielle: exposant les différentes méthodes pour determiner et mesurer les forces motrices, ainsi que le travail mécanique des forces*, vol. 2 (Liège: 1845), 15.

technical operations are grounded on passive materials, electrical energy, and, finally, electrically processed information. As a result, symbolic control operations were separated from the mechanical operations of the working machinery, or, in terms of electricity, low-current engineering of information processing was separated from high-current engineering of the working machine. Since the early days of cybernetics, matter has been conceived as a passive carrier of energy and information and has thus suffered an even sharper distinction from the symbolic, as Norbert Wiener put it: "Information is information, not matter or energy. No materialism which does not admit this can survive at the present day."[14] This triple separation is fundamental for digitally controlled and electronically driven technology.

Material is considered a neutral carrier both of external energy and information implemented in the physical gears. In the case of the robotic arm, the gearbox is assisted and replaced by the digital code. The kinetic chain is not mechanically constrained by the joints, but able to perform several degrees of freedom. Where in mechanical machines, the desired movement is controlled by the code executed by the mechanical constraints arranged by the connected devices, in digitally controlled machines, the code is completely separated from the material mechanics and has to be transmitted from outside in order to execute its programmed procedure. This obviously allows more flexibility in coding compared to the reduced mode in gear-controlled mechanisms.

The external character of electric code is also reflected within information theory: Claude Shannon's theory is based on the idea that information has to be transmitted from a sender to a receiver. The receiver is considered an empty entity; it contains no information about its own behavior and thus completely depends on the transmission of the external electrical code. Thus, the receiver is a passive material whose essence is to receive and obey the external code (fig. 5).

It is obvious how this structure and arrangement shapes the idea of materials as a neutral substance that depends on receiving energy and information in order to become active and programmable. Within this context, any self-activity of the material or some kind of external condition is seen as failure or disturbance and as such has to be strictly inhibited. Therefore, the device has to be isolated from possible impacts from outside—in the case of electrical machines, this requires above all dry conditions. Programmable material, which is often described within the context of active materials, is thus actually a passive material whose activity is the result of externally coded and activated operations.

Classical digital technology emphasizes information transmission and processing as a sequential flow. Therefore, hardware is based on matter used as a neutral carrier for the transmission of electrical signals that flow through matter: the material of switching circuits can be regarded as a material flowchart. In this way, symbolic logical

---

14 Norbert Wiener, *Cybernetics: Or Control and Communication in the Animal and the Machine* (Cambridge, MA: MIT Press, 1961), 132.

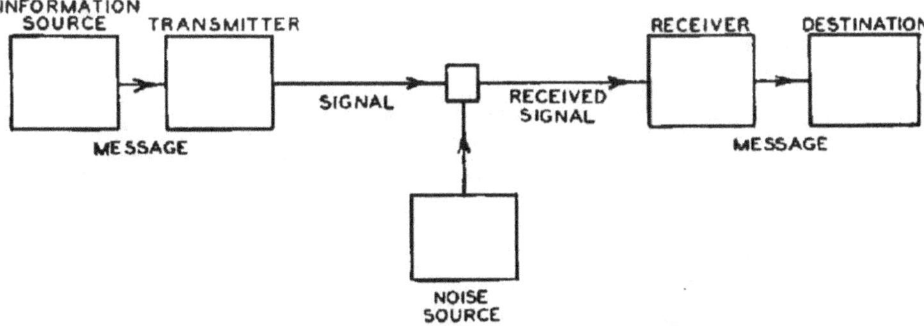

Fig. 5: Diagram of a general communication system, in Claude Elwood Shannon, "A Mathematical Theory of Communication," *The Bell System Technical Journal* 27 (1948), 381.

operations that over more than 2,000 years were processed on papyrus or paper could be "interpreted," as Claude Shannon did in his master's thesis of 1936, as electric switching circuits, and transferred from paper to electronic machines.[15] Information processing and transmitting is thus deeply shaped by these electrical conditions and relies on the inactive character of its material carrier.

Active Materials

This relationship between code and material is completely different in biomaterials (fig. 6). A comparison with a robot shows that the awn of an Erodium seed is—also in contrast to the sheet of paper—an extraordinary case of active cellulose structures. The material performs a strong torsion by drying and elongation through humidity, both as reversible operations.[16] The change of temperature and humidity between day and night triggers and fuels this activity every day, pushing the seed into the ground step by step. On earth, the sun floods the natural system day by day, altering the environmental conditions periodically. This change provides energy to the materials and makes the awn move. The clock-rate of this operation that triggers the inner activity of the material is low, in general two per day, and, accordingly, so is the resulting speed of the material's movement.

Bio-material is an open system and consists of heterogeneous structures and corresponding properties that depend on environmental conditions, mainly humidity and temperature. It is the environment that interacts with the material cellulose structure

---

15 Shannon's master's thesis of 1936 was published two years later as an article: Claude Elwood Shannon, "Symbolic Analysis of Relay and Switching Circuits [1938]," in *Collected Papers*, ed. Neil J. A. Sloane and Aaron D. Wyner (New York: IEEE Press, 1993), 474.
16 Michaela Eder et al., "Wood and the Activity of Dead Tissue," *Advanced Materials* 33, no. 28 (2021): 1–15, https://doi.org/10.1002/adma.202001412.

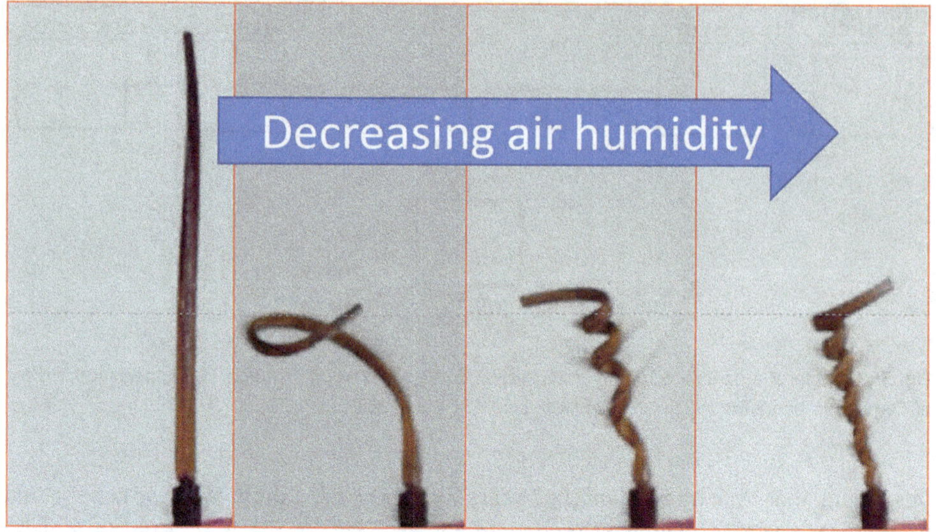

Fig. 6: Erodium awn.

composed of micro fibrils and cell structures with hydrophilic or hydrophobic inner surfaces. This is where Peter Fratzl's research on biomaterial's activity[17] jives with Jürgen Rabe's analysis of material activity driven by internal interfaces, where the attraction of water controls both their mechanical and electronic activities.[18]

The swelling or shrinking of the material as its fundamental activity is performed by the water that enters or leaves the hollow interstices of the cellulose structure. The material structure reacts to the changing environment or boundary condition of humidity and temperature; there is no sender or receiver, but a nested structure of many interconnected layers, where the special intrinsic geometric arrangement of the material defines and controls its performance. In plants, the actual performance depends on the inner geometry of the cellulose fibers whose properties can vary between strong stiffness and high flexibility.

In the case of wood used as a stiff construction material, this intrinsic activity of cellulose has to be minimized as much as possible. Plywood tries to silence this activity when different wood layers are glued together to a rigid staple. The directions of the inner possible activity of every layer are oriented in counter direction so as to mutually neutralize each other. It becomes obvious that passivation or passivity is not an ab-

---

https://doi.org/10.1002/adma.202001412.

18 Hu Lin et al., "Reversible Switching of Charge Transfer at the Graphene-Mica Interface with Intercalating Molecules," *ACS Nano* 14 (2020): 11584–604, https://doi.org/10.1021/acsnano.0c04144; Hu Lin et al., "Influence of Interface Hydration on Sliding of Graphene and Molybdenum-Disulfide Single-Layers," *Journal of Colloid and Interface Science* 540 (2019): 142–47, https://doi.org/10.1016/j.jcis.2018.12.089.

sence of activity, but the neutralization of forces to achieve a stable equilibrium that has the ostensible effect of passiveness. If stronger forces or humidity are applied, the stable structure may "explode" and lose its stability.

The difference between activity and passivity, stiffness and flexibility, however, is also a question of scale, according to which a process appears to be static, slow, or fast. The interaction between environmental conditions and materials structure is the performance, the material itself. The inner structure is organized in terms of interfaces, acting as a filter of its environment and providing the two necessary elements of a mechanical couple. Water that interacts with material surfaces that attract or repel it is a component as essential as the inner geometry and architecture of the material. The interplay between the fluid and the cellulose encodes the activity.

Wood as an exemplary bio-material shows the fitness of an open system in terms of specific functions, containing distributed agency and information embodied within the material structure. The material is not at all passive—it is composed of cellulose, water, and their interaction transformed into activity. All of its material structure is defined by this activity. If the material is taken out of its natural context, it loses its special function. This intrinsic interaction of biological materials depends on a completely different version of relating material, energy, and information that are all integrated into one and the same material structure.

And it is obvious that the analysis of this constellation requires a different approach. Compared to our technology in general, the awn—taken as an exemplary model system—shows a fundamental difference. There is a dynamic cellulose structure that interacts with its environment and thus produces the intrinsic code and energy necessary for the possible performance of a certain function.

Analog Code

It may be astonishing that a simple awn surpasses digital technology and is able to show that the field of biomaterials can be taken as a starting point for a material revolution and an "analog age" to come.[19]

Today, the enormous distance that defines the classical decisive gap between material and code, matter and mind can be transformed into the analysis of active material, where one can identify the same duality in its essential form: the material and the symbolic. Now, however, active material can be regarded as its intrinsic entanglement and radical fusion. From this perspective, biological material is an example of a different kind of symbolic operation or emergence. The material's structure is the code of its own operation, the motor of the activity and a sensor for being responsive, adaptive

---

19 In contrast to Alexander Galloway, it has to be emphasized that we did not yet arrive at all at an age of analog, but the real challenge is bring a future analog age into being. Alexander Galloway, "Golden Age of Analog," *Critical Inquiry* 48, no. 2 (2022): 211–32, https://doi.org/10.1086/717324.

and interactive with the environment that is not an additional but an essential element of the material's activity. This kind of machine is integrated—it is symbolic material operation. Material activity, information, and energy are intertwined within one and the same dynamical structure.

This new notion of active or symbolic material changes our classical understanding of material and code. Materials being regarded as different to code do not only symbolically represent but also physically execute their intrinsic information at the same time. This is a downright revolutionary analog device compared to digital coded machines. This is why, today, matter has to be examined as active, coded structures in order to reset the prevalent idea of algorithms and code within matter itself, thus reinventing the analog as a new code.

The material code, which now can be taken as the basic constellation for the analog code, is not a flow or chaotic process.[20] A symbolic system which is not discrete but continuous, material and symbolic at the same time is the very classical realm of geometry. Since Greek antiquity, geometric elements are considered continuous in contrast to discrete numbers and letters. And it was regarded as both symbolic and spatially extended, combining the symbolic and the material (fig. 7).

Analyzing more closely Euclid's *Elements*, one can see that geometric elements are not static objects, but operations and, in a certain sense, essential boundary conditions: the plane is a boundary of the 3D body, the line is a boundary of the plane, and the point is a boundary of the line. In this sense the elements combine presence and absence, the point the boundary of all boundaries, and as such an essential absence. Euclid's point is a σημεῖόν, a sign, a relationship, and as such a fundamental operation. Nevertheless, it exists due to its very constraint that makes its activity possible. Therefore, the point is an original difference and an activity that extends itself in terms of angles or lines.

The Euclidean point-sign is in the same way as the Epicurean atom not a minimal passive unity of space or matter but a fundamental geometric operation, παρέγκλισις (*parénklisis*), a swerving.[21] Geometry is an ancient analog operative system based on a small ensemble of devices such as point, angle, plumb line, gnomon, lever, spindle, or mesolabium, a system that was explicitly reset in Early Modernity by Leon Battista Alberti, Albrecht Dürer and Leibniz. All these constellations show that over long periods of time—which can be understood as past "analog ages"—the idea of geometric operations had been a basic feature. But still today, this operational character of geometry

---

20 There is a real need for a fundamental philosophical perspective on the analog. Only partially relevant: Brian Massumi, "On the Superiority of the Analog," in *Parables for the Virtual Movement: Affect Sensation* (Durham, NC: Duke University Press, 2002); also Galloway, "Golden Age of Analog."

21 According to Cicero, *De finibus bonorum et malorum: Libri quinque* 1 (Cambridge, UK: Cambridge University Press, 2010). See also Michel Serres, *Birth of Physics* (Manchester: Clinamen Press, 2000).

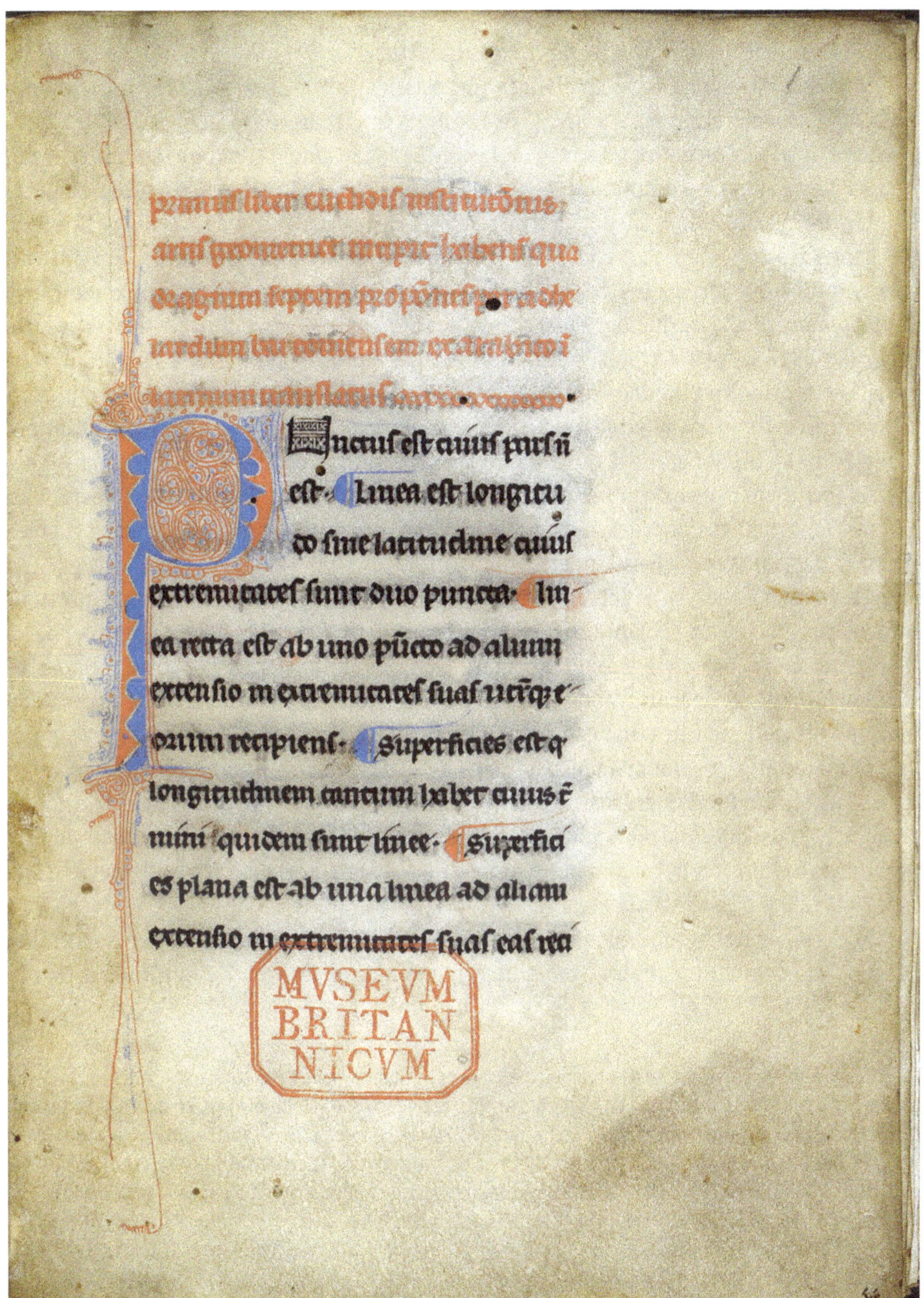

Fig. 7: Euclid, *Elementa* (translated by Adelard of Bath), ed. Johannes Campanus (Venice: Erhard Ratodolt, 1482).

is crucial for rethinking the analog as a symbolic operation that not only represents what it is referring to, but also performs it as such.[22]

The idea of analog code challenges the classical term of code based on alphanumeric or informational perspectives. In recent years, there have been huge efforts in the field of bio-informatics to go beyond a reductionist understanding of "information under the dominant computational form," as Pedro Marijuán put it in his seminal paper "Information and Life."[23] Erwin Schrödinger already argued in this way, when the gene was regarded as a code: "But the term code-script is, of course, too narrow. The chromosome structures are at the same time instrumental in bringing about the development they foreshadow. They are law-code and executive power—or, to use another simile, they are architect's plan and builder's craft—in one."[24] In a similar way, René Thom criticized the idea of linear code as reducing DNA's "topological complexity . . . and throw[ing] away almost all of its significance."[25] "A finite sequence of letters taken from an alphabet . . . is only one of the possible aspects of information; any geometric form whatsoever can be the carrier of information."[26] Thom refers precisely to geometric structures at work everywhere in the material structures, which becomes the starting point for a new understanding of code.

This kind of analysis is further developed in studies about the insufficient status of information theory to explain functional and interactive modes referring mainly to the transition from non-living to living, and from living systems to those capable of thought. Terrence Deacon's book *Incomplete Nature* tries to bridge these thresholds by symbolic potentialities that emerge within nested hierarchical structures.[27] My analysis of materials does not primarily address "life" or "consciousness," but focuses these questions on the fundamentally active, symbolic qualities of the material and its coded structure. Consequently, talking about active or symbolic material is not about adding additional dimension to matter, but making the absent potentialities visible beyond the mere presence of a material structure.

Material is not a substance, but a process and a relationship between something present and something absent, something that is not, or not yet. Deacon's focus on ab-

---

22 This genealogy of geometry as an analogue code is the object of my research over years and will be published soon in two volumes. For elements of this genealogy see: Wolfgang Schäffner, "The Point: The Smallest Venue of Knowledge in the 17th Century (1585–1665)," in *Collection Laboratory Theater: Sciences of Knowledge in the 17the Century*, ed. Helmar Schramm et al. (Berlin: De Gruyter, 2005); Wolfgang Schäffner, "Euklids Zeichen: Zur Genese des analogen Codes in der Frühen Neuzeit," in *Bildwelten des Wissens: 7, 2 Mathematische Forme(l)n* (Berlin: De Gruyter, 2010).
23 Pedro Marijuán, "Information and Life: Towards a Biological Understanding of Informational Phenomena," *TripleC* 2, no. 1 (2004): 6–19.
24 Erwin Schrödinger, *What Is Life* (Cambridge, UK: Cambridge University Press, 1992), 20.
25 René Thom, *Structural Stability and Morphogenesis: An Outline of a General Theory of Models* (Reading, MA: W. A. Benjamin, 1975), 157.
26 Ibid., 144–45.
27 Terrence Deacon, *Incomplete Nature: How Mind Emerged from Matter* (New York: W. W. Norton, 2012).

sence as a fundamental—and generally overlooked—feature in nature's processes is at the core of the question about symbolic material. In this sense, functions, adaptive processes, and potentialities as basic elements of the symbolic dimension of materials result from this structural absence inherent in the materials' geometric structure. The symbolic dimension is not an additional element but an intrinsic feature of active materials, and therefore the basic condition of an analog code.

It is helpful to remember again the robotic arm where function seems to be externally added by the code transmitted. However, in contrast, one can invert the classic idea of function. As an effect of constraints imposed on the machine that prevent it from doing everything possible, function is not an addition to its material and energy: "This negative view of function is easily recognized in the case of machine failure. When constraints break down, and previously restricted states or behaviors of the machine become possible, functionality is degraded."[28] Boundary conditions reduce the degrees of freedom and produce information.

Here, the dominant role of structures and morphologies becomes evident: points, point defects, holes, interfaces, hyperbolic structures, etc., geometric operations being the most active elements directed to something absent. Their relational character deeply intertwines structures and environments as boundaries of boundaries, where material, energy, and information are intrinsically related. This shift from substance to process is analyzed at the Cluster, which focuses on weaving, filtering, and cutting as intrinsic features of the material, as active processes producing relationships and boundaries that redefine the property of the material by interlacing the structure and its environment. This is where information and the symbolic emerge as an intrinsic process of materials.

Thus, the pure dichotomy of the symbolic and the material is transformed into an interrelated structure, where the material existence is defined in terms of constraints and absence, thus containing the symbolic. Overcoming this dichotomy, however, does not only change the idea of material from passive to active and symbolic materiality. Similarly, vice versa, the symbolic requires a different material character. The classical idea of ideal symbolic objects is in the same sense dependent on a silenced, passivated, and de-materialized material carrier. If such a de-materialization of material allowed a kind of fusion of the material and the symbolic, the seemingly extracted and abstracted quality of the graphic operation would now need to be conceived as active matter.

Looking back once more to the graphic surface of paper, one can say that it now makes sense as part of the materials' table: it is another starting point—alongside wood, seed capsules, or biofilms—that needs a fundamental figure-ground inversion for making its active matter evident. This is the place of my own research on geometric operations, on points, lines, and angles as active processes, as boundary conditions and constraints of degrees of freedom. The geometric point is a symbolic and spatial oper-

---

28 Terrence Deacon, "Shannon—Boltzmann—Darwin: Redefining Information (Part I)," *Cognitive Semiotics* 1, no. 1 (2013): 123–48, 129, https://doi.org/10.1515/cogsem.2007.1.fall2007.123.

ation, it is a fundamental difference and absence, placed at the core of the idea of active materials.

Outlook

My historical and theoretical genealogies of geometric operations as analog code are linked with the transversal importance of geometry as an operative system within active materials at MoA. Active matter has to be synthetically analyzed so as to make it visible as a decisive historical moment of an extraordinary coincidence in various disciplines, as some sort of contemporary *episteme* and order of things. Dealing with the reinvention of the analog goes far beyond a mere epistemic analysis, and requires feedback-looping speculative and historical epistemology with experimental practice and design projects.

On our interdisciplinary operation table, where this new order of things and analog code can be developed, we put different materials as model systems: seed capsules, biofilms, or felt, all of which consist of highly dynamic structures, mostly based on fiber structures. They become part of a new kind of epistemic endeavor that performs historical, epistemological, and intercultural practices within the science lab through corresponding experimentation of active materials and in the workshops for developing new design processes. We also analyze cultural practices related to active and symbolic materials, traditional practices within, and, above all, outside of Europe that can be taken as a deposit for technical strategies and solutions for symbolic material operations. As long as design is understood as transferring preconceived design ideas into matter, the material still plays the role of the passive recipient of the design idea. Design, especially in its modern form up to today's 3D printing, is a highly constructivist process relying on programming passive materials. This idealistic core of design is equal to the classical engineering process of implementation and thus part of the same strategy. This approach changes radically when we take the material's intrinsic activity as the starting point for design processes. Thus, result- and shape-oriented design is transformed into an adaptive, process-oriented procedure that includes mutual adaption of material and building process with its environment, comparable to traditional cultural practices developed over long periods of time, and, above all, to natural processes of biological growth. Instead of implementing human intelligence into the world, the thorough analysis of active material is finally directed toward a new kind of design, able to merge technical and natural processes in a nondestructive manner.

Authors' Biographies

Frank Bauer enquires into methods and ontologies of computational design and manufacturing in interdisciplinary and experimental formats. His ongoing dissertation engineers operative, instrumental, and material extensions of digital modeling and making processes in computational art production. He applies notions of affordance, persistence, and activity in these workflows and embeds and probes them in industry. His teaching efforts are situated within the BA/MA programs at Universität der Künste Berlin and Technische Universität Berlin as well as in the Design for Manufacture MArch at The Bartlett School of Architecture (University College London). He attempts to bridge the broader agendas of Matters of Activity outwards and shift discourses from possibility to constraint spaces of technology.

Bastian Beyer's research deals with fiber-based materials in the design context. He studied architecture in Munich (University of Applied Sciences) and Berlin (Universität der Künste Berlin). After his studies, he received a Marie Curie Research Fellowship in the ArcInTex program and received his doctorate from the Royal College of Art in London. In his dissertation, he investigated structural and reactive properties of biofilms on fiber-based substrates and explored new manufacturing methods for bio-based materials. At Matters of Activity, he currently works together with Regine Hengge, Skander Hathroubi, and Iva Rešetar on cellulose-based biofilms and their application for design and architecture.

Horst Bredekamp is professor of art and visual history at Humboldt-Universität zu Berlin and senior co-director of Matters of Activity. In 2000, he founded the project The Technical Image at the Hermann von Helmholtz Center for Cultural Techniques. He was a permanent fellow of the Wissenschaftskolleg from 2003 to 2012 and directed the Centre for Advanced Studies in the Humanities and Social Sciences project Picture Act and Embodiment funded by the German Research Association (DFG). From 2012 to 2018, he co-directed the Cluster of Excellence project Image Knowledge Gestaltung along with Wolfgang Schäffner. Bredekamp was also one of the founding directors of Humboldt Forum, Berlin.

Maxime Le Calvé is an anthropologist of art and science and currently a postdoctoral research associate at Matters of Activity. He trained in general ethnology in Paris Nanterre and has a PhD in social anthropology and theater studies from EHESS Paris and Freie Universität Berlin. He has published on the ethnographic study of atmospheres (Exercices d'ambiances, 2018), on performance art, music, Berlin, brains, and ethnographic training. He is also the co-curator of the exhibitions Field/Works in Lisbon (2020 – 2021), Stretching Materialities (Berlin, 2021 – 2022), and currently the participant exhibition Sketching Brains (Charité, Berlin).

Emile De Visscher has been a practice-based design researcher at Matters of Activity since 2019. Trained as an engineer (Université de Technologie de Compiègne) and designer (Royal College of Art, London), he obtained a PhD in the SACRe doctoral program from Université Paris Sciences & Lettres and École des Arts Décoratifs Paris in 2018. His research focuses on the invention of small-scale production tools at the intersection of engineering, design, material sciences, and performing arts. He is the editor in chief of the design research journal Obliquite, and co-curated the exhibition Material Legacies with Michaela Büsse at Kunstegewerbemuseum Berlin in 2022.

Myfanwy Evans is a scientist applying geometric ideas in the study of a wide variety of entangled structures. She is professor for applied geometry and topology at the Institute for Mathematics at the University of Potsdam. She aims to create robust geometric tools to deepen the analyses of complex and hierarchical materials, shedding light on the important role of geometry in the structure-function relationship in materials.

Mohammad Fardin Gholami graduated in polymer science with a Bachelor of Engineering and a Master of Science. As part of his doctoral thesis, he focuses on carbon-based 2D nanomaterials and their polymeric functionalization, primarily using scanning probe microscopy and spectroscopy methods. Fardin is a project leader of Cutting at Matters of Activity, where he is developing tools for a micro- and nanometer-size surgical operation room for biological matters as well as high-precision cutting and manipulation of materials.

Lorenzo Guiducci obtained a PhD in physics and science of biomaterials at the University of Potsdam for the research performed at the Biomaterials Department at the Max Planck Institute for Colloids and Interfaces in Golm. He was a researcher and teacher at the Cluster of Excellence project Image Knowledge Gestaltung at Humboldt-Universität zu Berlin. As a postdoc at Matters of Activity, he aims to explore structure-function relationships in both biological role models and mechanical metamaterials using both engineering tools (FE simulations and modeling) and design methods including traditional fabrication practices (such as weaving).

Regine Hengge studied biology and obtained her doctorate at Universität Konstanz. After a post-doctoral phase at Princeton University, she completed her habilitation in microbiology and molecular genetics at Universität Konstanz. She has been a professor of microbiology at Freie Universität Berlin (1998–2013) and Humboldt-Universität zu Berlin (since 2013). Going through the scales from molecules to macroscopic morphogenesis, she studies regulation and signal transduction in environmental stress adaption and biofilm formation of bacteria. She received the Gottfried Wilhelm Leibniz Prize and an ERC Advanced Investigator Grant and is an elected member of the German National Academy Leopoldina, EMBO and the American Academy of Microbiology.

Heidi Jalkh is an experimental designer, design educator, and researcher based in Buenos Aires, Argentina. Trained in industrial design, she is a specialist in the logic and technique of form, holds a master's degree in interdisciplinary research, and received the Humboldt Innovation Prize in 2022. Her professional practice as a designer consists of many interests, including the design and manufacture of bio-inspired and bio-fabricated materials, interdisciplinary research, and craft-based processes. As leader of the research group Sistemas Materiales, she focuses on making significant connections between biology, design, and engineering, aiming to achieve a higher level of integration between form, behavior, and material.

Yoonha Kim is an anthropologist interested in the relationship between technology and locality in the age of the planetary. As a pre-doctoral researcher in the project Object Space Agency as well as Centre for Anthropological Research on Museums and Heritage (CARMAH) at Humboldt-Universität zu Berlin, she looks into East Asian cosmology and its relation to technical objects through the constantly reimagined Korean sartorial heritage. From filmmaking to design workshops, she explores multimodal forms of ethnographic fieldwork, which include co-curating the exhibition Stretching Materialities (2021–2022, Berlin) and co-organizing the Stretching Senses School.

Claudia Mareis is an expert for design as well as cultural history and theory. Since 2021, she has been professor of design and history of knowledge at the Department of Cultural History and Theory at Humboldt-Universität zu Berlin. Her research interests include histories, theories and methodologies of design, knowledge cultures in design, cultural history of creativity, design and eco-material politics. From 2013 to 2021, she directed the Institute for Experimental Design and Media Cultures (IXDM) as well as the Critical Media Lab at the FHNW Academy of Art and Design in Basel. At IXDM, she has built up a pioneering interdisciplinary research group converging design, media arts, anthropology, historical studies, and technology. As co-director of Matters of Activity, she has co-lead the Cluster with Horst Bredekamp, Peter Fratzl, and Director Wolfgang Schäffner since 2019.

Sabine Marienberg studied philosophy and Romance studies in Munich, Berlin, and Perugia and completed her doctoral degree on sign actions in the thought of Giambattista Vico and Johann Georg Hamann at Freie Universität Berlin. From 2014–2018, she led the research project Symbolic Articulation, where she investigated the relation of language and images. After a subsequent visiting professorship in the interdisciplinary program Diversity of Knowledge at Humboldt-Universität zu Berlin, she joined the research project Symbolic Material at Matters of Activity. Her work focuses on the conceptual and terminological framework of the Cluster from a philosophical and linguistic perspective.

Natalija Miodragović is an architect whose interdisciplinary and experimental work starts from an understanding of art and space as vehicles for social change. Her work focuses on the perception and understanding of space and lightweight, flexible, and textile structures. As a researcher, she has taught at the Institute for Architecture-Related Art at Technische Universität Braunschweig, Weißensee School of Art and Design Berlin, and the Frankfurt University of Applied Sciences. She works in cooperation with artists and researchers and co-authored a series of projects and exhibitions, such as the Serbian Pavilion EXPO 2010, dreidreidrei Organ for Zionskirche Berlin, and artist Tomas Saraceno.

Thomas Ness holds the Cluster Professorship for Embodied Interaction at the Product Design Department of Weißensee School of Art and Design Berlin. With a background in designing physical interactions and focusing on exploration and prototyping as design practice, he has done work for international clients ranging from automotive and tech industries to educational and cultural institutions.

Léa Perraudin is a media theorist and speculative material scholar and a postdoc at Matters of Activity. Her research interests include (media) infrastructures of the Anthropocene as well as practices of knowing and making in the realm of environmental humanities and queer feminist technology studies. Perraudin is currently working on a media theory of phase transitions by investigating the material ties of media infrastructures in contemporary technocapitalist environments through situated phenomena of transience, dispersal, abundance, and solidification. Perraudin holds a PhD in media studies from University of Cologne. She co-leads the experimental laboratory for knowledge exchange and speculative design "CollActive Materials" funded by the Berlin University Alliance (2022–2024).

Jürgen P. Rabe has been professor of the physics of macromolecules at the Department of Physics at Humboldt-Universität zu Berlin since 1994. His research aims to correlate the structure and dynamics of molecular systems at interfaces with mechanical, electronic, optical, and (bio)chemical properties, from the molecular to the macroscopic length and time scales. He is also the founding director of the Integrative Research Institute for the Sciences – IRIS Adlershof at Humboldt-Universität zu Berlin.

Felix Rasehorn is a practice-based researcher with a background in product and interaction design from Weißensee School of Art and Design Berlin. He works as a research assistant at Matters of Activity and is a PhD candidate at Technische Universität Berlin. In his dissertation, he works with bio-mimetic design approaches and collaborates with natural scientists to sustainably, technically, and economically co-develop prototypes that envision preferable futures.

Khashayar Razghandi is a material scientist and design teacher working at the intersection of natural sciences, humanities, and design, between Max Planck Institute of Colloids and Interfaces, Department of Biomaterials, and the Cluster of Excellence project Matters of Activity. He is engaged with active new materialism, creating interdisciplinary educational co-creation spaces, and materials sustainability. Recent publications include "Rethinking Active Matter: Current Developments in Active Materials, Active Materials" with Mohammad Fardin Gholami, Lorenzo Guiducci, and Susanne Jany, in Active Materials (De Gruyter, 2021) and "Rethinking Materials Paradigm: Towards an Active Understanding of Gestalt," in Design, Gestaltung, Formatività (Birkhäuser, 2022).

Iva Rešetar studied architecture at the University of Belgrade and at the Städelschule in Frankfurt am Main. She has been a practicing architect in the field of digital and experimental design, and has held several research positions, among others at the Akademie Schloss Solitude, in the ArcInTexETN, and at Universität der Künste Berlin. In her research and teaching, she explores fluid and marginalized materialities in architecture through experimental and interdisciplinary ways of designing and making. Her interests include histories and aesthetic practices of thermodynamics around issues of climate adaptation, energy, and the environment in architecture, design processes between the analogue and the digital, and textile architectures, understood as temporary and more-than-human.

Nina Samuel is an art and science historian and curator with a PhD from Humboldt-Universität zu Berlin. Her thesis The Shape of Chaos investigates visual epistemologies in the field of complex dynamics and fractal geometry and drawing as a mode of thinking. She held various research positions, among others with Technisches Bild (HU), Embodied Information – "Lifelike" Algorithms and Cellular "Machines" (Freie Universität Berlin), and at the Bard Graduate Center in New York City. Samuel has received scholarships and research grants from the Fulbright Program, eikones Basel, and the Max Planck Institute for the History of Science. Before coming to Matters of Activity, she was program director of the museum-centered PhD program "PriMus – Promovieren im Museum" at Leuphana University Lüneburg and is now a project lead of Object Space Agency.

Wolfgang Schäffner is a historian of science and media technologies and has been professor of cultural history of knowledge at the Department of History and Theory of Culture at Humboldt-Universität zu Berlin since 2009. From 2012–2018, he was the Director of the Hermann von Helmholtz Center for Cultural Techniques and Director of the Cluster of Excellence project Image Knowledge Gestaltung at Humboldt-Universität zu Berlin. He is a permanent guest professor and director of the Walter Gropius Program at the Faculty of Architecture, Design and Urbanism at Universidad de Buenos Aires, and head of the German-Argentinian MA program Open Design. Since 2019, he is co-director of Matters of Activity.

Nelli Singer is a textile designer with a focus on active materials research. She holds an MA in textile and surface design from the Weissensee School of Art and Design Berlin and has been part of Matters of Activity since September 2020. Her interests lie in innovative material structures and the experimental process of their design. Her creative work is characterized by an interdisciplinary approach involving design, art, architecture, natural sciences, and technology. Her work has been shown at Humboldt Forum under the heading Active Curtain and the exhibition Stretching Materialities at Tieranatomisches Theater, Berlin.

Mareike Stoll holds a PhD in German studies from Princeton University (2015). Her first book titled *ABC der Photographie: Photobücher der Weimarer Republik*—for which she was awarded a research and publication grant by the Deutsche Gesellschaft für Photographie (DGPh)—was published with Walther König in 2018. At Weißensee School of Art and Design Berlin and Matters of Activity she is a research associate, currently working on her second monograph about contemporary children's picturebook as site of visual and emotional literacy and artistic practices as "Making Sense by Hand."

Daniel Suárez is an architect and researcher interested in innovative materials, design interfaces, and technology processes for architectural production. He holds an MSc in architecture and urban design from the Polytechnic University of Madrid. In his research, Daniel explores textile-informed material systems on a large scale, combining fluid interactions of human craftsmanship and vernacular design with computational design processes and digital fabrication. He was part of the ArcInTexETN network and worked as associated researcher at Universität der Künste Berlin. Currently, he is part of the Structural Textiles project at Matters of Activity, examining relations between active materiality, bio-based materials, digital instructions, and geometry formation processes.

Kolja Thurner is an art and visual historian. He studied art history, German studies, and philosophy at Eberhard-Karls University, Tübingen, and art and visual history at Humboldt-Universtiät zu Berlin. His research focuses on the early modern history of double imagery as well as on the material and symbolic interplay of chance, amorphism, and figuration in landscape painting. His PhD project Patinir and the Pioneers of Seeing investigates the visual culture of double images in sixteenth-century Antwerp art. Since 2017, he has worked as a research assistant to Horst Bredekamp at Matters of Activity.

Hanna Wiesener is a designer, researcher, and design consultant. With a background in product design and cultural studies, she works as a research associate at Matters of Activity in the projects Filtering and Robotic-Assisted Surgery. At Designfarm, the design-in-tech accelerator at Weißensee School of Art and Design Berlin, she mentors teams in turning their design prototypes into meaningful concepts.

Clemens Winkler is a design researcher at Matters of Activity, whose work relates environmental concerns and social, scientific, and technological development. His current projects explore epistemic foundations of atmospheric control, especially on experiencing, notating, archiving, and speculating in experimental exhibition formats with his research group Object Space Agency. In his current position as a guest professor at the University of Performing Arts Ernst Busch in the MA studio Spiel und Objekt, he is setting up a playful framework for negotiations on current societal concerns, energy, and pollution governance, post-fossil futures, and digital media use on the theater stage.

Image Credits

Bauer, Kim

Fig. 1, 2, 4:   Frank Bauer, Yoonha Kim, Matters of Activity

Fig. 3:   Courtesy of Antiquariat Norbert Haas

Beyer, Rešetar

Fig. 1, 9:   Photo: Michaela Eder, Max Planck Institute of Colloids and Interfaces, Potsdam

Fig. 2, 5, 8, 10: Photo: Michelle Mantel, Matters of Activity

Fig. 3:   Table originally published in Dietrich Fengel and Gerd Wegener, *Wood: Chemistry, Ultrastructure, Reactions* (Berlin/New York: De Gruyter 1983), 46. © 1983 De Gruyter

Fig. 4:   Bastian Beyer, Matters of Activity

Fig. 6, 7:   Photo: Marius Land

Fig. 11:   Cécile Bidan, Max Planck Institute of Colloids and Interfaces, Potsdam

Bredekamp

Fig. 1:   Courtesy of Harvard Art Museums/Busch-Reisinger Museum, The Willy and Charlotte Reber Collection, Louise Haskell Daly Fund. Photo © President and Fellows of Harvard College, 1995.431.6

Fig. 2:   Public domain, Courtesy of Ronald Feldman Fine Arts, New York

Fig. 3:   Courtesy of Horst Bredekamp

Fig. 4:   © Hessisches Landesmuseum Darmstadt, VG Bild-Kunst, Bonn 2024; photo: Wolfgang Fuhrmannek. Courtesy Hessisches Landesmuseum Darmstadt

Fig. 5:   Estate Joseph Beuys, © VG Bild-Kunst, Bonn 2024, photo: Jürgen Müller-Schneck, Berlin

Fig. 6, 7:   Estate Joseph Beuys, © VG Bild-Kunst, Bonn 2024, photo: Carolin Tisdall, London

Fig. 8:   Estate Joseph Beuys, © VG Bild-Kunst, Bonn 2024, photo: Gwen Phillips, New York

Fig. 9, 10:   © Trustees of the British Museum

Fig. 11:   Münzkabinett, Staatliche Museen zu Berlin—Stiftung Preußischer Kulturbesitz, 18221653 (R. Saczewski)

Fig. 12:   Bayerische Staatsbibliothek München, 4 L.eleg.m. 176, S. 313, urn:nbn:de:bvb:12-bsb10522866-7

Fig. 13:   Patrimonio Nacional

Fig. 14   Loan Stichting Museum Boijmans Van Beuningen (former Koenigs collection), photo: Rik Klein Gotink

Fig. 15:   Museum Kurhaus Kleve—Ewald Mataré-Sammlung, Kleve, Germany

De Visscher

Fig. 1, 3:   Courtesy of Palta Studio

Fig. 2, 4 – 8:   Courtesy of *.able journal*

Guiducci, Jalkh

All figures: Lorenzo Guiducci, Heidi Jahlkh, Matters of Activity

Hengge, Evans

Fig. 1:	Susanne Herbst, Regine Hengge
Fig. 2, 3:	Diego O. Serra, Regine Hengge, Matters of Activity
Fig. 4:	Myfanwy Evans, Matters of Activity
Fig. 5:	Regine Hengge, Matters of Activity

Le Calvé

All figures:	Maxime Le Calvé, Matters of Activity

Mareis

Fig. 1, 3, 4, 8:	Courtesy of Smithsonian Libraries and Archives
Fig. 2:	Heritage Images / Oxford Science Archive / akg-images
Fig. 6:	Alamy Stock Foto / Florilegius
Fig. 7:	Royal Collection Trust / © His Majesty King Charles III 2023
Fig. 9:	Library of Congress Prints and Photographs Division Washington, DC
Fig. 10:	Alamy Stock Foto / gameover
Fig. 12:	Internet Archive / University of Toronto

Miodragović, Singer, Suarez, Gholami, Sauer

Fig. 1A:	Skander Hathroubi, Bastian Beyer, Matters of Activity
Fig. 1B:	Myfanwy Evans, Roland Roth
Fig. 1C, 6:	Nelli Singer, Weißensee Kunsthochschule Berlin
Fig. 1D:	Natalija Miodragović, Nelli Singer, Daniel Suárez, Weißensee Kunsthochschule Berlin
Fig. 2A:	Lindsey Webb
Fig. 2B:	Nelli Singer, Daniel Suárez, Weißensee Kunsthochschule Berlin
Fig. 3:	Dinesh Valke, Mimbres Corredor, Michaela Eder, Max Planck Institute of Colloids and Interfaces, Potsdam, Michelle Mantel, Matters of Activity
Fig. 4:	Lara Ladik, Natalija Miodragović, Nelli Singer, Daniel Suárez, Matters of Activity
Fig. 5A–D:	Nelli Singer, adapted by Mohammad Fardin Gholami, Matters of Activity

Ness, Rasehorn, Wiesener

All figures:	Felix Rasehorn, Matters of Activity

Perraudin

Fig. 1, 2, 4:	Léa Perraudin, Matters of Activity
Fig. 3:	Gina Caison

Rabe

Fig. 1:	© 2012 American Chemical Society
Fig. 2, 3:	© 2015 American Chemical Society
Fig. 4, 5:	© 2018 Elsevier Inc.
Fig. 6:	© 2020 American Chemical Society
Fig. 7:	Christian Kassung, Jürgen P. Rabe, Matthias Staudacher, José D. Cojal González, Matters of Activity